Vegetation Mapping

Biogeography Research Group
Symposia Series

Vegetation Mapping
Edited by **Roy Alexander** and **Andrew C. Millington**

Vegetation Mapping

From Patch to Planet

Published on behalf of the Biogeography Research Group of the Royal Geographical Society with the Institute of British Geographers by

Roy Alexander

Environment Research Group, Department of Geography, Chester College, UK

and

Andrew C. Millington

Department of Geography, University of Leicester, UK

JOHN WILEY & SONS, LTD

Chichester · New York · Weinheim · Brisbane · Singapore · Toronto

Copyright ©2000 by John Wiley & Sons Ltd,
Baffins Lane, Chichester,
West Sussex PO19 1UD, England

National   01243 779777
International   (+ 44) 1243 779777
e-mail (for orders and customer service enquiries): cs-books@wiley.co.uk
Visit our Home Page on http://www.wiley.co.uk
or http://www.wiley.com

Other Wiley Editorial Offices

John Wiley & Sons, Inc., 605 Third Avenue,
New York, NY 10158-0012, USA

WILEY-VCH Verlag GmbH, Pappelallee 3,
D-69469 Weinheim, Germany

Jacaranda Wiley Ltd, 33 Park Road, Milton,
Queensland 4064, Australia

John Wiley & Sons (Asia) Pte Ltd, 2 Clementi Loop #02-01
Jin Xing Distripark, Singapore 129809

John Wiley & Sons (Canada) Ltd, 22 Worcester Road,
Rexdale, Ontario M9W 1L1, Canada

Library of Congress Cataloging-in-Publication Data

Vegetation mapping: from patch to planet / edited on behalf of the
Biogeography Research Group of the Royal Geographical Society with
the Institute of British Geographers by Roy Alexander and Andrew C.
Millington.
    p. cm. — (Biogeography Research Group symposia series)
    Proceedings of a conference held in May, 1995.
    Includes bibliographical references.

ISBN 0-471-96592-8
    1. Vegetation mapping Congresses.   I. Alexander, Roy W.
    II. Millington, A. C.   III. Royal Geographical Society (Great
    Britain).   Biogeography Research Group.   IV. Institute of British
    Geographers.   V. Series.
    QK63.V445 2000
    333.95′3—dc21   99-39850
                                                     CIP

British Library Cataloguing in Publication Data

A catalogue record for this book is available from the British Library

ISBN 0-471-96592-8

Typeset in 10/12pt Times from the author's disks by Vision Typesetting, Manchester
Printed and bound in Great Britain by Biddles Ltd, Guildford and King's Lynn

This book is printed on acid-free paper responsibly manufactured from sustainable forestry,
in which at least two trees are planted for each one used for paper production.

Contents

List of Contributors

Dr Roy Alexander Environment Research Group, Department of Geography, Chester College, Parkgate Road, Chester, CH1 4BJ, UK

Dr Richard Armitage School of Geography, Centre for Earth and Environmental Science Research, Kingston University, Kingston upon Thames, Surrey, KT1 2EE, UK

Raul Arquepino Centro Universitario Estudios Medio Ambiente y Desarollo (CUEMAD), Universidad Mayor de San Simon, Casilla 2121, Cochabamba, Bolivia

Dr Peter Atkinson Department of Geography, University of Southampton, Highfield, Southampton, SO17 1BJ, UK

Clare Billington Formerly with the World Conservation Monitoring Centre, 219 Huntingdon Road, Cambridge, CB3 0DL, UK

Dr Iain Brodie Department of Biology, Anglia Polytechnic University, East Road, Cambridge, CB1 1PT, UK

Dr Anne Brookes EOS, Broadmede, Farnham Business Park, Farnham, Surrey, GU9 8QT, UK

T.A. Brown Institute of Terrestrial Ecology, Edinburgh Research Station, Bush Estate, Penicuik, Midlothian, EH26 0QB, UK

Dr Paddy Coker Department of Environmental Sciences, University of Plymouth, Drake Circus, Plymouth, PL4 8AA, UK

Margaret Cruickshank School of Geography, Queen's University, Belfast, BT7 1NN, UK

Robin Fuller Institute of Terrestrial Ecology, Monks Wood, Huntingdon, Cambridgeshire, PE17 2LS, UK

Mike Furse Institute of Freshwater Ecology, River Laboratory, East Stoke, Dorset, BH20 6BB, UK

Dr David Green Centre for Remote Sensing and Mapping Science, Department of Geography, University of Aberdeen, Elphinstone Road, Aberdeen, AB9 2UF, UK

Dr Richard Gulliver Carraig Mhor, Imeravale, Port Ellen, Isle of Islay, Argyll, PA42 7AL, UK (Associate of the Department of Geography, University of Glasgow)

Stephen Hartley Department of Biology, University of Leeds, Leeds, LS2 9JT, UK

William Hickin Leicester Environmental Remote Sensing Unit (LERSU), Department of Geography, University of Leicester, University Road, Leicester, LE1 7RH, UK

Shazad Jehangir   Punjab Forestry Service, Islamabad, Pakistan

E.W.G. Jones   Department of Geography, University of Cambridge, Downing Place, Cambridge, CB2 3EN, UK

Simon Jones   Leicester Environmental Remote Sensing Unit (LERSU), Department of Geography, University of Leicester, University Road, Leicester, LE1 7RH, UK

Professor Martin Kent   Department of Geographical Sciences, University of Plymouth, Drake Circus, Plymouth, PL4 8EE, UK

Dr Roberto Lázaro   Estación Experimental de Zonas Aridas (CSIC), General Segura, 1, 04001, Almería, Spain

E.C. Mackey   Environmental Audit Branch, Scottish Natural Heritage, 2 Anderson Place, Edinburgh, EH6 5NP, UK

Professor Andrew Millington   Department of Geography, University of Leicester, University Road, Leicester, LE1 7RH, UK

Dr R. Milne   Institute of Terrestrial Ecology, Edinburgh Research Station, Bush Estate, Penicuik, Midlothian, EH26 0QB, UK

Dr Juan Puigdefábregas   Estación Experimental de Zonas Aridas (CSIC), General Segura, 1, 04001, Almería, Spain

Neil Quarmby   I.S. Ltd, Atlas House, Simonsway, Manchester, M22 5PP, UK

Eliane Reid   National Strategy, Scottish Natural Heritage, Caspian House, Clydebank Business Park, Clydebank, Glasgow, G81 2NR, UK

Paula Spearman   Otley College, Charity Lane, Otley, Ipswich, Suffolk, IP6 9DY, UK

Dr Roy Tomlinson   School of Geography, Queen's University, Belfast, BT7 1NN, UK

G.J. Tudor   Environmental Audit Branch, Scottish Natural Heritage, 2 Anderson Place, Edinburgh, EH6 5NP, UK

Dr Ruth Weaver   Department of Geographical Sciences, University of Plymouth, Drake Circus, Plymouth, PL4 8EE, UK

Dr Jane Wellens   Leicester Environmental Remote Sensing Unit (LERSU), Department of Geography, University of Leicester, University Road, Leicester, LE1 7RH, UK

Barry Wyatt   Institute of Terrestrial Ecology, Monks Wood, Huntingdon, Cambridgeshire, PE17 2LS, UK

Preface

During the closing decades of the twentieth century we have seen an increase in public awareness of environmental issues and with this has come an increased demand for reliable environmental data at a range of spatial scales. Over the same period, the developing technologies of remote sensing and geographical information systems have had a significant impact on the ways in which many types of environmental data are acquired, analysed and reported in map form. This combination of needs and technologies has led to considerable changes in the science of vegetation mapping since the publication of Kuchler's seminal work on the subject in 1967.

This book has its origins in a conference with the same title run by the Biogeography Research Group of the Royal Geographical Society with the Institute of British Geographers (RGS–IBG) in May 1995. That conference aimed to present a review of contemporary vegetation mapping in terms of both methodology and application at a range of spatial scales. This book has the same aim and comprises a series of peer-reviewed contributions which is a mix of those originating from the conference and those which have been contributed subsequently to fill some gaps of both scale and approach. Thus what is contained represents a snapshot of vegetation mapping in the closing years of the century and millennium, together with a review of developments over the past three decades and the establishment of a research agenda.

Given the breadth of the field of vegetation mapping in terms of techniques, applications and scales, the chapters of this book could have been organised in a variety of ways. Following the subtitle, from patch to planet, we have chosen to take spatial scale as the organising criterion, thus placing chapters concerned with techniques together with those reporting applications within the same broad domains of scale. The volume opens with a review of the main data sources for vegetation mapping in which Barry Wyatt considers the use of data gathered from ground, air and space in relation to the purposes for which the maps are produced. Section 2 consists of 11 chapters reporting techniques, issues of error and applications using data from ground, air and space at scales ranging from the site to the local. Section 3 contains a further five chapters dealing with techniques and applications at regional and continental scales. The final chapter returns to the overall theme, looking at developments in vegetation mapping in recent years and developing a research agenda for the future.

We are grateful to the many people who helped in producing this book. In particular we wish to thank the contributors and reviewers, the committee of the Biogeography Research Group for encouragement to compile this volume, Chester College for the provision of resources to assist in production, Ruth Pollington for redrawing some of the diagrams and staff at Wiley, especially Sally Wilkinson, Isabelle Strafford, Louise Portsmouth and Karin Fancett, for their ready assistance at all stages of the production cycle.

Roy Alexander
Andrew C. Millington

Section 1

INTRODUCTION

1 Vegetation Mapping from Ground, Air and Space – Competitive or Complementary Techniques?

B. K. WYATT

Institute of Terrestrial Ecology, Monks Wood, Huntingdon, UK

The relative contributions of ground observation techniques and remote sensing from air and space are reviewed in relation to the requirements of ecological survey and mapping. Four complementary approaches are identified, covering: (i) use of remotely-sensed data to aid field survey design; (ii) use of ground observations to calibrate the models used to interpret remotely-sensed data; (iii) validation of airborne or spaceborne observations against reference data collected on the ground; (iv) integration of data from field observations and remote sensing to generate new information. These approaches are considered in the context of survey and mapping of vegetation and land cover and in relation to the use of Earth observation to make quantitative estimates of surface properties and fluxes of material and energy. It is concluded that no single approach is optimal in all circumstances. Efficient land survey and mapping programmes, particularly when these cover extensive areas, will take advantage of the full suite of available data acquisition methods.

INTRODUCTION

The era of operational Earth observation from space could be said to have begun in 1972 with the launch of Landsat 1, or ERTS as it was then known (Freden and Gordon, 1983). From that time, began also a period of more or less open competition between the practitioners of traditional ground-based mapping and the advocates of the new space technology (Fuller, 1981). Traces of this survive, even now the 'new' space systems are approaching their quarter century. There are several reasons for differences of opinion as to the relative merits of the two approaches. Remote sensing was clearly 'oversold' in its infancy. It has subsequently been demonstrated that even the relatively crude Landsat MSS is capable of generating useful information, particularly in areas which are poorly mapped from the ground (e.g. Fox *et al.*, 1983; Nelson *et al.*, 1987; Pickup *et al.*, 1993). But it is certainly true that space-based Earth observation was slow to live up to the ambitious claims that the aerospace lobby made in those early years. Conversely, the mapping community – particularly those from a biological background – tended to be deterred by the technological complexity of space-based digital imagery and by the new demands it placed on them. Until very recently, the number of published articles in mainstream ecological journals which mentioned the use of remote sensing could be counted on the fingers of one hand (Fuller *et al.*, 1989).

In contrast, aerial photography, which only came to maturity as a result of technology developed during the Second World War, has never been the cause of similar scepticism. Often, users rationalise this by making unfavourable comparisons between the spatial resolutions achievable from air photographs and space imagery. However, it is also true that an analogue photograph presents fewer conceptual challenges than a digital image and that much of the information it contains can be extracted without the need for expensive computer systems; these considerations may also have a bearing on the relative rates of acceptance of the two technologies.

Meanwhile, there has emerged a growing requirement for ecological surveillance, sometimes in response to statutory obligations, more usually to meet the needs of planning, environmental management and research (e.g. Commission of the European Communities, 1992; Meyer-Roux and King, 1992; Barr *et al.*, 1993). These applications are increasingly demanding, especially in regard to the spatial extent and the repeat frequency required and the need to access remote and difficult terrain, and these demands are difficult to satisfy by other means than by the use of remote sensing. When Sir Dudley Stamp undertook the first national land use survey in the 1930s (Stamp, 1962), he, for one, would surely have been happy to supplement the family saloon, which was his principal survey platform, with an image processor and access to comprehensive national Landsat Thematic Mapper (TM) coverage. By the same token, remote sensing is now increasingly regarded as an indispensable tool for any extensive survey of vegetation or land cover. However, many of the applications mentioned in the previous paragraph require information at a level of detail that cannot be achieved from remote sensing alone. Consequently, there is growing interest in the integration of data acquired from satellite and/or airborne survey with ground observations (Bolstad and Lillesand, 1992; Barr *et al.*, 1993; Brondizio *et al.*, 1996).

SURVEY CHARACTERISTICS

Ground-based survey, airborne and space-borne remote sensing may be regarded as complementary sources of data for ecological mapping precisely because they each have particular strengths and weaknesses (see Table 1.1).

Ground-based survey is labour intensive, requires high levels of skill (certainly for purposes of vegetation mapping) and is difficult or expensive when applied to extensive areas, other than by adopting a sample-based approach. Field observation presents special problems in inaccessible areas and difficult terrain. Unless special attention is given to quality control, it can be difficult to ensure consistency in identification and mapping of vegetation and land cover units. In compensation, given the requisite skills and resources, ground survey, over localised areas, can provide much more detail about the land surface, especially its vegetation cover and spatial variability, than airborne or space-based systems are ever likely to deliver. Furthermore, field survey is the only means we have to observe land use and the history of land management that is often the dominant factor in determining present land cover. Field survey can also be used to obtain information about sub-surface conditions (e.g. soils and hydrology) that are detectable from remote sensing only with difficulty. Ground survey is not depend-

Table 1.1 Summary of strengths and weaknesses of different survey techniques in vegetation mapping

	Ground survey	Aerial photography	Digital airborne remote sensing	Space-borne remote sensing
Resolution limits	< 1 m	1–10 m	1–20 m	5 m– > 1 km
Potential areal extent	Constrained by resources	$km^2 \times 10^2$	$km^2 \times 10^2$	Global
Features mapped	Land cover, land use	Land cover, land use inferred from context	Land cover	Land cover
Discrimination	Individual species and finer	Structural vegetation types	Structural vegetation types	Broad land cover classes
Revisit frequency	Constrained by resources	Resource and weather dependent	Resource and weather dependent	Daily to *ca.* 20 days but also weather dependent
Recording medium	Usually analogue	Analogue	Digital	Digital
Interpretation methods	Human observation	Usually human interpretation	Automated image processing or photo-interpretation	Automated image processing or photo-interpretation
Investment	Low	Moderate	High	Very high
Costs per unit area	High	Moderate	Moderate	Low

ent on the weather, but frequent revisits in order to study change are normally impractical if the area of interest is at all extensive.

Airborne methods, which include both photography and digital imagers, offer a cost-effective solution to the need for continuous recording over extensive areas, but at the cost of reduced capacity for discrimination. Information is usually derived from aerial photographs by expert human interpretation, which means that context, shade, texture and other characteristics may be called on in the mapping task. It is often possible to *infer* land use from the context in which a given land cover category appears – for example, grassland in an urban setting is likely to be there for amenity or recreational purposes. These sorts of inferences are much more difficult to implement in automatic digital image processing.

Digital airborne scanners differ from *space-based systems* mainly in their typical levels of spatial resolution and in the operational characteristics of the platforms on which they are mounted. Satellites provide continuous or repeated overflights, but with the disadvantage that the instant of data acquisition over a given target is predetermined by the orbital pattern, and frequently coincides with cloud or other adverse atmospheric conditions (Fuller *et al.*, 1994b). Aircraft can be more opportunistic, but cannot hope to provide the global coverage of satellite systems. Imaging radars, both airborne and satellite-mounted, have the capacity to overcome the cloud problem, but no operational radar system is optimised for observing vegetation, and we still have a great deal to learn about the interpretation of radar imagery from terrestrial targets.

It is possible to apply photointerpretation to analogue products from digital remote sensing, and a number of important land surveys, such as the CORINE Land Cover Project (Heyman *et al.*, 1993) employ this approach. Increasingly, however, digital image processing methods are becoming the norm. Multivariate classification in the spectral or temporal domains is the most commonly used technique (Schowengerdt, 1983). Remote sensing, precisely because it observes from a distance, is capable of detecting pattern at scales which may not be obvious from the ground. Examples of this include the use of aerial survey in archaeology and the observation of large-scale patterns of disturbance which can be an important factor in controlling ecological succession (Hall *et al.*, 1991; Marchetti *et al.*, 1995; Suffling, 1995; Acevedo *et al.*, 1996).

Because image classification in remote sensing is usually applied to individual pixels, it cannot readily be used to identify objects in the landscape or to detect contextual relationships. However, as we shall see later, digital imagery may be interpreted in ways other than classification; for example, features in regions of the spectrum outside the visible wavelengths can be used to compute measures such as vegetation indices which can provide information which is not readily measured, even from ground survey.

COMPLEMENTARY APPROACHES

Data on vegetation and land cover acquired from ground, air and space may be used to complement each other at four different stages of the mapping operation:

1. Airborne and satellite data may be exploited in the design of field sampling procedures.
2. Ground reference data are required to calibrate the models needed to interpret remotely-sensed data. The most common instance of this is in the selection of training areas for supervised classification, but the same principle is employed, for example, in calibrating regression models to predict vegetation cover or leaf area index from remotely-sensed vegetation indices.
3. Ground reference data or airborne data are needed to validate information from satellite remote sensing.
4. Data from field survey may be combined with remotely-sensed data to create new information.

Remote Sensing in the Design of Field Mapping

A particular strength of remote sensing is its ability to provide an overview that would be difficult to achieve from ground survey alone. This characteristic can be very helpful in preparing and executing survey and mapping operations in the field and numerous examples could be quoted. For instance, it is becoming commonplace in planning any extensive survey, particularly when it covers relatively poorly mapped regions, to take into the field hard copy prints of suitable remotely-sensed imagery, preferably geo-corrected, for use as ortho-maps.

A series of countrywide ecological surveys conducted in Britain by the Institute of Terrestrial Ecology (ITE) from 1978, culminating in Countryside Survey-1990 (Barr *et al.*, 1993), were carried out by intensive field observations within a stratified sample of 1 km cells within the National Grid. Field observations, including land use, land cover, and vegetation type and condition were transferred to 1:10 000 Ordnance Survey map sheets. The method was comparatively easy to apply in well-structured landscapes, but in the uplands and in other more featureless terrain, delineation of the boundaries of natural vegetation communities proved time-consuming and imprecise. In the 1990 survey, aerial photography was acquired in advance of the field campaign, and these features were identified by photointerpretation and transferred to the field recording maps. Field surveyors were then able to use this information to orient themselves and to record the precise composition of the regions identified from remote sensing.

The National Countryside Monitoring Scheme, begun by the Nature Conservancy Council in the late 1980s (Nature Conservancy Council, 1987) and continued subsequently by Scottish Natural Heritage, uses current and historic aerial photographs to observe and measure change in the rural landscape (see Mackey and Tudor, Chapter 13). The scheme employs aerial photographic interpretation, supplemented by field checking, within a stratified random sampling framework, in which the strata were defined by classification of Landsat data. In this case, we have an example of space-based remote sensing being used to ensure the statistical integrity of an aerial photographic survey.

Ground Data for Model Calibration

This embraces everything from the training of supervised classifiers to the validation of complex radiative models of vegetation canopy reflectance.

Image classification essentially involves the transformation of remotely sensed measures of spectral radiance to information about the composition of the land surface to which a given user community can relate. In the case of vegetation scientists, the principal aim is to map the distribution of vegetation in terms of its floristic composition, its structure, productivity or some other key characteristic of interest. The assumption is that radiometrically unique categories can be distinguished which correspond exactly with the classes of interest. The validity of this assumption largely determines the success and accuracy of the classification.

In order to match radiometric classes with land cover, reference data must be obtained, representative of each of the target categories. These data may be used either *post hoc*, for example to interpret the clusters generated by an unsupervised classification (e.g. Belward *et al.*, 1990) or, more usually, *propter hoc*, as training data in a supervised classification (e.g. Fuller *et al.*, 1994a). Training data are almost always acquired from ground observations. They may have been collected specifically for the purpose or sometimes it is possible to make use of existing reference data sets or (less reliably) to rely on individual knowledge of the terrain to be classified.

Kershaw and Fuller (1992) draw attention to the importance for classification accuracy of a statistically sound basis for the selection of training data. Satisfactory allocation of the training sites to their appropriate classes does not necessarily guarantee a high classification accuracy for the overall image. It is important to ensure that reference data independent of the training set are available as a basis for validation. Training data must be fully representative of the scene as a whole. On the other hand, it is important to minimise variability *within* training sites. Taken together, these two points suggest that training sites should be well distributed across the scene and that heterogeneous areas, as far as possible, should be avoided. If both criteria cannot be satisfied, it may be possible to subdivide a highly variable class into two or more sub-classes. In the case of the ITE Land Cover Map, it was necessary to define in excess of 80 sub-classes (to take into account, for example, variations caused by topography) in order to construct the final 25 target categories (Fuller *et al.*, 1994a).

Classification of remotely-sensed data is thus already an important means of acquiring qualitative information about land cover. However, a more ambitious, and therefore longer-term goal for Earth observation is to generate quantitative and spatially explicit estimates of important biophysical variables. The ecological variables which are, in principle, accessible from remote sensing, include rates of photosynthesis and evapotranspiration, soil moisture, surface temperature, leaf area index, proportional vegetation cover and standing biomass. Estimation of these variables from remotely-sensed radiance requires the development, calibration and validation of a whole suite of models, ranging from simple statistical regressions to physically based models describing the interception, absorption and reflection of radiation by plant canopies (Woodcock *et al.*, 1994; Myneni *et al.*, 1995b).

Vegetation indices are among the simplest and most widely used tools for making quantitative estimates of properties of plant canopies. All green plants exhibit a spectral response which is characterised by strong absorption by chlorophyll and other photosynthetic pigments at 500–650 nm and high reflectance in the near-infrared region from 800 to 1250 nm, due to light scattering within leaf cells (see Figure 1.1). Vegetation indices are designed to emphasise these features, typically by computing the arithmetic difference or the ratio between measured reflectance in the critical regions (Myneni *et al.*, 1995a; Leprieur *et al.*, 1996). Various indices have been proposed, and have been shown to be well correlated with vegetation cover, leaf area index and absorbed photosynthetically active radiation. When integrated across the growing season, vegetation indices show strong correlations with biomass. Several indices are now in quite common use for purposes of vegetation monitoring, ranging from observations of seasonality in vegetation at global scales (Townshend and Justice, 1986) to regional and local applications for drought management in arid regions (Prince, 1991). Figure 1.2 provides a good example of how a vegetation index may be calibrated against measurements made in the field to allow regional-scale estimation of standing biomass from remotely-sensed imagery (Stewart *et al.*, 1987).

Although vegetation indices have proved to be a useful means of deriving quantitative information about vegetation canopies from remote sensing, and although, as we have shown, there is a theoretical basis for their use in this way, the relationships between these indices and the agronomic variables of interest are essentially statistical ones. Increasingly, it is coming to be recognised that diagnostic estimates of vegetation cover and condition from Earth observation data require a different approach, in which the detailed mechanisms of interactions between radiation and the vegetation canopy are understood and modelled. The development of such mechanistic models of

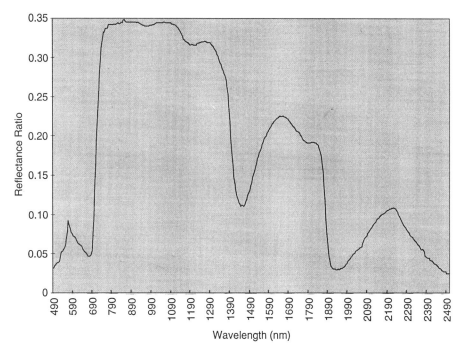

Figure 1.1   Spectral reflectance for typical vegetation

canopy reflectance and backscatter is especially hungry for ground reference data, in order to initialise and calibrate the models. This is particularly true in the case of forest ecosystems. Here, land cover, measured in terms of floristic composition, may be functionally less important, and may influence the remotely-sensed response less than do canopy density and structure. Considerable research effort is therefore currently being directed to the modelling of relationships between forest canopy structure and bi-directional reflectance (Kimes *et al.*, 1986; Bartlett *et al.*, 1988; Nilson and Peterson, 1991). In Figure 1.3, data on tree size, shape and distribution, collected in the field, have been used to simulate how the structure of tropical forests can determine the extent and patterns of shading. The results of these simulations provide inputs to a model which predicts canopy hemispherical reflectance, and model outputs are being compared with remotely-sensed estimates of forest reflectance from Landsat TM and from the dual-angle Along-Track Scanning Radiometer (ATSR) to validate the models and, ultimately, to provide a means of mapping forest structure over large areas (Gerard *et al.*, 1998; Gerard and North, 1997).

For reliability in any of these model predictions, robust ground reference data are an essential requirement for calibration purposes. Reliability here refers not just to the absolute accuracy of the ground observations, but also to their spatial integrity. It must be possible to relate the ground observations to the corresponding image location, and the data must be representative of the variability encountered within the resolution element of the image. These are far from trivial criteria, and they become more difficult to satisfy as the resolution element of the remotely-sensed data increases (Roesch *et al.*, 1995; Schreuder *et al.*, 1995).

(a)

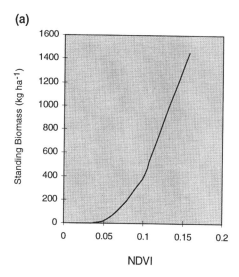

(b)

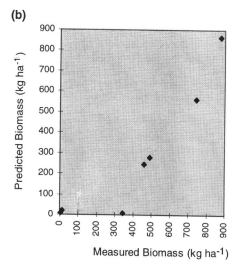

Figure 1.2 Calibration of remotely-sensed vegetation index against ground measurements of standing biomass in African rangeland. (a) Calibration plot – measured biomass vs. remotely-sensed NDVI. (b) Validation data – standing biomass predicted from remotely sensed NDVI vs. measured biomass

Ground Data for Validation of Remotely Sensed Products

Thankfully, the time when maps or land cover statistics from remote sensing were presented without validation is largely past. It is now accepted practice to provide estimates of quality, usually by comparing remotely-sensed output with data collected on the ground. These validation data are commonly presented in the form of corre-

(a)

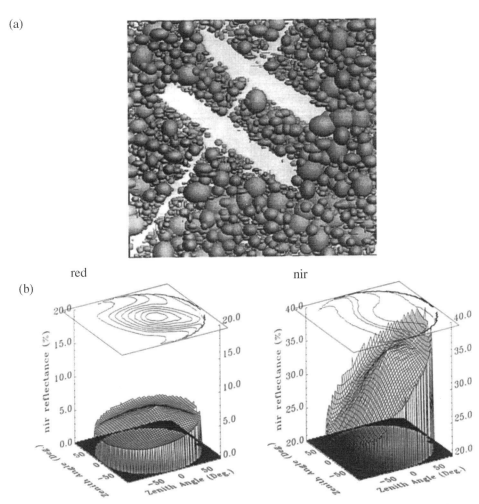

red   nir

(b)

Figure 1.3 Modelling of tropical forest canopy reflectance from ground observations of tree size, number and distribution. (a) Nadir view of modelled canopy based on data from 1 ha forest plot; (b) hemispherical reflectance diagram, ATSR-2 red channel; (c) hemispherical reflectance diagram, ATSR-2 near-infrared channel

spondence matrices or some overall index of goodness of match, such as the kappa statistic (Hudson *et al.*, 1987; Næsset, 1996).

The ITE Land Cover mapping project (Fuller *et al.*, 1994a) provides a good illustration of the use of reference datasets to check and validate output. Early proving was carried out semi-quantitatively against the results of photointerpretation. This demonstrated generally good correspondence between the TM classification and the aerial photointerpretation, with most mis-classification concentrated along parcel boundaries (presumably because of difficulties with mixed pixels). The initial national classifications were assessed by sampling on a field-by-field basis along pre-planned transects. Finally, rigorous comparisons were undertaken against independent data, first using maps compiled by field observers in Countryside Survey-1990, covering 512 1 km squares nationally, and, secondly, using data from photointerpretation of a sub-sample of the 1 km squares.

The results provide interesting insights into the difficulties of validating datasets of this type. Absolute correspondence between the three surveys was between 50 and 60%, though there was extreme variability between classes. The fundamental problem is the difficulty of arriving at 'truth', since there are substantial known errors associated with all three approaches. ITE undertook quality control checks of the performance of its field recorders, which suggests that the field data were internally consistent at a level of about 85 to 90%. However, there were undoubtedly significant differences in class definitions between the three surveys (e.g. in their treatment of wetlands and in the boundary conditions between managed and unmanaged grassland) as well as differences due to probable changes in land cover between the time of the individual surveys. In retrospect, the validation appears to have raised as many questions as it answered. The best evidence suggests that, after accounting for these differences in methodology, the results are consistent with an overall accuracy of 80–85% in the land cover map and 90–95% in the field data, giving a correspondence range of 70–80% (Fuller et al., 1998).

Integration of Data from Ground Survey and Remote Sensing

The examples considered so far have covered the complementary use of data from ground survey and remote sensing in ways which enhance one or the other. But there are also applications in which the final product benefits from synergies between the different data sources. Examples of such applications feature elsewhere in this volume. Milne and Brown (Chapter 15) describe the use of remotely-sensed data on land cover in association with measurements of the carbon content of soils and vegetation to generate national maps of carbon pools for use in assessing policy options for ameliorating climate change impacts. Similar approaches, combining remotely-sensed land cover with ground-based data on soil type and water flow, are being used to model catchment hydrology and the leaching of agricultural chemicals into water supplies.

Countryside Survey-1990 provides a particularly potent vehicle for exploiting the advantages of the parallel approaches; remote sensing gives synoptic coverage, but with relatively poor discrimination; the field samples can provide much detail but predictions about precise geographical distributions are poor. As an example, the Land Classification records East Anglia in five land classes, distributed in uniform zones from west to east across the region (Figure 1.4a). On the basis of field survey, it is possible to say a great deal about the typical land uses and their ecological associations within each land class. Conversely, the Land Cover Map reveals a great deal of local variability – for example, Thetford Forest and the coastal conurbations (Figure 1.4b), but the remotely-sensed data are incapable of providing the detailed land use information accessible from field survey. Nevertheless, we can combine the spatial qualities of the land cover map with the fine land use detail of the field survey. For example, the land classification tells us the proportional cover of individual crops within the arable land classes, while the satellite map can indicate much more precisely the geographical

(a)   (b)

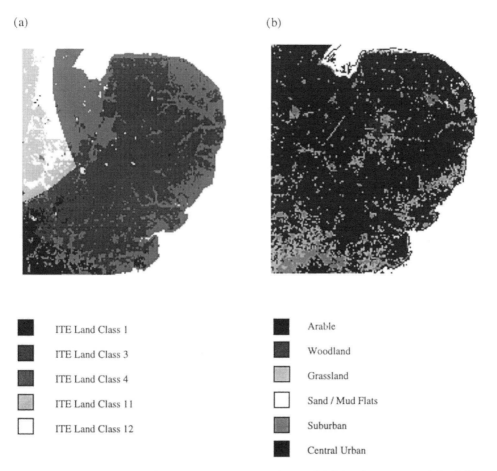

■ ITE Land Class 1	■ Arable
■ ITE Land Class 3	■ Woodland
■ ITE Land Class 4	▨ Grassland
▨ ITE Land Class 11	☐ Sand / Mud Flats
☐ ITE Land Class 12	▨ Suburban
	■ Central Urban

Figure 1.4 Countryside Survey-1990 – comparison of (a) land strata used in field campaign and (b) land cover classes from supervised Landsat TM classification

distribution of arable land. In combination, the two datasets can therefore be used to map the regional variability of individual land use categories.

CONCLUSIONS

This observation brings me neatly back to my starting hypothesis. Ground survey, airborne methods and satellite remote sensing each have their strengths and weaknesses. But it is rarely helpful to advocate one approach to the exclusion of the others. Vegetation mapping and ecological survey are sufficiently challenging tasks that we should exploit the complementary strengths of all the techniques that are available to us in order to carry them out as effectively and as efficiently as possible.

REFERENCES

Acevedo, M.F., Urban, D.L. and Shugart, H.H. (1996) Models of forest dynamics based on roles of tree species. *Ecological Modelling*, **87**(1–3), 267–284.

Barr, C.J., Bunce, R.G.H., Clarke, R.T., Fuller, R.M., Furze, M.T., Gillespie, M.K., Groom, G.B., Hallam, C.J., Hornung, M., Howard, D.C., *et al.*, (1993) *Countryside Survey 1990: Main Report*. Department of the Environment, London.

Bartlett, D.S., Hardisky, M.A., Johnson, R.W., Gross, M.F., Klemas, V. and Hartman, J.M. (1988) Continental scale variability in vegetation reflectance and its relationship to canopy morphology. *International Journal of Remote Sensing*, **9**(7), 1223–1241.

Belward, A.S., Taylor, J.C., Stuttard, M.J., Bignal, E., Matthews, J. and Curtis, D. (1990) An unsupervised approach to the classification of semi-natural vegetation from Landsat Thematic Mapper data. A pilot study on Islay. *International Journal of Remote Sensing*, **11**(3), 429–445.

Bolstad, P.V. and Lillesand, T.M. (1992) Rule-based classification models: flexible integration of satellite imagery and thematic spatial data. *Photogrammetric Engineering and Remote Sensing*, **58**(7), 965–971.

Brondizio, E., Moran, E., Mausel, P. and Wu, Y. (1996) Land cover in the Amazon estuary: linking of the Thematic Mapper with botanical and historical data. *Photogrammetric Engineering and Remote Sensing*, **62**(8), 921–929.

Commission of the European Communities (1992) Council Directive 92/43/EEC of 21 May 1992 on the conservation of natural habitats and of wild fauna and flora. *Official Journal of the European Communities*, **L**(206), 7–50.

Fox, L., Mayer, K.E. and Forbes, A.R. (1983) Classification of forest resources with LANDSAT data. *Journal of Forestry*, **81**(5), 283–287.

Freden, S.C. and Gordon, F. (1983) Landsat satellites. In: R.N. Colwell (ed.), *Manual of Remote Sensing*. American Society of Photogrammetry, Falls Church, Virginia, pp. 517–570.

Fuller, R.M. (ed.) (1981) *Ecological Mapping from Ground, Air and Space*. Institute of Terrestrial Ecology, Abbots Ripton, Huntingdon, UK.

Fuller, R.M., Jones, A.R. and Wyatt, B.K. (1989) Remote sensing for ecological research: problems and possible solutions. *15th Annual Conference of the Remote Sensing Society*, University of Bristol, pp. 155–164.

Fuller, R.M., Groom, G.B. and Jones, A.R. (1994a) The Land Cover Map of Great Britain: An automated classification of Landsat thematic mapper data. *Photogrammetric Engineering and Remote Sensing*, **60**(5), 553–562.

Fuller, R.M., Groom, G.B. and Wallis, S.M. (1994b) The availability of Landsat TM images of Great Britain. *International Journal of Remote Sensing*, **15**(6), 1357–1362.

Fuller, R.M., Barr, C.J. and Wyatt, B.K. (1998) Countryside survey from ground and space: different perspectives, complementary results. *Journal of Environmental Management*, **54**(2), 101–126.

Gerard, F.F. and North, P.R. (1997) Analysing the effect of structural variability and canopy gaps on forest BRDF using a geometric–optical model. *Remote Sensing of Environment*.

Gerard, F.F.G., Wyatt, B.K., Millington, A.C. and Wellens, J. (1998) The role of data from intensive sample plots in the development of a new method for mapping tropical forest types using satellite imagery. In: F. Dallmeier and J.A. Comiskey (eds.) *Forest Biodiversity Research, Monitoring and Modeling: Conceptual Background and Old World Case Studies*. Man and the Biosphere Series, Volume 20. UNESCO, Paris, pp. 141–158.

Hall, F.G., Botkin, D.B., Strebel, D.E., Woods, K.D. and Goetz, S.J. (1991) Large-scale patterns of forest succession as determined by remote sensing. *Ecology*, **72**(2), 628–640.

Heyman, Y., Steenmans, C., Croisille, G. and Bossard, M. (1993) *CORINE Land Cover – Technical Guide*. Commission of the European Communities, Luxembourg.

Hudson, W.D., Ramm, C.W.W. and Baumer, G.M. (1987) Correct formulation of the Kappa coefficient of agreement. *Photogrammetric Engineering and Remote Sensing*, **53**(4), 421–422.

Kershaw, C.D. and Fuller, R.M. (1992) Statistical problems in the discrimination of land cover from satellite images: a case study in lowland Britain. *International Journal of Remote Sensing*,

13(16), 3085–3104.

Kimes, D.S., Newcomb, W.W., Nelson, R.F. and Schutt, J.B. (1986) Directional reflectance distributions of a hardwood and pine forest canopy. *IEEE Transactions on Geoscience and Remote Sensing*, **GE-24**(2), 281–293.

Leprieur, C., Kerr, Y.H. and Pichon, J.M. (1996) Critical assessment of vegetation indices from AVHRR in a semi-arid environment. *International Journal of Remote Sensing*, **17**(13), 2549–2563.

Marchetti, M., Ricotta, C. and Volpe, F. (1995) A qualitative approach to the mapping of post-fire regrowth in Mediterranean vegetation with Landsat TM data. *International Journal of Remote Sensing*, **16**(13), 2487–2494.

Meyer-Roux, J. and King, C. (1992) Agriculture and forestry. *International Journal of Remote Sensing*, **13**(6,7), 1329–1341.

Myneni, R.B., Hall, F.G., Sellers, P.J. and Marshak, A.L. (1995a) The interpretation of spectral vegetation indices. *IEEE Transactions on Geoscience and Remote Sensing*, **33**(2), 481–486.

Myneni, R.B., Maggion, S., Iaquinta, J., Privette, J.L., Gobron, N., Pinty, B., Kimes, D.S., Verstraete, M.M. and Williams, D.L. (1995b) Optical remote sensing of vegetation: modeling, caveats and algorithms. *Remote Sensing of Environment*, **51**(1), 169–188.

Næsset, E. (1996) Use of the weighted Kappa coefficient in classification error assessment of thematic maps. *International Journal of Geographic Information Systems*, **10**(5), 591–604.

Nature Conservancy Council (1987) *Changes in the Cumbrian Countryside*. First report of the National Countryside Monitoring Scheme. Research and Survey in Nature Conservation No. 6, Nature Conservancy Council, Peterborough, UK, 39 pp.

Nelson, R., Horning, N. and Stone, T.A. (1987) Determining the rate of forest conversion in Mato Grosso, Brazil, using Landsat MSS and AVHRR data. *International Journal of Remote Sensing*, **8**(12), 1767–1784.

Nilson, T. and Peterson, U. (1991) A forest canopy reflectance model and a test case. *Remote Sensing of Environment*, **37**, 131–142.

Pickup, G., Chewings, V.H. and Nelson, D.J. (1993) Estimating changes in vegetation cover over time in arid rangelands using Landsat MSS data. *Remote Sensing of Environment*, **43**(3), 243–263.

Prince, S.D. (1991) Satellite remote sensing of primary production: comparison of results for Sahelian grasslands 1981–1988. *International Journal of Remote Sensing*, **12**(6), 1301–1311.

Roesch, F.A., Van Deusen, P.C. and Zhu, Z. (1995) A comparison of various estimators for updating forest area coverage using AVHRR and forest inventory data. *Photogrammetric Engineering and Remote Sensing*, **61**(3), 307–311.

Schowengerdt, R.A. (1983) *Techniques for Image Processing and Classification in Remote Sensing*. Academic Press, London.

Schreuder, H.T., LaBau, V.J. and Hazard, J.W. (1995) The Alaska four-phase forest inventory sampling design using remote sensing and ground sampling. *Photogrammetric Engineering and Remote Sensing*, **61**(3), 291–297.

Stamp, L.D. (1962) *The Land of Britain: its Use and Misuse*. Longman, London.

Stewart, J.B., Barrett, E.C., Milford, J.R., Taylor, J.C. and Wyatt, B.K. (1987) Estimating rainfall and biomass for the pastureland zone of the West African Sahel. *Proceedings of the International Astronautical Federation Congress*, Brighton, UK.

Suffling, R. (1995) Can disturbance determine vegetation distribution during climate warming? A boreal test. *Journal of Biogeography*, **22**(2–3), 501–508.

Townshend, J.R.G. and Justice, C.O. (1986) Analysis of the dynamics of African vegetation using the normalized difference vegetation index. *International Journal of Remote Sensing*, **7**(11), 1435–1445.

Woodcock, C.E., Collins, J.B., Gopal, S., Jakabhazy, D., Li, X., Macomber, S., Ryherd, S., Harward, V.J., Levitan, J., Wu, Y., *et al.* (1994) Mapping forest vegetation using Landsat TM imagery and a canopy reflectance model. *Remote Sensing of Environment*, **50**(3), 240–254.

Section 2

SITE AND LOCAL SCALE

2 Cover Distribution Patterns of Lichens, Annuals and Shrubs in the Tabernas Desert, Almería, Spain

R. LÁZARO,[1] R. W. ALEXANDER[2] and J. PUIGDEFÁBREGAS[1]
[1]*Estación Experimental de Zonas Aridas (CSIC), Almería, Spain*
[2]*Environment Research Group, Department of Geography, Chester College, UK*

The Tabernas Desert, in semi-arid south-east Spain, supports a sparse and patchy vegetation of shrubs, tussock grasses, annuals and terricolous lichens. Both the topography and the vegetation exhibit sharp variations over fine spatial scales. Random sampling of the cover of each of the main plant life-forms (lichens, annuals and shrubs) stratified by landforms in each of the three main lithologies present, was carried out in order to establish the spatial relationships between life-form types and between life forms and environment. More than 500 samples were taken, in each of which a number of topographic, soil and surface feature variables were recorded. Geomorphology and topography were found to exert strong controls on most of the environmental variables measured, so association between landforms and plant life-form covers exists. The three life forms exhibited broadly the same preferences for environmental variables, but were differently demanding. This would explain the spatial segregation into patches, mainly between lichens and higher plants, and would also suggest temporal segregation, with lichens being the first colonisers. The lack of association between lichens and a number of erosion-related variables supports the suggestion of their role as primary colonisers but does not necessarily indicate that they are more resistant to erosion than the other life forms; rather that they are more able to colonise and survive in the most infertile locations. Occurrence of lichens on some slopes with a southerly orientation suggests that they are not limited simply by aspect but by the degree of erosion, which tends to be higher on south-facing slopes. The main factors controlling lichen, annual and shrub covers are established, and thus the basis is provided for mapping vegetation within the area in future.

PREAMBLE

Studies aimed primarily at enhancing our understanding of the ecology of patchy vegetation in small but diverse sites involve the collection, analysis and interpretation of sample data sets in order to identify patterns and trends in the distribution of the vegetation. Such studies frequently result in the explanation of distribution patterns rather than in the production of vegetation maps (cf. Chapter 18). The results, however, provide important information on vegetation–environment linkages that can enable the later production of predictive vegetation maps based upon ancillary data sources such as aerial photographs, remotely-sensed imagery and digital elevation models.

In this chapter we describe a study of the distribution pattern of vegetation types (based on life form) in the intensely dissected terrain of the Tabernas Desert in semi-arid south-east Spain. The scale of variation in both the topography and the vegetation make synoptic mapping difficult in such terrain but the identification of major vegetation–environment associations from detailed field sample data can inform a subsequent mapping exercise utilising secondary data sets with a coarser resolution.

INTRODUCTION

The Tabernas Desert is located some 20 km north of the city of Almería and 5 km south-west of the village of Tabernas (Figure 2.1). The area contains some of the most extensive badlands in south-east Spain. These badlands are cut into an uplifted sequence of Upper Miocene (Tortonian) marls, shales and turbidites with occasional interbedded sandstone bands, and occur at the juxtaposition of maximum uplift at the western end of the Tabernas basin with the downfaulted Rioja corridor (Alexander *et al.*, 1994). The basin uplift continued through the Pliocene and Quaternary leading to the developments of a landscape dominated by deep dissection. The area is thermo-mediterranean semi-arid with a mean annual precipitation of 218 mm (Lázaro and Rey, 1991; Solé-Benet and Alexander, 1996). The combination of climate, erosion and many centuries of human activity has led to the development of a sparse scrub vegetation consisting of shrubs and tussock grasses with an, often extensive, lichen crust at the soil surface. The deep dissection has created a complex suite of landforms with much variation in slope angle, length and aspect. Many steep, south-facing slopes are devoid of vegetation.

Work by Lázaro and Puigdefábregas (1994) and Lázaro (1995) has demonstrated that certain characteristics of the vegetation, particularly floristic composition, are significantly associated with landforms in these badlands. This chapter explores the spatial dimension of these relationships by examining the cover distribution of each physiognomic stratum of the vegetation in relation to a number of environmental variables. Although data relating to all vegetation strata are examined here, emphasis is placed on the terricolous lichens (considering them as a physiognomic group rather than in floristic terms) as they achieve high cover values in this environment.

Lichens play an important role in the area, particularly in terms of stabilisation mechanisms, slope hydrology (Solé-Benet *et al.*, 1997) and vegetation succession (Lázaro, 1995). They have been studied in relation to their effect on infiltration and runoff (Alexander and Calvo, 1990) and in the context of all of the vegetation (Alexander *et al.*, 1994). The latter paper established relationships between vegetation types and environmental features at the site. As the analysis was based on samples containing floristic data on both lichens and higher plants however, the role of lichens cannot be separately determined. Life-form data collected as part of the same investigation have been re-examined and these indicate that the different physiognomic groups appeared to show different patterns of occurrence in relation to slope facets. While

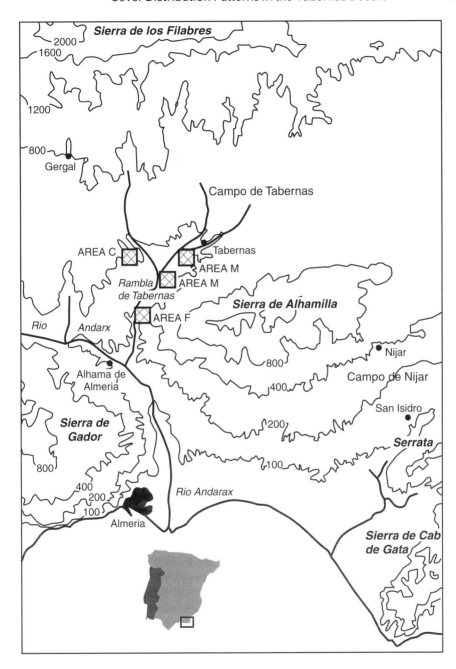

Figure 2.1 Location of the sample areas

lichens showed no marked preference for stable or eroded sites at the scale of sampling (3 × 3 m plots), shrubs were most abundant on the most stable sites and grasses/ annuals occupied an intermediate position (occurring with greater abundance on both stable and moderately stable slopes than on eroded slopes). This would seem to indicate that lichen cover is restricted due to light competition on more stable sites and may be equally restricted in the most eroded sites by the erosion processes themselves. In relation to slope gradient, shrub and grass/annual cover were greatest on the weakest slope angles whereas lichen cover values were higher on moderate and steep slopes. Data on orientation were inconclusive as few samples were collected on south-facing slopes due to their being largely devoid of any vegetation.

These and other observations suggest that the lichens of the Tabernas badlands form recognisable *sinusiae*, often more or less segregated from the rest of the vegetation (i.e. areas in which lichens are the dominant vegetation), and the role of these patches and their dynamics seem important (Lázaro, 1995), but have not been fully studied. In order to address the question, 'where is the terricolous lichen crust in the Tabernas desert?', it is thus essential to examine not only the distribution of the lichens but also that of the other main vegetation strata from which they are frequently segregated. Lázaro (1995) found therophyta and chamaephyta to be the most abundant life forms in the area accounting, between them, for 90% of the higher plant species (55% were therophyta and the rest mainly dwarf shrubs and perennial herbs). Lichens accounted for 25% of the total flora and hence the three 'strata' considered here are dwarf shrubs, annuals and lichens.

From previous work in the area (cited above) and field observations the following hypotheses can be formulated:

1. Lichens are the first colonisers.
2. Cover of the different plant forms (lichens, annual and shrubs) is associated with the landforms.
3. There is a certain degree of spatial segregation according to the dominant life form. This is due, at least in part, to the influence that cover by higher plants exerts on the lichens. Thus it is necessary to study the cover distribution of all main life forms in order to understand lichen cover distribution.
4. Although the majority of slopes receiving most insolation are devoid of lichens, indeed devoid of any vegetation, lichens (sometimes with higher cover values) do occur on slopes with a southerly orientation and in more or less horizontal and open places. Thus the cause of the frequent absence of lichens on south-facing slopes must be other than orientation alone.

METHODS

Three sampling areas were selected to represent the main lithologies present (Figure 2.1). Area 'M' represents the most extensive lithology, highly bioturbated, deep-water marls of the Chozas Formation of Tortonian age, and includes the badlands to the south and south-east of the Rambla de Tabernas (UTM 30S WF 4996); the same area as that investigated by Alexander *et al.* (1994). Area 'F' contains alternating marls and sandstones, and includes the catchment of the Rambla del Aguilón and surroundings (UTM 30S WF 4894); it is generally orientated to the west. Limestone and calcareous sandstones, corresponding to the Upper Miocene shoreline (IGME, 1975, 1983), occur in area 'C', which consists mainly of a small catchment orientated toward the east, and flowing into the Rambla de Lanujar (UTM 30S WF 4897). In each lithology we distinguished the landforms (or parts of landform, according to Peterson, 1981) that occurred using the typology shown in Table 2.1.

Two sets of vegetation samples were collected from each of the three sampling areas. The samples were located at random, stratified by landform, and at least 16 were taken from each landform of each sampling area. In each sample, the orientation and slope angle were measured, and cover values of shrubs, annual plants, lichens and stones were estimated (as a percentage, with the aid of a rigid and graduated metal square of 0.5 m side for the first set of samples and of 1 m side for the second). The first sampling included 152 samples (at least four in each landform of each area) of 1.5 × 1.5 m, in which the regolith thickness was approximated with a 1.8 m metal stake, and the upper 7 cm of the soil were taken to the laboratory for analysis. In the second sampling 352 samples of 2 × 3 m were taken, in which orientation, slope angle and cover values were recorded and regolith thickness, the intensity of erosive processes and the frequency of surface microforms (including salt deposits) were estimated by eye. The intensity of erosion by splash, concentrated runoff, diffuse runoff, mass movement and piping was recorded on a three-point scale using 1 for low intensity, 2 for medium and 3 for high intensity. Frequency of the surface microforms and salts was recorded in a similar fashion, using 1 for low frequency, 2 for medium and 3 for high frequency. In assessing erosion processes and surface microforms, all uncertainties between classes 1 and 2 or 2 and 3 were resolved as class 2. The surface microforms and salt deposits recorded were:

- Cracks
- Drop holes (holes of 0.5–2.0 cm diameter and a few millimetres depth, in lichens or crust, seemingly caused by raindrop impact)
- Micro clearings without crust (areas with no biotic or mineral crust, of 4–10 cm diameter and uncertain origin, set within a more extensive mat of lichen and/or cyanobacterial crust)
- Rills
- Micro mass movement marks
- Micro gullies
- Pipes
- Pedestals
- Castles (similar to pedestals but wider and capped by lichens rather than stones; in shape and size they often resemble a caramel cream, with more or less furrowed sides; castles require the existence of, at least, a layer of silt over the marls and, where they are abundant, they produce a crusted 'pop-corn' surface)
- Micro escarpments
- Salts (white patches of salt on the soil surface formed by evaporation)
- Gypsum (as either protruding veins or individual clasts)
- Micro deposits (areas, normally of a few square decimetres in extent, with evident accumulation of sediments from micro mass movement or micro gullies upstream)

Table 2.1   Landform typology

Landform	Label	Description
Control slope	CS	The nearest slope unaffected by the present drainage net
Headwaters	HW	Upper part of a catchment, upstream of the gully incisions
Stable slope	SS	Little erosion; Calcic Regosol
Semi-stable slope	MS	Eroded but original surface remaining
Eroded slope	ES	Slopes with the highest sediment output, or which have been strongly eroded previously and have a shallow regolith. Regosols or Leptosols
Eroded divide	ED	Top of an interfluve between contiguous gullies when practically nothing of the original surface between the gullies remains
Residual divide	RD	Top of an interfluve between contiguous gullies still having the original surface, at least *c.* 2 m wide, because the gullies have not yet met
Lower and upper frontal facet	LF, UF	Where parallel gullies flow into another of a superior order, the interfluve frequently forms a small, more or less triangular slope: the front interfluve facet, which marks the original position of the general slope. LF is the lower half of the facet and UF the upper half
Stable pediment	SP	Basal surface with a slight slope generated by deposition and with weak erosive processes. Deep Calcic Solonchaks. Micropediments of SS are not considered SP but are included in SS
Eroded pediment	EP	Pediment with evidence of actual processes either aggradational or degradational affecting more than 50% of its surface

For each soil sample, pH, electrical conductivity, $CaCO_3$, organic C, C/N ratio (from organic C and total N) and total S (to approximate the gypsum) were determined; the proportions of organic N and organic S in the soil were considered irrelevant.

For each environmental variable measured on a quantitative scale (six for the soil, plus slope angle and stone cover) a series of analyses of variance (one-way ANOVA) was performed. This was done using lithology and landform as factors, as well as both factors at the same time when possible, to show possible interactions. Because only four landforms (CS, SS, MS and ES) were represented in all three areas, the comparison between areas was made considering only the samples corresponding to those four landforms. ANOVA comparing landforms within lithologies were then carried out, including all samples from each area. Finally, as areas C and M had eight landforms in common, further ANOVA were carried out comparing areas C and M including those eight landforms.

Using data from the first set of samples, 10 contingency tables were evaluated by χ^2

to verify whether a significant association exists between landforms (into each area and as a whole, that is, including the samples from all three areas in each landform) or lithologies and the variables treated as qualitative: orientation and regolith thickness. To verify relationships between the different plant forms and the landforms, a Kruskal–Wallis test was carried out using the data from the second (larger) set of samples, on each plant form (lichens, annuals and shrubs), in which landforms were treated as an independent variable or factor.

Binary Discrimination Analyses (BDA) (Strahler, 1978) were used to examine the relationships between vegetation and environmental characteristics. Vegetation cover data were reduced to cover classes. BDA requires at least five samples from each variable and also that each value of the factor or independent variable has at least five samples. In order to retain the maximum amount of information from the data collected, a different number of cover classes was used for each of the different life-form groups (see Table 2.2) and the maximum number of cover classes permitted by the analysis was used for each group. Six BDA were carried out (two for each plant form considered: one per set of samples), using the vegetation cover classes as a factor and, as variables, the data (reduced to binary level) on soil, topography, regolith thickness and stone cover, erosion processes and surface microforms and salts. Reduction to binary level was achieved by breaking down the numeric range of each continuous variable into categories based upon its frequency distribution (see Table 2.3). The landforms were not included in the BDA to prevent possible redundancies. BDA was also carried out by Lázaro (1995) with the same variables, but using landform as a factor, to verify whether the free selection in the field of the landforms was too subjective, or whether the landforms were consistent with real environmental variation. The result was that the landforms are significantly associated with almost all variables, the only exceptions being two corresponding to surface microforms.

The BDA first produces a contingency table for each variable and all the factor values. Each table is evaluated for its significance by G statistics (which are similar to χ^2). In a second step BDA calculates Haberman's D for each variable and factor class, which are standardised residuals generated from G. The sign of D indicates whether the relationship between the variable and the factor is direct (+) or inverse (−). This relationship becomes stronger the higher the absolute value of D.

Both BDA made for each plant form have 10 variables in common and gave very consistent results. To avoid redundancies and save space, BDA results have been synthesised in one table per plant form, with each synthetic table including:

Table 2.2 Cover classes used in Binary Discriminant Analyses

Life-form	Cover classes (%)							
Lichens	0	$>0\leq2$	$>2\leq8$	$>8\leq20$	$>20\leq35$	$>35\leq55$	$>55\leq75$	>75
Annuals	0	$>0\leq2$	$>2\leq8$	$>8\leq20$	$>20\leq40$	$>40\leq65$	>65	
Perennials	0*	$>0\leq2$	$>2\leq8$	$>8\leq20$	$>20\leq35$	>35		

* Only in the first set of samples.

Table 2.3 Derivation of binary variables from the original numeric and rank variables

Numeric/rank variable	Label	Units	Binary category	Label	Range/ threshold
Topographic and soil variables					
Slope angle	SA	degrees	Weak slope	WEA	$< 20°$
			Moderate slope	MOA	$\geq 20° < 40°$
			Steep slope	STA	$\geq 40°$
Orientation	OR	degree	North orientation	NOR	$315°–45°$
			East orientation	EOR	$46°–134°$
			South orientation	SOR	$135°–225°$
			West orientation	WOR	$226°–314°$
Regolith thickness	RT	cm	Shallow regolith	SRT	≤ 75
			Medium regolith	MRT	$> 75 \leq 150$
			Deep regolith	DRT	> 150
Stone cover	SC	%	$> 25\%$ stone cover	S25	If $> 25\% = $ '1'
			$> 50\%$ stone cover	S50	If $> 50\% = $ '1'
Rock/large stone outcrop	RC	%	$> 50\%$ outcrop	R50	If $> 50\% = $ '1'
pH	pH		pH	pH	If $> 7.032 = $ '1'
Electrical conductivity	EC	mS cm^{-2}	Electrical conductivity	EC	If $> 1.829 = $ '1'
Calcium carbonate	CC	% weight/weight	Calcium carbonate	CC	If $> 19.81\% = $ '1'
Organic carbon	OC	% weight/weight	Organic carbon	OC	If $> 0.40\% = $ '1'
Carbon/nitrogen ratio	C/N	weight/weight	Carbon/nitrogen ratio	C/N	If $> 3.56 = $ '1'
Total sulphur	TS	% weight/weight	Total sulphur	TS	If $> 0.426\% = $ '1'
Erosion processes					
Splash	SPL	1–3 scale	Splash	SPL	If $> 1 = $ '1'
Concentrated runoff	CRF	1–3 scale	Concentrated runoff	CRF	If $> 1 = $ '1'
Diffuse runoff	DRF	1–3 scale	Diffuse runoff	DRF	If $> 1 = $ '1'
Mass movement	MMO	1–3 scale	Mass movement	MMO	If $> 1 = $ '1'
Piping	PIP	1–3 scale	Piping	PIP	If $> 1 = $ '1'
Surface microforms and deposits					
Cracks	CRA	1–3 scale	Cracks	CRA	If $> 1 = $ '1'
Drop holes	DHO	1–3 scale	Drop holes	DHO	If $> 1 = $ '1'
Micro clearing without crust	MCC	1–3 scale	Micro clearing without crust	MCC	If $> 1 = $ '1'
Rills	RIL	1–3 scale	Rills	RIL	If $> 1 = $ '1'
Micro mass movement marks	MMM	1–3 scale	Micro mass movement marks	MMM	If $> 1 = $ '1'
Micro gullies	MGU	1–3 scale	Micro gullies	MGU	If $> 1 = $ '1'
Pipes	HOL	1–3 scale	Pipes	HOL	If $> 1 = $ '1'
Pedestals	PET	1–3 scale	Pedestals	PET	If $> 1 = $ '1'
Castles	CAS	1–3 scale	Castles	CAS	If $> 1 = $ '1'
Micro escarpment	MES	1–3 scale	Micro escarpment	MES	If $> 1 = $ '1'
Salts	SAL	1–3 scale	Salts	SAL	If $> 1 = $ '1'
Gypsum	GYP	1–3 scale	Gypsum	GYP	If $> 1 = $ '1'
Deposit	DEP	1–3 scale	Deposit	DEP	If $> 1 = $ '1'

- the results corresponding to the variables included only in the first sampling (from the original first table),
- the results of the variables included only in the second sampling (from the original second table),
- and, in the case of 10 topographic variables included in both samplings, always the results from the second, because of its larger number of samples.

Finally, to synthesise the environmental variables, a Principal Component Analysis (PCA) was carried out for each life form, i.e. from each synthetic BDA result table, using Haberman's D values obtained from the BDA, the vegetation cover classes as individuals or samples and the variables significantly related to the vegetation in each case. The PCA were carried out using the SYN-TAX IV computer package (Podani, 1991).

RESULTS

Values of quantitative environmental characteristics from the first sampling (those which included soil data) are shown in Table 2.4. The quantitative variables were averaged acording to several criteria, and subjected to analysis of variance (ANOVA) to verify whether the landforms and lithologies are significantly different and whether there are interactions between lithology and landform. Table 2.5 summarises ANOVA results.

In the 10 contingency tables examining the association of landforms or lithologies with orientation and regolith thickness, all associations were significant at $P \leq 0.001$, except:

- landform of Area F/regolith thickness ($P \leq 0.01$),
- landform of Area M/regolith thickness ($P \leq 0.005$), and
- lithologies (areas)/regolith thickness ($P \leq 0.0025$).

From the second set of samples, quantitative values were recorded only for slope angle and stone cover. The corresponding ANOVA results showed no significant difference in average slope angle for the areas (although areas C and M differed in terms of slope angle at a low level of significance), but significantly different ($P < 0.001$) average slope angles for the landforms, both within each area and as a whole (i.e. including in each the samples from all three areas). For stone cover, the areas as well as the landforms (both within each area and as a whole) were significantly different at $P < 0.001$, though areas C and M were not significantly different from one another. Contingency tables constructed from the qualitative variables in the second set of samples (orientation, regolith thickness and erosion processes) were evaluated by χ^2. This involved 15 tests including, for each variable, areas, landforms within each area, and landforms as a whole (i.e. including the samples from all three areas in each). All associations proved significant ($P \leq 0.001$) with only the relationship between land-forms of area C and regolith thickness being significant at a lower level ($P \leq 0.01$).

The results of the Kruskal–Wallis tests, carried out to assess hypothetic associations among life forms and landforms, are shown in Table 2.6. Landforms were found to be

Table 2.4 Means and standard deviations of the quantitative variables (pH, electrical conductivity, calcium carbonate, organic carbon, carbon/nitrogen ratio, total sulphur, slope angle and stone cover) by landforms, including data from all three areas

Variable		pH	EC	CC	OC	C/N	TS	SA	SC
CS	Mean	7.834	0.344	20.04	0.703	5.202	0	24.1	37.9
	S.D.	0.429	0.526	7.436	0.293	1.27	0	6.2	27.5
HW	Mean	6.673	2.504	20.03	0.359	3.282	0.716	41.0	6.0
	S.D.	0.602	0.608	5.132	0.095	0.571	0.736	5.3	7.2
LF	Mean	7.133	1.714	19.69	0.443	3.871	0.586	29.2	16.8
	S.D.	0.686	0.956	6.07	0.114	0.559	0.57	7.9	28.5
UF	Mean	6.932	1.893	20.53	0.451	4.103	0.895	26.2	6.2
	S.D.	0.687	0.793	5.084	0.108	0.814	1.02	7.9	10.9
ED	Mean	6.313	2.44	31	0.298	2.228	1.586	26.8	36.0
	S.D.	0.225	0.367	3.225	0.111	0.857	0.769	9.9	32.4
RD	Mean	7.768	0.801	29.49	0.337	3.815	0.066	30.5	90.7
	S.D.	0.693	0.959	0.97	0.104	0.993	0.147	4.9	9.9
SP	Mean	7.668	0.442	11.49	0.426	4.452	0.09	7.0	3.3
	S.D.	0.634	0.523	0.996	0.036	0.362	0.124	5.3	2.0
EP	Mean	6.945	1.998	19.48	0.396	3.381	0.434	16.1	11.6
	S.D.	0.562	0.913	5.474	0.092	0.951	0.496	11.5	16.9
ES	Mean	6.643	2.446	26.98	0.275	2.454	0.556	44.1	26.5
	S.D.	0.596	0.947	6.981	0.096	0.876	0.755	5.9	33.1
SS	Mean	7.814	0.445	19.19	0.566	4.387	0.091	33.0	9.5
	S.D.	0.524	0.673	4.41	0.147	0.818	0.252	10.0	12.3
MS	Mean	7.052	1.465	27.94	0.355	3.189	0.264	34.9	45.8
	S.D.	0.632	0.932	9.867	0.099	0.895	0.357	9.2	35.6

significantly different ($P < 0.001$) in terms of the cover of each plant form considered (lichens, annuals and shrubs).

Tables 2.7 to 2.9 give a summary of the results of the several Binary Discriminant Analyses (BDA) carried out to examine the relationships between vegetation cover classes and environmental characteristics. Some erosive processes and surface micro-forms (splash, mass movement, micro gullies, marks of micro mass movement, micro escarpment, pedestals, drop holes and gypsum), as well as moderate slope angle, have not shown significant relationships to lichen cover (Table 2.7). Similarly, certain variables do not show significant relationships to annual or shrub cover (Tables 2.8, 2.9) but are retained in the tables so that comparisons can be made (though they were removed prior to the subsequent PCA analyses). As the variables piping, pipes and salts occurred in less than five samples it was not possible to include them in the BDA and they are not shown in the tables of results.

The PCA carried out for each of the synthetic tables (Tables 2.7, 2.8 and 2.9) show that only three factors account for an important amount of the total variance in all cases (Table 2.10). The component scores, the variance of variables accounted for by each factor and the component correlations of each variable, allow ecological interpretation of the main factors. There are small differences in the nature of the main factor depending on the life form: lichen cover is favoured more by easterly orientation than by organic matter, whereas organic matter is more important for annuals and also for shrubs (for which easterly orientation is in the negative part of factor 1).

Table 2.5 Significance of each ANOVA. In the first group the three sampling areas are compared using both factors at the same time, but including only four landforms, because only four are common to the three areas. Areas C and M have eight landforms in common and the last group compares these two areas with the data from eight landforms. The central group uses only the landform factor, in each area. Quantitative variables are the same as in Table 2.4

Group	Factor	pH	EC	CC	OC	C/N	TS	SA	SC
Litho-Landf C, F & M	Litho	0.0053	0.0646	0.0000	0.4168	0.4092	0.0013	0.0007	0.0000
4 common landf	Landf	0.0046	0.0000	0.0217	0.0001	0.0007	0.0000	0.0000	0.1410
	Litho & Landf	0.7000	0.5681	0.0018	0.4941	0.3299	0.1355	0.0226	0.0000
Landf into each Litho	C Landf	0.0046	0.0004	0.2329	0.0000	0.2489	0.6087	0.0000	0.0518
	F Landf	0.0011	0.0000	0.0000	0.0000	0.0000	0.0143	0.0000	0.0435
	M Landf	0.0132	0.0000	0.2940	0.0013	0.0078	0.8204	0.0002	0.0000
Litho-Landf C & M	Litho	0.2327	0.3391	0.0000	0.0000	0.3468	0.1244	0.8574	0.0010
8 common landf	Landf	0.0005	0.0000	0.6185	0.0000	0.0033	0.0159	0.0000	0.1829
	Litho & Landf	0.3179	0.1597	0.3134	0.1659	0.5173	0.2667	0.0000	0.0002

Table 2.6 Results of Kruskal–Wallis tests on life-form with landform associations

Life-form	d.f.	N	H	p
Lichens	9	352	124.510	0.0000
Annuals	9	352	114.209	0.0000
Shrubs	9	352	182.038	0.000

However, schematically, factor 1 is of a similar nature in the three PCA. It represents a gradient between the most stable, topographically favoured sites with more developed soil and the most unstable or eroded sites with unfavourable topography and shallow, more or less saline regolith, which is often cracked (Figure 2.2). For the three life forms, the greater covers are in the positive part of factor 1 (but not always in strict order) and, the lesser, in the negative part.

Factors 2 and 3 do not seem to be similar for the different life forms, and do not have a clear ecological interpretation in any of the PCA.

DISCUSSION

The differences among the three sampling areas are due mainly to the characteristics of area F. It has an average of 31.2% $CaCO_3$, while area M has 23.9% and area C only 13.4%. Similarly, the stone cover is 58% in area F, but only 14% in area M and 7% in area C. Also, although a band of sandstone at or close to the surface makes some parts of area F more stable than the other areas, its northern slopes are generally less stable and have lower vegetation cover, due to their situation on scarp rather than dip slopes. Despite this singularity of area F, there are greater overall differences in environmental characteristics between landforms than between lithologies.

Lithology and landform show interactions in three variables: $CaCO_3$, slope angle and stone cover ($CaCO_3$ only when all three areas are compared, but the other two variables also when areas C and M are compared). The model of the relationships of these variables with the landforms differs depending on the lithology.

Because it is easily leached, the surface distribution of $CaCO_3$ in the Tabernas badlands generally shows an inverse pattern to infiltration capacity. Thus landforms with high rates of runoff have more $CaCO_3$ in the surface layer. As infiltration depends not only on slope angle, local factors, such as the particle size and vegetation, may change the pattern of $CaCO_3$ in the landforms in area M. The pattern of slope angle varies among the three sample areas due to the existence of eroded pediments in area C, but, in general, it increases with erosion. The pattern of stone cover varies with the lithology because of the greater abundance of sandstone in area F.

Slope orientation and regolith thickness are related to both landforms and sample areas (lithologies). Association of regolith thickness with sample areas is logically due to area F. The association of orientation with sample areas hinders discrimination of its effect from that of lithology. Thus the results for orientation are valid only within each sample area.

Table 2.7 Synthesis of results of Binary Discriminant Analyses based on cover classes of lichens.
* significant at p < 0.1; ** significant at p < 0.05; *** significant at p < 0.01; (NS) not significant

Variable	Freq.	G	Signif.	% cover							
				0	>0 ≤ 2	>2 ≤ 8	>8 ≤ 20	>20 ≤ 35	>35 ≤ 55	>55 ≤ 75	>75
WEA	108	18.53	***	−2.50	−1.67	−1.13	0.41	0.84	1.95	2.26	0.34
MOA	144	4.50	(NS)	0.18	−1.37	1.34	0.29	0.92	−0.29	−0.7	0.02
STA	67	28.59	***	2.68	3.62	−0.33	−0.84	−2.11	−1.91	−1.77	−0.42
NOR	135	34.46	***	−2.21	−3.99	−1.60	1.72	1.22	1.52	1.50	2.61
EOR	55	31.78	***	−2.63	−1.86	−1.42	−1.63	1.05	1.88	2.1	2.73
SOR	53	43.33	***	2.07	3.89	1.79	0.32	−2.27	−1.84	−2.21	−2.75
WOR	76	32.20	***	3.09	2.87	1.55	−0.83	−0.36	−1.82	−1.67	−3.05
SRT	25	59.40	***	7.60	1.03	1.05	−1.92	−1.49	−1.83	−1.77	−2.32
MRT	45	16.29	**	0.42	1.60	0.99	−0.18	1.59	−2.03	0.18	−2.37
DRT	249	53.09	***	−5.29	−2.01	−1.51	1.39	−0.38	2.89	1.00	3.50
S25	32	43.28	***	−2.32	−2.29	2.08	3.11	2.17	2.25	−1.27	−2.61
S50	72	78.29	***	1.48	4.53	3.37	1.00	−1.23	−2.53	−3.28	−4.30
R50	15	26.24	***	−0.37	1.98	3.93	−1.46	−0.13	−1.39	−1.35	−1.77
pH	46	21.24	***	−1.22	−3.41	−0.24	1.14	−0.16	0.59	2.38	1.55
EC	67	41.48	***	3.37	2.99	1.21	−1.12	−1.40	−0.32	−3.59	−1.63
CC	78	29.35	***	2.77	−1.31	−0.70	0.98	1.86	0.94	0.10	−3.67
OC	52	26.30	***	−0.40	−2.99	−1.79	0.13	0.76	−0.53	3.59	1.63
C/N	54	31.07	***	−2.89	−0.87	−1.34	0.46	−0.63	−1.45	3.44	2.56
TS	41	24.34	***	4.03	−0.13	0.70	−1.49	0.14	0.82	−2.42	−0.86
SPL	134	6.65	(NS)	−1.79	0.03	0.12	−1.04	1.26	0.83	0.80	−0.08
CRF	68	72.39	***	3.09	5.93	1.64	−1.73	−1.62	−3.26	−1.81	−3.42
DRF	22	12.68	*	0.70	1.98	0.07	0.26	1.13	−1.00	−1.65	−1.58
MMO	20	3.78	(NS)	−0.70	1.15	0.25	−0.99	0.43	0.60	−0.81	−0.20
CRA	75	49.27	***	5.40	3.28	−0.34	−1.20	−2.32	−0.94	−1.20	−3.00
DHO	28	5.16	(NS)	−1.11	−0.80	0.77	−0.82	0.67	1.22	0.72	−0.35
MCC	19	16.96	**	−1.45	−1.07	−1.75	−0.19	1.41	2.96	0.03	0.54
RIL	55	65.68	***	3.47	5.54	0.77	−1.63	−0.64	−2.39	−2.28	−3.64
MMM	12	7.51	(NS)	−1.14	0.44	−1.37	0.52	1.22	−0.30	−0.23	0.80
MGU	5	8.01	(NS)	−0.73	−1.13	−0.88	0.56	−0.64	2.09	0.71	0.21
PET	30	10.61	(NS)	0.77	0.47	1.73	0.84	−0.19	−0.18	−1.32	−2.05
CAS	45	15.80	**	−2.33	−1.22	2.41	−0.18	0.98	0.55	0.18	−0.21
MES	25	10.84	(NS)	−0.96	2.07	−0.18	−1.92	0.09	0.84	−0.4	−0.09
GYP	10	10.36	(NS)	−1.04	−0.81	2.55	−0.19	0.30	0.93	−0.04	−1.43
DEP	26	35.11	***	0.39	5.51	−0.26	−1.32	0.03	−1.87	−1.14	−2.37

Table 2.8 Synthesis of results of Binary Discriminant Analyses based on cover classes of annuals.
* significant at $p < 0.1$; ** significant at $p < 0.05$; *** significant at $p < 0.01$: (NS) not significant

Variable	Freq.	G	Signif	\% cover						
				0	>0 ≤ 2	>2 ≤ 8	>8 ≤ 20	>20 ≤ 40	>40 ≤ 65	>65
WEA	108	19.28	***	−1.91	−2.95	2.17	1.93	−0.34	1.48	−0.03
MOA	144	13.62	**	−2.43	1.75	0.31	−1.49	−0.43	−0.60	1.07
STA	67	35.86	***	5.19	1.30	−2.90	−0.42	0.91	−0.99	−1.28
NOR	135	28.85	***	1.58	−4.33	1.11	1.39	1.86	0.21	2.89
EOR	55	16.03	**	−1.22	−2.88	1.84	1.94	0.52	0.90	−1.13
SOR	53	24.05	***	0.86	3.99	−0.99	−1.94	−2.00	−1.51	−1.10
WOR	76	22.30	***	−1.50	4.08	−2.05	−1.63	−0.87	0.27	−1.38
SRT	25	33.72	***	6.33	1.81	−2.61	−1.21	−0.20	−0.98	−0.72
MRT	45	5.48	(NS)	0.01	1.28	−0.81	−1.07	1.07	−0.49	−1.00
DRT	249	28.22	***	−4.12	−2.25	2.38	1.68	−0.77	1.05	1.31
S25	32	24.47	***	−2.51	0.47	1.24	1.64	1.45	−1.78	−1.78
S50	72	43.93	***	−1.44	5.74	−1.15	−2.51	−2.58	−1.82	−1.34
R50	15	23.68	***	−0.59	4.24	−1.99	−1.77	−1.25	−0.75	−0.55
pH	46	41.44	***	−2.15	−3.38	−1.76	2.18	1.14	2.19	3.69
EC	67	47.88	***	3.09	3.66	1.35	−2.56	−2.11	−1.85	−3.32
CC	78	18.97	***	3.00	0.31	0.07	0.29	−1.09	−1.73	−1.73
OC	52	14.33	**	−1.42	−2.46	0.05	0.18	2.11	1.11	1.85
C/N	54	26.81	***	−3.22	−1.91	−0.06	0.01	2.92	1.74	1.74
TS	41	20.95	***	3.39	0.94	0.66	−1.28	−1.49	−0.58	−2.12
SPL	134	16.83	***	−1.50	0.09	2.58	−0.08	−1.26	−1.01	−2.10
CRF	68	19.98	***	2.34	3.08	−1.58	−1.21	−1.51	−1.01	−1.29
DRF	22	7.68	(NS)	−0.73	1.65	0.32	−1.58	0.00	−0.92	−0.67
MMO	20	11.34	*	2.46	1.69	−1.16	−0.82	−1.46	0.39	−0.64
CRA	75	30.97	***	4.83	0.79	−1.33	−1.58	1.46	−1.15	−1.37
DHO	28	7.85	(NS)	−0.83	0.76	1.18	−1.41	0.31	−1.05	−0.77
MCC	19	12.67	**	−0.67	−0.41	3.06	−0.74	−1.42	−0.85	−0.62
RIL	55	16.61	**	1.81	2.77	−2.60	−0.85	0.00	−0.73	−1.13
MMM	12	10.23	(NS)	3.49	0.23	−1.03	0.80	−1.12	−0.67	−0.49
MGU	5	8.30	(NS)	−0.34	−1.21	−1.13	0.21	2.42	2.04	−0.31
PET	30	5.93	(NS)	−0.86	−0.39	0.47	0.01	0.49	2.07	−0.80
CAS	45	8.98	(NS)	−1.08	0.32	1.19	−0.64	0.51	−1.37	−1.00
MES	25	11.44	*	2.06	0.55	1.03	−0.65	−1.65	−0.98	−0.72
GYP	10	6.54	(NS)	−0.48	0.21	1.60	−1.43	0.10	−0.61	−0.44
DEP	26	23.91	***	−0.80	3.63	−0.11	−2.37	−1.68	−1.01	−0.74

Table 2.9 Synthesis of results of Binary Discriminant Analyses based on cover classes of shrubs. * significant at p < 0.1; ** significant at p < 0.05; *** significant at p < 0.01; (NS) not significant

Variable	Freq.	G	Signif	% cover				
				>0 ≤ 2	>2 ≤ 8	>8 ≤ 20	>20 ≤ 35	>35
WEA	108	18.42	***	−3.59	−0.05	2.72	1.06	−0.34
MOA	144	9.76	**	−1.99	−0.14	−0.67	1.24	2.31
STA	67	48.57	***	6.59	0.23	−2.34	−2.75	−2.43
NOR	135	16.94	***	−2.36	−1.64	0.14	3.22	1.47
EOR	55	8.23	*	1.24	1.46	−0.15	−1.76	−1.55
SOR	53	7.01	(NS)	1.03	0.39	0.73	−0.85	−2.00
WOR	76	4.34	(NS)	0.74	0.26	−0.68	−1.43	1.41
SRT	25	11.37	**	3.23	−2.20	−0.82	0.38	−0.20
MRT	45	9.16	*	1.73	0.61	−0.06	−2.45	−0.05
DRT	249	12.85	**	−3.55	0.92	0.58	1.82	0.17
S25	32	13.55	**	−2.10	0.14	2.47	0.74	−0.58
S50	72	21.78	***	1.02	2.65	−0.72	−1.21	−3.05
R50	15	2.85	(NS)	−0.61	−0.99	0.39	0.29	1.51
pH	46	40.83	***	−3.34	−2.62	3.45	1.79	3.17
EC	67	45.82	***	3.00	2.69	−3.73	−2.04	−2.85
CC	78	12.64	**	−1.24	0.76	−0.62	1.19	−1.70
OC	52	19.85	***	−2.12	−0.42	1.02	1.40	2.85
C/N	54	27.73	***	−2.31	−0.16	1.13	1.91	2.76
TS	41	22.96	***	0.32	2.08	−2.21	−1.19	−1.82
SPL	134	42.25	***	2.85	4.04	−2.24	−2.73	−3.62
CRF	68	14.74	***	2.69	1.02	−0.44	−2.07	−1.99
DRF	22	4.55	(NS)	−1.27	0.49	1.60	−1.05	0.00
MMO	20	15.31	***	3.57	−0.66	−0.28	−2.11	−0.66
CRA	75	60.57	***	7.15	0.36	−2.19	−3.47	−2.67
DHO	28	12.08	**	−0.72	2.61	0.32	−1.48	−1.75
MCC	19	9.37	*	0.78	1.52	−1.86	0.45	−1.42
RIL	55	11.34	**	2.74	0.82	−1.21	−1.37	−1.55
MMM	12	4.98	(NS)	0.50	0.13	0.91	−1.61	−0.09
MGU	5	3.62	(NS)	1.17	0.40	−0.14	−1.03	−0.71
PET	30	22.26	***	4.44	0.21	−2.19	−2.12	−0.49
CAS	45	28.80	***	2.54	2.33	−0.83	−3.30	−1.73
MES	25	18.27	***	0.60	2.73	−0.32	−2.38	−1.65
GYP	10	11.59	**	3.29	−0.81	−0.20	−1.47	−1.02
DEP	26	3.34	(NS)	−1.06	0.78	1.04	−0.26	−0.97

Within each sample area, the landforms are significantly related to both orientation and regolith thickness, as well as to slope angle and many of the soil characteristics. This fact explains the particular spatial distribution of the vegetation, which shows discontinuous patches and, consequently, also the pattern of erosion, which overall looks like a negative of the vegetation pattern. Therefore, the Tabernas badlands, in spite of their limited extent, are very rich in microhabitats due to their complex geomorphology.

Southerly, westerly, easterly and northerly orientations, respectively, show positive associations with progressively higher classes of cover by both lichens and annual

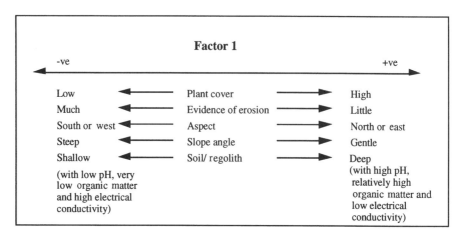

Figure 2.2 Schematic interpretation of the environmental gradient represented by PCA factor 1

plants, suggesting variations in water availability with aspect. Although there are lichens on the south-facing slopes, cover is rarely more than 25%. Shrub cover is not significantly associated with southerly or westerly orientations, but a northerly orientation is favourable and an easterly orientation, relatively unfavourable. However the closest positive association with a northerly aspect is in the 20–35% cover class; therefore shrub covers over 35% are limited by other factors that cannot be replaced by favourable orientation. An examination of the samples with more than 35% shrub cover shows that in most cases they correspond to the control slope (CS), a certain number to the lower part of SS and, some to LF or SP. Thus, for shrubs to attain more than 35% cover, a favourable orientation is insufficient; time and a minimum contributing area to increase soil moisture are also necessary. Dense patches of shrubs occur in concave sections of slopes and often extend upslope above concavities.

The effects of slope angle and regolith thickness are similar for all life forms. Steep slopes and thin regolith are associated with low cover or bare soil, while the highest covers are related to deep regoliths and medium or weak slopes. Slopes steeper than 40° seem, in general, to limit covers of more than 8%, in lichens as well as higher plants. However annual plants are more demanding than lichens, and shrubs more demanding than annuals. This trend is reflected in the BDA tables (Tables 2.7, 2.8 and 2.9), which show that shrubs have no significant association with four erosion-related variables, annuals with seven and lichens with eight. Thus the influence of erosion is greatest on shrubs and least on lichens. The observed segregation of lichen *sinusiae* on relatively steep slopes is likely to be caused by competition and supports the suggestion that terricolous lichens are the first colonisers.

Stone cover of between 25 and 50% favours colonisation by vegetation, probably acting as a mechanism of soil stabilisation (Calvo *et al.*, 1991; Alexander *et al.*, 1994), but greater cover by stones limits colonisation as the stones take up more space. In this situation lichens reach no more than 55%, annuals 40% and shrubs 35% cover. Stone cover values greater than 50% are strongly associated with lower vegetation cover,

Table 2.10 Synthesis of results of Principal Component Analyses based on BDA results: variance accounted for by each factor in each analysis

	Cumulative percentage of eigenvalues		
	Factor 1	Factor 2	Factor 3
Lichens	64.73	75.77	85.44
Annuals	55.13	75.32	91.12
Shrubs	67.87	87.18	95.28

limiting lichens to 20%, annuals to 2% and shrubs to 8%. A sandstone stratum beginning to surface has the same effect on annuals, limits lichens to 8% but shows no association with shrubs, the roots of which frequently cross the sandstone stratum through the cracks.

The absence of, or only slight cover by, either lichens or annuals is associated with a high concentration of gypsum (total sulphur), electrical conductivity and calcium carbonate, very little organic matter and pH lower than 7, while high cover by both layers of vegetation is associated with the opposite conditions. However, lichens endure salts better. The different relationships with the variables, particularly the soil chemical variables, displayed by the different classes of lichen cover suggest that there are several lichen communities, more or less associated with the levels of cover. Shrubs and annual plants have similar relationships with soil chemical variables, except for calcium carbonate, where association with shrub cover is not sufficiently clear.

Piping is both infrequent and localised in the Tabernas badlands. This process was recorded in less than five samples and, as a consequence, eliminated from all analyses. The erosion processes most associated with lichen cover are the formation of rills and deposition of sediments. Evidence of erosion by sheet flow has a significant relationship only with the lichens, with medium and low covers. Neither mass movement nor splash are significantly associated with the classes of lichen cover. However, this result should not be taken at face value. Few samples were located on the upper parts of eroded slopes, where most mass movement is concentrated, precisely because there are no lichens nor higher plants to sample in such locations. On the other hand, the marks of drop impact can normally only be seen for sure on the lichen or pre-lichen (cyanobacterial) crust. But statistically the drop impact is homogeneously or almost homogeneously distributed, even if it does not produce the same effect in all sites. It is only the holes in the lichen crust which have an irregular distribution. Thus all classes of lichen cover are more or less related to splash and lichens seem unaffected by drop impact. This cannot be entirely the case as drop impact has the effect of producing holes in lichens, but it seems that the lichen crust as a whole is resistant and can increase cover against splash erosion. With annual plants, sediment deposition and rill formation are also very important (associated positively with low cover), but mass movement is less significant. The positive association of low cover of annual plants with drop impact is probably, in reality, with high cover of lichens; the higher plants protect the lichens from the drops, and only where the cover of annuals is low and that of lichens is high is it easy to see drop impact marks in the lichen crust.

Mass movement, drop impact (the same consideration as that with annuals) and also rill formation are positively associated with low covers of shrubs, but there is no significant relationship between shrubs and sedimentation.

The general trend in the relationships between the different erosion processes and the life forms is for positive association with low plant cover values and negative association with high plant cover values. However the individual cover class most closely associated with any particular process varies depending on the process and the life form involved. The fact that only two erosion processes have an important relationship with lichen cover does not necessarily lead to the conclusion that lichens are more resistant to erosion. Lichens are competitively displaced to places that are more eroded than those occupied by the higher plants and thus, if extensive as well as low cover by lichens is found at relatively eroded sites, they will not show a significant relationship with erosion processes. This situation in the relatively more eroded sites shows that lichens are able to colonise and survive in the driest and most infertile locations, but not that lichens are more resistant to erosion. Observational evidence suggests that the terricolous lichen crust is, in fact, quite fragile. For example, the marks of the shoes or similar persist for a number of years and the very abundance of lichens itself demonstrates the absence of livestock.

The surface microform most closely associated with lichens (negatively to cover) is crack formation. The same is true of the annuals, though they are also related to micro rills; but dense covers by annuals are not strongly associated with any microform. These relationships are probably indirect. The soil surface cracks when it is bare, so the frequency of cracks or micro rills is proportional to the amount of bare soil and, as a result, is associated with low covers. The protective role of vegetation can be seen here: the lichen crust cracks only occasionally; cracks are rare in sites with high lichen or annual cover; in fact, lichen cover of more than 8% already shows negative association with cracks.

Several microforms are significantly related to the cover of shrubs: cracks and pedestals, in particular, show strong relations. The association is with low covers (0–2% and 2–8% classes) and the samples involved correspond to 'resistant' (Lázaro, 1995) vegetation. (Lázaro (1995), defines 'resistant' plants as those which are specialists in surviving in eroded and unstable sites, almost without soil, producing unappreciable changes in the environment.) This indicates that shrubs are found in less eroded sites than annuals (and annuals in less eroded sites than lichens); so, as a result, evidence of erosion is associated with low shrub cover.

The lack of association between lichens and pedestals, as well as several field observations (e.g. fast lichen colonisation on the vertical surface of a soil profile), could indicate that the growth rate of lichens is faster than might be anticipated in such a dry climate.

The PCA results (Table 2.10) suggest strong control by geomorphology and topography over most of the environmental variables, since almost all variables must be interrelated as the first factor always accounts for between 55 and 68% of total variance. These results also support the interpretation that the three life forms have broadly the same preferences, though they are differently demanding, since factor 1 is essentially the same in the three cases.

The partial spatial segregation among life forms observed in the field and in the data from Alexander *et al.* (1994) would indicate that the cover of each, and particularly that of the lichens, depends not only on the environmental variables but also on the cover of the other life forms. This spatial segregation would appear to be due to competition, mainly between lichens and higher plants, since the latter intercept both light and dew. A linear regression between annual and lichen covers shows significant association at $P < 0.05$, but between shrubs (which produce greater interception) and lichens there is no significant association. This seemingly anomalous result is interpreted as being due to the sampling scale: the sample plots are of several square metres and thus can contain medium or high cover of lichens and, simultaneously, high cover of shrubs, since lichens do not occur beneath shrubs but are often abundant in the spaces between them. Further work is required to provide a clearer explanation of the vegetation mosaic (at scales between a few and some tens of square metres) in which significant covers of annuals and shrubs, and at times lichens, occur simultaneously, but with little or no overlap or stratification.

This detailed analysis of life form–environment relationships thus highlights the difficulties of mapping vegetation in such complex terrain but, at the same time, provides a basis from which mapping can be approached using ancillary data such as low level (airborne) imagery, aerial photographs and digital elevation models. Further work is in progress both to investigate this approach to mapping the distribution patterns of life forms and to examine the distribution and structure of plant communities in the area using floristic data.

CONCLUSIONS

The results of these analyses demonstrate the rich variety of microhabitats that exist in these badlands as a consequence of the complex geomorphology. They also provide an explanation for the distribution patterns exhibited by the main physiognomic types of vegetation in terms of the topographic, edaphic and erosional characteristics of each site. The landforms are significantly different in terms of the lichen, annual and shrub vegetation that they support. Further, the results provide at least a conceptual map of the distribution pattern of the lichen crust.

Spatial segregation occurs both between the discontinuous, patchy vegetation and zones of erosion, and also, though more lightly, among the dominant life form types within the vegetation patches. All life forms seem ultimately to prefer the same environmental conditions and, where favourable conditions occur, shrubs become the dominant vegetation albeit with relatively low cover values in this extreme environment. Annuals occupy an intermediate position in terms of site stress, with lichens displaced to dry and infertile sites and the vacant patches between shrubs and annuals in the better sites. However, in the most extreme sites erosion dominates and there are no, or scarcely any, lichens, only more or less isolated individuals of 'resistant' higher plants such as *Salsola genistoides* and *Moricandia foetida*.

The evidence presented indicates that lichens are the first colonisers in that they are able to colonise and survive in the most infertile locations, but not that lichens are more resistant to erosion; in fact, the terricolous lichen crust seems quite fragile. Their occurrence on some slopes with a southerly aspect suggests that it is not aspect alone which limits them but rather the degree of erosional activity, which tends to be higher on south-facing slopes. The main erosion processes restricting lichen cover are mass movement and concentrated runoff, whereas lichens are able to increase cover in the presence of splash erosion.

After (relatively) stable sites with limited erosion processes or gentle slope gradients and a minimum depth of regolith (associated with low erosion), organic matter, basic pH, a favourable orientation (to north and east, in order to have light during dew persistence) and, probably, the cover of higher plants, are the main factors controlling the lichen cover. The different relationships that classes of lichen cover have with the environmental variables, particularly the soil chemical variables, suggest that there are several lichen communities, more or less associated with the kinds of cover. The higher plants similarly require a low erosion rate, but also a minimum degree of soil development and the greater moisture availability afforded by slope concavities and other locations favoured by increased runoff. Overall the shrub stratum is slightly more demanding than the annual one but, in comparison with the lichens, the differences are small, and often unclear, between typical sites with dense annual cover and those with relatively dense shrub cover.

ACKNOWLEDGEMENTS

The research for this chapter was carried out as part of three consecutive collaborative research projects funded by Spanish CICYT, through its 'Plan Nacional de I + D' (projects NAT 89-1072-C06-04, AMB93-0844-C06-01 and AMB95-0986-C02-01) as well as others funded by the European Community: MEDALUS III, (CE)ENV4-CT95-0118, (CSIC) 0523/FF. We are grateful to Dr Gabriel del Barrio for a program to compute the Binary Discriminant Analysis, Dr Gerardo Sánchez for the map in Figure 2.1, and to the two referees for their many constructive comments on an earlier version of the manuscript.

REFERENCES

Alexander, R.W. and Calvo, A. (1990) The influence of lichens on slope processes in some Spanish badlands. In: J.B. Thornes (ed.) *Vegetation and Erosion*. Wiley, London.

Alexander, R.W., Harvey, A.M., Calvo, A., James, P.A. and Cerda, A. (1994) Natural stabilisation mechanisms on badland slopes: Tabernas, Almería, Spain. In: A.C. Millington and K. Pye (eds) *Environmental Change in Drylands: Biogeographical and Geomorphological Perspectives*. Wiley, Chichester.

Calvo, A., Harvey, A.M., Paya-Serrano, J. and Alexander, R.W. (1991) Response of badland surfaces in SE Spain to simulated rainfall. *Cuaternario y geomorfologia*, **5**, 3–14.

IGME (1975) *Mapa geológico de España 1:50000. Hoja 1030 (23–42), Tabernas*. Servicio Publicaciones Ministerio Industria, Madrid.

IGME (1983) *Mapa geológico de España 1:50000. Hoja 1045 (23–43), Almería*. Servicio Publicaciones Ministerio Industria, Madrid.

Lázaro, R. (1995) Relaciones entre vegetación y geomorfología en el área acarcavada del Desierto de Tabernas. Unpublished PhD thesis, Valencia, 244 pp. + annexes.

Lázaro, R. and Puigdefábregas, J. (1994) Distribución de la vegetación terofítica en relación con la geomorfología en áreas acarcavadas cerca de Tabernas, Almeŕa. *Monografías Flora y Vegetación Béticas*, **7/8**, 127–154.

Lázaro, R. and Rey, J.M. (1991) Sobre el clima de la provincia de Almería (SE Ibérico): Primer ensayo de cartografía automática de medidas anuales de temperatura y precipitación. *Suelo y Planta*, **1**(1), 61–68.

Peterson, F.F. (1981) *Landforms of the Basin and Range Province. Defined for Soil Survey*. Nevada Agricultural Experiment Station, Technical Bulletin, 28. University of Nevada, Reno.

Podani, J. (1991) SYN-TAX IV. Computer programs for data analysis in ecology and systematics. In: E. Feoli and L. Orlóci (eds) *Computer Assisted Vegetation Analysis*. Kluwer Academic Publishers, The Netherlands, pp. 437–452.

Solé-Benet, A. and Alexander, R. (1996) Contemporary processes in the Tabernas Basin, SE Spain. In: A.E. Mather and M. Stokes (eds) *2nd Cortijo Urra Field Meeting, SE Spain: Field Guide*. University of Plymouth, UK.

Solé-Benet, A., Calvo, A., Cerdá, A., Lázaro, R., Pini, R. and Barbero, J. (1997) Influence of micro-relief patterns and plant cover on runoff related processes in badlands from Tabernas (SE Spain). *Catena*, **31**, 23–38.

Strahler, A.H. (1978) Binary discriminant analysis: a new method for investigating species–environment relationships. *Ecology*, **59**(1), 108–116.

3 The Use of Microscale Field Mapping in a Study of the Ox-eye Daisy (*Leucanthemum vulgare* L.) as a Component of Wild Flower Meadows

P.J. SPEARMAN,[1] **R.W. ALEXANDER,**[2] **and I.D.S. BRODIE**[3]
[1] *Department of Biology, Anglia Polytechnic University, Cambridge, UK (Present address: Otley College, Suffolk, UK)*
[2] *Environment Research Group, Department of Geography, Chester College, UK*
[3] *Department of Biology, Anglia Polytechnic University, Cambridge, UK*

The sowing of wild flower seed on former arable land has become more common in recent years. Concern has been expressed about the ecotypic variability and performance differences of commercial seed mixtures. Permanent quadrats were established at three sites sown with wild flower seed mixes and individual *Leucanthemum vulgare* plants were tagged and monitored throughout the 1994 growing season. Data analysis was aided by digital mapping of the quadrats. Sites had different densities of plants present, with the oldest site having the lowest mean number of plants per square metre. With only one exception, seedlings and mature plants showed a clumped pattern of dispersion, although there was no association between clumps of seedlings and clumps of mature plants. The persistence of mature plants was also found to differ between sites and between areas of different seed type on an individual site. The variation shown here may support current concerns about the variable performance of seed from different ecotypes used in commercial wild flower seed mixtures.

INTRODUCTION

Many ecologists are concerned with the fate of individuals of a species over time. This chapter reports research involving the ox-eye daisy, *Leucanthemum vulgare* L. This species is a native perennial of the British Isles, occurring on a range of soil types, but being more successful in drier areas such as chalk grasslands and meadows (Howarth and Williams, 1968). Such habitats have suffered major decline (e.g. the loss of an estimated 95% of unimproved grasslands, Anon., 1984) since the Second World War as a result of the intensification of modern agriculture.

In recent years the reseeding of areas of formerly arable land, roadside verges and land adjacent to developments has become more common and, as a result, there has been a rise in demand for wild flower seed. It is believed that this has resulted in the sale

Vegetation Mapping: From Patch to Planet. Edited by Roy Alexander and Andrew C. Millington.
© 2000 John Wiley & Sons Ltd.

of non-native seed (Akeroyd, 1992, 1993; Clover, 1994; Rothschild, pers. comm., 1995). Reseeding provides a diverse sward of vegetation in a relatively short period of time and the seed mixture content can be tailored to suit the area, in terms of suitable species for the district and establishment on land with certain environmental conditions. Mixtures generally contain 80% grasses and 20% herbs. Grasses in seed mixes have to be bred to conform with Ministry of Agriculture, Fisheries and Food (MAFF) regulations and there is a tendency for many non-native grasses to be used, which are chosen for their low competitiveness. Herbs are chosen for their chances of establishment, likelihood to be present naturally in the area, and appearance. *Leucanthemum vulgare* is often included as it produces a large number of good-sized flowers within one year of sowing and flowers even more profusely in the second year, thus giving a young sward a quick show of flowers (Wells, 1987).

Concern has been expressed about the possible sowing of non-native seed under the misconception of its being of 'wild' origin, giving rise to plants that are not indigenous to the area, or in some cases the country, in which they are growing. Akeroyd (1992) suggests that on Magog Down, Cambridgeshire, foreign ecotypes of a number of species are present. These include a robust agricultural strain of *Achillea millefolium* (yarrow), *Chrysanthemum segetum* (corn marigold) in a variety of unexpected colours, and three ecotypes of ox-eye daisy, *L. vulgare* $(2x = 2n = 18)$, *L. vulgare* $(4x = 2n = 36)$ and, *L. vulgare* × *superbum* $(12x = 2n = 108)$. The diploid variety is native to Britain, is smaller and is less hairy than the other two. The tetraploid seems to be common in seed mixtures. It is similar to the native ecotype but is slightly larger in size, is more hairy and has a greyer appearance. The most notable distinguishing difference is that it flowers three to four weeks later than the diploid. *Leucanthemum vulgare* × *superbum*, the dodecaploid, is a garden variety, the shasta daisy. It is much larger, and has a longer flowering period, but being polyploid and a cultivar it is less likely to reproduce successfully (P.D. Sell, pers. comm., 1995).

In order to investigate the presence, distribution and relative abundance of native and non-native ecotypes, a study of individuals of *L. vulgare* was established at three sites, in each of which the performance, morphology, growth and phenology of plants were examined. Initial field data collection led to a number of other investigations being established.

The creation of simple quadrat maps acts primarily as a display tool allowing the comparison of a plant's growth form type within quadrats at different survey times. Site differences also become more apparent through the spatial display of the data and the maps serve as the first level of distribution analysis. Given that the vegetation swards at all sites were relatively 'new', the oldest being six years old, and that they have been artificially seeded using methods designed to achieve an even seed distribution, the regular spacing of plants would be a reasonable expectation. Conversely, sites established from the seed bank and undisturbed for a longer time span may be expected to be more likely to show irregular patterns of distribution (Goldsmith *et al.*, 1986). The use of digital maps means that monthly samples can be compared in order to visualise distribution patterns and their changes over time. Such mapping also facilitates the overlay of growth form data with other variables measured at the sites and hence the analysis of associations between plant performance and environmental variation.

METHODS

Site details are summarised in Table 3.1 and their locations are shown in Figure 3.1. Permanent quadrats were established at each of the three sites. The quadrats were mostly 1×1 m, though on sites where the density of mature individuals of *L. vulgare* was low ($< 5 \, \text{m}^{-2}$), quadrats were expanded to encompass enough plants to justify the monitoring programme. The boundaries of the permanent quadrats were marked using metal tubing buried in each of the four corners. A quadrat with 100 equal subdivisions (10×10 cm grids in the case of 1×1 m quadrats) was used to give precise coordinates for each plant's location. This grid was precisely positioned during each survey by anchoring it to the metal markers. Data on the percentage cover (subjective estimate) of all species present, light intensity at canopy and ground level and canopy height were recorded for each whole quadrat. Individuals of *L. vulgare* were then identified and specific information recorded. Larger plants were tagged, for ease of recognition on subsequent surveys, using coloured electrician's tags placed around the basal rosette. Record sheets were tailored to the growth stage of plants and the time of survey, but included the following items:

- Plant location coordinates
- Colour of tag
- Plant growth form type[1]
- Growth environment[2]
- Plant height
- Presence of buds, flowers or seed heads
- Grazing, disease, or evidence of wilting

Surveys of permanent quadrats were carried out monthly throughout the growing season, and the results recorded on a spreadsheet. Results were then summarised in a grid form and transferred to the Map II Map Processor package (Pazner *et al.*, 1989), where summary maps of individual quadrats were created. Maps were then available for comparison.

The density of mature plants and seedlings (number of individuals m^{-2}) was calculated as a monthly mean for all quadrats at a site. Analysis of the association between the location of mature plants and the location of seedlings was carried out by means of the χ^2 test of association. This was performed at two scales: (i) Using the 10×10 cm (or equivalent) grid cell divisions of the quadrat and (ii) by grouping blocks of four adjacent 10×10 cm cells to form 20×20 cm cells.

The nature of the spatial distribution of plants within quadrats was analysed using

---

[1] Growth form type: Plants that were small and showed no evidence of leaf scars were recorded as seedlings. Other plants were 'mature' and were categorised in accordance with the number of stems attached to the basal rosette. During analysis the number of stems on mature plants was discounted for ease of interpretation of results.

[2] Growth environment: This used a subjective scale developed for this research which is as follows:
  1 = Plant clear of neighbours, at least at the top, < 10% cover by neighbours.
  2 = Plant with neighbours in contact but not over-topped, 11–30% cover by neighbours.
  3 = Plant directly shaded by neighbours, 31–50% cover by neighbours.
  4 = Plant significantly shaded by neighbours, 51–75% cover by neighbours.
  5 = Plant under canopy of neighbours, > 75% cover by neighbours.

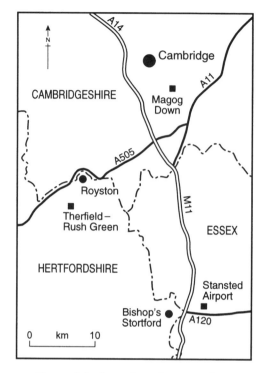

Figure 3.1   Location of survey sites

the index of dispersion (the variance:mean ratio). An index of dispersion of 1.0 implies random distribution, less than 1.0, over-dispersion and greater than 1.0, contagion or clumping. Persistence of mature plants is particularly important in the maintenance of a stable plant population. Persistence values were calculated as the percentage of mature plants present in March (or May for Magog Down) which remained present in August.

RESULTS

Populations show considerable between-site variation in the density of both mature plants and seedlings, with Magog Down exhibiting the highest values and Stansted Airport the lowest (Figure 3.2). The Stansted site as a whole exhibits a rather patchy distribution of *L. vulgare* with most patches having very low numbers but one being particularly highly populated.

Statistical analysis of mapped data proved difficult due to the low numbers of individual plants encountered in some quadrats and the consequent high frequency of zero as a category. However, trends can be established allowing between-quadrat and between-site comparisons to be made over the six month monitoring period.

Examples of quadrat maps for each site, over time, are shown in Figure 3.3. Calculated dispersion indices exceed 1.0 in all but one case (Magog Down, Quadrat 4,

Table 3.1   Summary details of survey sites

Site name	Magog Down	Rush Green, Therfield	Stansted Airport
OS Grid Ref	Sheet No: 154 485585	Sheet No: 166 397344	Sheet No: 167 525216
Description of location	South of the City of Cambridge on the Magog Hills	Through Therfield village, towards A10, a small area attached to a large modern house, before North End Farm	On the main airport site in Essex adjacent to the A120 road to Dunmow
Date of sowing seed	Spring 1991	Spring 1991	Autumn 1989
Seed mixture supplier	Bookers and Emorsgate	Emorsgate	Emorsgate and Miriam Rothschild
Cutting dates	Cut October 1993 August 1994 Topped September 1994	Cut autumn 1993 Not cut 1994 Cut early 1995	Cut early spring 1994 Cut November 1994
Grazing by domestic animals	Grazed late summer onwards by sheep, only lower part of the site	No grazing	No grazing
1994 survey period	May onwards	March onwards	March onwards
Area survey (m^2)	5	5	11
Nature of quadrats	Permanent	Permanent	Permanent
Degree of grazing by rabbits	None, rabbit-proof fence surrounds site	Site subject to some grazing	Site subject to a lot of grazing
Other comments	The site is heavily used, especially during the summer months	Site is small, 0.46 ha. and surrounded by farmland, the site is rarely used	Site has no public access and is 1 km from the runway of the airport

July) indicating that clumping of individuals occurs at all sites and throughout the growing season. Clumping is visually most apparent at moderate densities such as those found at Therfield (Figure 3.3a). Indices of dispersion at this site (range 6.369–156.51) were greater than at either of the others (Stansted: 5.76–63.408; Magog Down: 14.49–51.46).

Examination of the time sequence of quadrat maps (Figure 3.3) reveals the relative stability of the mature plant populations as compared to seedling populations. The latter show higher recruitment and death rates. This observation is supported by the density data shown in Figure 3.2, where, on all sites, monthly fluctuations in the total number of mature plants present are relatively low. Mature plants suffer a steady decline over the season on all sites except Stansted where numbers are much lower and fluctuate throughout the season. Persistence of mature plants is highest at Therfield (mean = 66%) and lowest at Stansted (mean = 45.75%). Magog Down has the

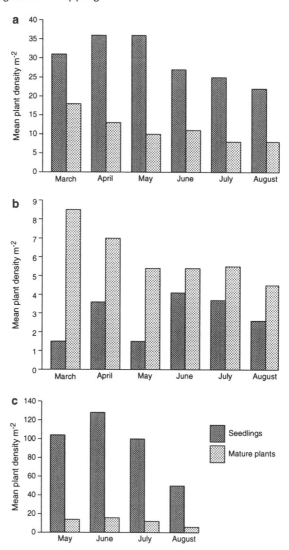

Figure 3.2 Mean plant density (no. m^{-2}) at each of the survey sites during the 1994 growing season. (a) Therfield, Rush Green; (b) Stansted Airport; (c) Magog Down

greatest variability in persistence values between quadrats (10% mean survival of mature plants within quadrats 1 and 3 and 83% mean survival within quadrats 2, 4 and 5). Persistence values are less easy to apply to seedlings as recruitment and mortality of individuals occur throughout the growing season and over time scales shorter than the sampling intervals, thus making data difficult to interpret.

Maps of seedlings show the establishment of pattern over time. For example, Quadrat 2, Therfield (Fig. 3.3 a (i)) has a clump of seedlings establishing in and around cell 1,4 (origin at lower left, column 1, row 4) by April, when seedling numbers reach a

(a)

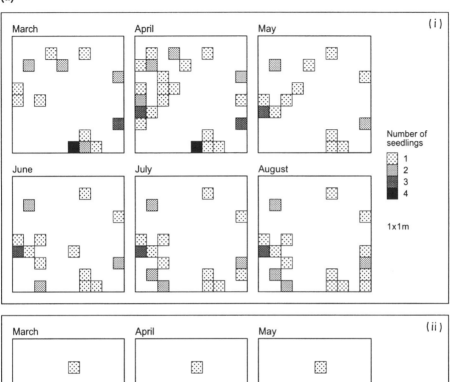

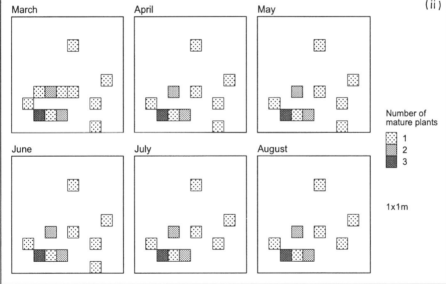

Figure 3.3 Examples of quadrat maps of seedlings (i) and mature plants (ii) at the three survey sites during the 1994 growing season. (a) Therfield, Rush Green; (b) Stansted Airport; (c) Magog Down.

(b)

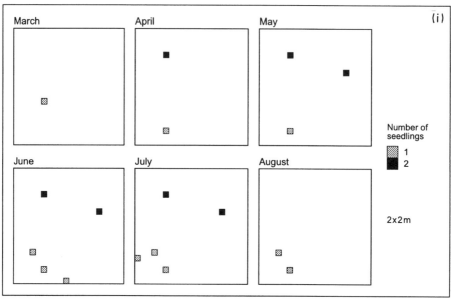

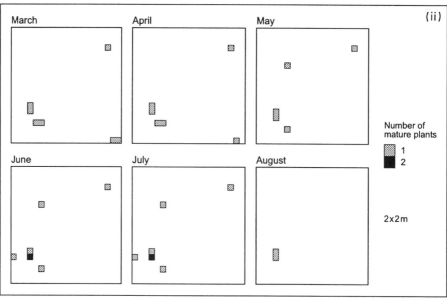

(c)

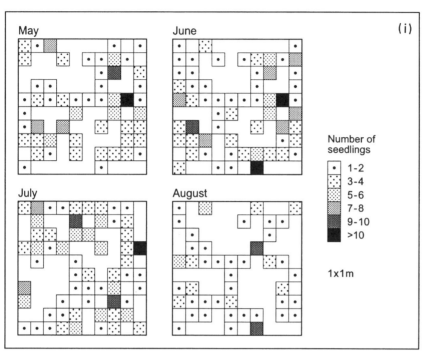

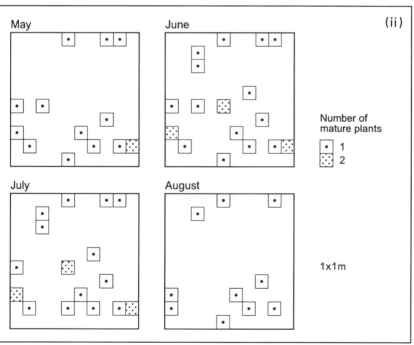

peak at this site. This pattern is reinforced over the following months with a group of seedlings surviving through to August in that area of the quadrat.

The maps in Figure 3.3 show differences in the pattern of clumping exhibited by mature plants and seedlings. Both seedlings and mature plants show clumped patterns of distribution but there is little or no association between the clump locations of the two age classes, i.e. clumps of seedlings are not found with clumps of mature plants at this scale of mapping. This lack of association is clear from the visual presentation of quadrat maps and is also supported by χ^2 analysis, although the results of the χ^2 tests should be treated with caution due to the low numbers of plants counted in many quadrats/cells. χ^2 values show much variation with only two significant results ($P < 0.05$), one a positive and the other a negative association.

DISCUSSION

The mapping of *L. vulgare* allows the visualisation of the variation in density both between sites and over time. It also gives an indication of trends in the dynamics of reseeded sites. Stansted, the oldest site, has the lowest density of *L. vulgare* (mean = 8.91 plants m^{-2}), while Magog Down has the highest density (mean = 107.6 plants m^{-2}), with Therfield being intermediate (mean = 49.2 plants m^{-2}). The low density of plants at Stansted compared to the other sites supports the work of Wells *et al.* (1986) who found *L. vulgare* to be abundant for the first two years after the sowing of wild flower seed, then to suffer a substantial decline in dominance on a site, behaving as a short-lived perennial. In general, density also decreases towards the end of the growing season in all quadrats at all sites, partly reflecting the temporal patterns of seedling germination. Peak germination occurs between April and June at Therfield and Magog Down, while the smaller numbers of seedlings at Stansted show a tendency for recruitment later in the growing season.

The lack of association between the location of clumps of seedlings and clumps of mature plants is clear from the visual presentation of quadrat maps and is also borne out by the χ^2 analysis. Mature plants have higher variance:mean ratios than seedlings suggesting that plants which persist in a community do so at preferred locations. All sites show clumping at a scale larger than the sample size. Given that seeding methods are aimed at regular dispersal of seed, the establishment of clumps at the sample size and larger scales suggests that plants persist preferentially in patches of suitable habitat. The establishment of clumps can be observed during the growing season as clusters of seedlings are recruited and spatially variable mortality reinforces the pattern (see Figure 3.3 a(i)).

The persistence of plants throughout the growing season is an indicator of population stability. Seedling dynamics are well accepted as being highly variable in a number of species with high germination and mortality rates (Harper, 1977; Silvertown, 1990). However, mature plants would be expected to show less fluctuation in numbers. Histograms (Figure 3.2) confirm the expected trend with lower variation in the density of mature plants than seedlings. Persistence values of mature plants serve as an indicator of this feature. Therfield has the highest mean persistence value (66%) followed by Magog Down (53.8%) and Stansted which has the lowest (45.75%). The highest within-site variability in persistence is seen at Magog Down, the possible reasons for which are discussed below.

The differences in density, distribution and persistence between sites may be due to site age (i.e. time since sowing), but if age were the sole controlling factor in plant dynamics then Magog Down and Therfield would be expected to show similar levels. Data from Stansted (the 'oldest' site) support a trend of declining *L. vulgare* numbers with time and the continuation of monitoring on Magog Down may reveal a similar trend.

Ecotypic variation may be another factor affecting the population dynamics of *L. vulgare*. The Stansted sward was established using seed from two seed merchants, one of which obtained seed from an ancient meadow of SSSI status, which therefore has a high probability of being from a native ecotype. Magog Down is considered to have three ecotypes present and was sown with seed from two suppliers, whilst the seed for Therfield came from a single source. Variations in the performance of different ecotypes may explain the variations in population density and survival which are observed at Magog Down. Seeds from the two sources of supply were spread onto different parts of the site. Quadrats 1 and 3 are located in an area sown with seed from one source, whilst quadrats 2, 4 and 5 are in an area sown predominantly with seed from a second. Plant density is lower in quadrats 1 and 3 (means of 78.25 and 88.75 m^{-2}, respectively) than in 2, 4 and 5 (means of 91.5, 78.25 and 151 m^{-2}, respectively). Also dispersion indices and quadrat maps suggest greater clumping and lower persistence in quadrats 1 and 3. No plants, either mature or seedlings, survived through to August in quadrat 1, and only 20% of mature plants and 21% of seedlings survived in quadrat 3. By contrast, 71–90% of mature plants and 51–77% of seedlings survived in quadrats 2, 4 and 5. These differences may indicate that the concerns expressed by Akeroyd (1993) on the varied success of different ecotypes are well founded.

Variations in population dynamics may also be attributable to site differences. Preliminary soil analyses show the Therfield site to have a clay soil, the Stansted site to have a clay loam, and Magog Down a chalky loam. In general, the moisture content of the soils follows the same sequence with Therfield being the wettest site and Magog Down the driest. Howarth and Williams (1968) noted that *L. vulgare* prefers drier soils but can grow on a variety of soil types. Variability of soil moisture also occurs within sites due to the topographic position of quadrats; for example, quadrat 4 at Stansted lies close to a ditch, making it wet at all times. Such differences in local environmental factors could account for within-site variations in population dynamics.

CONCLUSION

Preliminary monitoring has shown differences between sites sown with seed mixtures from different seed merchants. Density, dispersion and persistence of *L. vulgare* plants were found to vary both between sites and over the growing season. Microscale mapping and monitoring have proved useful in the visualisation and analysis of these variations in population abundance and dynamics, but the precise roles of causal factors remain uncertain. Further research will concentrate on comparing the performance of plants grown from seed from different sources using both glasshouse and field trials, and continued monitoring at Magog Down will provide further field data. This combination of strategies will, hopefully, yield a clearer insight into the relative importance of the factors controlling the population dynamics of *L. vulgare* in wild flower meadows.

ACKNOWLEDGEMENTS

We would like to thank the following for their co-operation and assistance: Mr Winterflood, The Magog Trust; Mr Ogden, Stansted Airport; Mrs Penny Anderson, Penny Anderson Associates; Mr Bell, Hertfordshire Countryside Management Services; the Conservators of Therfield. In addition, thanks are due to the technicians of the Biology Department, Anglia Polytechnic University, for assistance in the field.

REFERENCES

Akeroyd, J. (1992) A remarkable alien flora on the Gog Magog Hills. *Nature in Cambridgeshire*, **34**, 35–44.

Akeroyd, J. (1993) Seeds of destruction. *Natural World*, **39**, 26–27.

Anon. (1984) *Nature Conservation in Great Britain*. Nature Conservancy Council, Shrewsbury, UK.

Clover, C. (1994) Seeds of doubt in meadow. *The Daily Telegraph*, Monday 27 June, p. 6.

Goldsmith, F.B., Harrison, C.M. and Morton, A.G. (1986) Chapter 9. In: P.D. Moore and S.B. Chapman (eds) *Methods in Plant Ecology*. Blackwell, Oxford, pp. 437–524.

Harper, J.L. (1977) *Population Biology of Plants*. Academic Press, Oxford.

Howarth, S.E. and Williams, J.T. (1968) Biological flora of the British Isles, *Chrysanthemum leucanthemum* L. *Journal of Ecology*, **56**, 585–595.

Pazner, M., Kirby, K.C. and Thies, N. (1989) *Map II Map Processor*. Wiley, Chichester.

Silvertown, J. (1990) *Introduction to Plant Population Ecology*. Longman, Harlow, UK.

Wells, T.C.E. (1987) The establishment of floral grasslands. *Acta Horticulturae*, **195**, 59–69.

Wells, T.C.E., Frost, A. and Bell, S. (1986) *Wild Flower Grasslands from Crop Grown Seed and Hay Bales*. Nature Conservancy Council, Peterborough, UK.

4 Mapping Grassland at the Whole Site Level, with Special Reference to the Impact of Agricultural Improvement

R. GULLIVER

Associate of the Department of Geography, University of Glasgow, UK
Address for correspondence: Carraig Mhor, Imeravale, Port Ellen, Isle of
Islay, Argyll, Scotland

The whole of a 5 × 5 km area in central Leicestershire (England) was surveyed in 1996–70 and in 1982. Species rich grassland sites were visited at intervals thereafter, and the rate of loss of unimproved grassland determined. These two initial surveys resulted in the recognition of those plant species which are indicative of unimproved grassland.

The presence of low embankments within some of the farmed fields allowed the impact of grassland management intensification to be assessed by comparing a set of more or less unimproved banks with the body of the associated field. At some other locations the banks had been partly improved, and at some they had been largely eliminated. In a sample (vignette) area the following attributes of banks and their associated fields were mapped to illustrate the impact of intensification: (a) vascular plant complement, (b) the number of species indicative of unimprovement, (c) the number of species particularly associated with certain grassland plant communities, (d) the nature of the plant community, and (e) the number of anthills created by *Lasius flavus*, (these structures are destroyed by ploughing, mechanical mowing and silage making).

The collection of data at a whole site level allows the full range of species present to be recorded, and used in conservation assessment. A further advantage of the technique is that it allows the site characteristics to be investigated and mapped at the geographical level at which they were being managed.

The relationships revealed in this sub-regional study clearly demonstrate the nature and speed of change in this habitat. They also indicate the need for conservation activity, for example at the regional and national levels.

INTRODUCTION

The intensification of agriculture since the Second World War has led to a major loss of wildlife habitats in the countryside (Shoard, 1980; Bowers and Cheshire, 1983; Pye-Smith and Rose, 1984; Harvey, 1997). For hedgerows, farm ponds and ancient woods this process has produced measurable declines in the quantity of the resource (e.g. Relton, 1972; Pollard *et al.*, 1974; Beresford and Wade, 1982; Barr *et al.*, 1986; Gulliver, 1986; Spencer and Kirby, 1992; Barr, 1994). Qualitative changes have also taken place. Grasslands, for example, may have the appearance of having changed little or not at all, but the wildlife and conservation interest has often declined markedly over this period (Nature Conservancy Council, 1984) due to the intensifica-

Vegetation Mapping: From Patch to Planet. Edited by Roy Alexander and Andrew C. Millington.
© 2000 John Wiley & Sons Ltd.

tion of management. This may take one of two forms: (i) the conversion of permanent (unploughed) grassland to short or medium-term leys, or to resown 'permanent' grassland or (ii) the 'improvement' of permanent grassland by the implementation of one or more of the following management practices: drainage, fertilisation, herbicide use, liming, adoption of a more intense grazing or defoliation regime.

The use of field survey techniques allows these subtle changes to be detected. Detailed mapping of selected study areas (vignettes), as employed in this investigation, provides a clear insight into the processes involved in changes in grassland habitat quality. Previous studies, using different techniques, have indicated that very major changes have occurred. Examining the nature of the surviving grassland, in a range of sample areas, the Nature Conservancy Council (1984) concluded that only 5% of lowland neutral grassland (including hay meadows) now had any significant wildlife interest. Fuller (1987) analysed the data gathered during grassland surveys carried out between 1932 and 1984, and estimated that semi-natural pastures (excluding rough grazings) in lowland England and Wales occupied 0.2 million hectares, 4% of the 1984 grassland area, and 3% of its area in 1932. Some unimproved permanent grassland has been 'lost' by changes to other land uses, especially arable (Fuller, 1987; Green, 1989).

THE STUDY AREA – SP69NW

The 1:10 560 Ordnance Survey map sheet SP69NW, 5 × 5 km in extent, and located in central Leicestershire, UK, was selected for detailed investigation for the following reasons:

1. The area consisted largely of grassland farming which was likely to be subject to change due either to intensification, or to conversion to a different form of land use.
2. The area was well supplied with public footpaths.

The countryside is rolling, with boulder clay on the higher ground, Lower Lias Clay on the lower slopes, and alluvium in the valleys. Fields are small or medium sized (see later), and many of the boundaries are hedges. The River Sense runs from east to west in the southern part of the area, its valley having been utilised during both canal and railway construction. Seven of the component 1 × 1 km squares are occupied wholly or partly by suburban development at Wigston and Oadby. Six small villages or hamlets occur in the remaining 18 'rural' 1 × 1 km squares. No other major rural developments or attributes, e.g. large estates, reservoirs, quarries or airfields, are present.

METHODS

A ground-based survey of the whole area was undertaken between 1966 and 1970. Land use (Coleman and Maggs, 1968) was recorded together with the presence of four plant species indicative of unimproved grassland – *Cardamine pratensis*, *Lychnis flos-cuculi*, *Primula veris* and *Rhinanthus minor*. Land use in 1982 was determined using aerial photography. Ground surveys of all known unimproved grassland sites and

Table 4.1 The vascular plants used as indicators of Species Rich Permanent Grassland, SPRG, in SP69NW. Scientific and English names follow Stace (1991). Selection based on cumulative field experience

Scientific name	English name	Code (see notes)
Agrimonia eupatoria	Agrimony	
Briza media	Quaking-grass	D
Cardamine pratensis	Cuckooflower	A1
Carex flacca	Glaucous Sedge	D
Cirsium acaule	Dwarf Thistle	
Festuca pratensis	Meadow Fescue	
Filipendula ulmaria	Meadowsweet	
Galium verum	Lady's Bedstraw	
Helictotrichon pubescens	Downy Oat-grass	
Leontodon hispidus	Rough Hawkbit	
Lynchnis flos-cuculi	Ragged-Robin	A1
Pimpinella saxifraga	Burnet-saxifrage	
Plantago media	Hoary Plantain	
Primula veris	Cowslip	A1,D
Rhinanthus minor	Yellow-rattle	A1,D
Sanguisorba minor	Salad Burnet	
Sanguisorba officinalis	Great Burnet	
Silaum silaus	Pepper-saxifrage	
Succisa pratensis	Devil's-bit Scabious	

Notes:
A1 These four species were used in the initial identification of SRPG in 1966–70. Other plants in the list are the A2 species. D Differentials of the Centaureo–Cynosuretum cristati and of MG5.
General note: Further species were associated with unimproved grassland, but occurred at a very low frequency.

adjacent areas were undertaken in 1982, 1985, 1990 and 1993. Lists of vascular plant species were produced at 38 permanent grassland fields and 16 banks, and quadrat data were also gathered. During this phase of fieldwork, a further set of 15 species indicative of unimproved grassland was identified (Table 4.1). The first four indicators had fairly conspicuous blooms and were moderately widespread in unimproved grassland sites. They were given the code A1 to distinguish them from the second group of 15 (the A2 species), which tended to be less widespread and/or less conspicuous. Permanent grassland fields were categorised as:

(a) Highly Species Rich Permanent Grassland – HSRPG – 44 or more species per site: five or more A species (i.e. A1 plus A2), and at least one A1 species.
(b) Moderately Species Rich Permanent Grassland – MSRPG – with less than 44 species per site, and at least one A1 species (the mean of 10 sites was 33 species).
(c) Other Species Rich Permanent Grassland – OSRPG – with either (i) A2 but no A1 species present, or (ii) with no A species present at all but an overall species count above 44 (one site only).
(d) Permanent Grassland – PG – no signs of reseeding, no A species present and an overall species count below 44.

The SRPG designation was obtained by pooling the MSRPG and HSRPG categories.

Table 4.2 Loss of Species Rich Permanent Grassland, SRPG, in SP69NW; using all SRPG detected in 1966–70 as a baseline

Year	MSRPG area		HSRPG area		SRPG area	
	ha	%	ha	%	ha	%
1966–70	44.78	100	15.63	100	60.41	100
1982	15.93	35.6	9.15	58.5	25.08	41.5
1985	7.80	17.4	5.83	37.3	13.63	22.5
1990	6.68	14.9	4.49	28.7	11.17	18.5
1993	6.68	14.9	4.49	28.7	11.17	18.5

MSRPG: Moderately Species Rich Permanent Grassland.
HSRPG: Highly Species Rich Permanent Grassland.
SRPG: Total Species Rich Permanent Grassland.

The characteristics of the surviving 'ridge and furrow' were sketched onto the base map. 'Ridge and furrow' consists of linear strips of land, variable in width from one site to another, but often around 10 m wide, created during the mediaeval period to improve soil drainage for arable crops. The land was ploughed in the same manner every year, with the soil being turned upwards to the centre of the strip. The long-term result was a convex ridge grading down to a shallow depression on each side (Rackham, 1986). Providing the land is not ploughed after the abandonment of strip agriculture, these formations persist through to the present day.

RESULTS

Land Use Change

Species-rich grassland declined markedly between 1966–70 and 1982, the greatest loss being in the MSRPG category (Table 4.2). Land use at the two single dates 1966 and 1982 for the total area and for the purely rural area (18 1 × 1 km squares) is presented as Table 4.3. Land use for each component 1 × 1 km square was measured separately, and these data were then used to calculate the coefficient of variation for each land use category.

Survey data from 1966–70 and 1982 had been plotted on 1:10 560 maps. The SRPGs in part of SP69NW are reproduced as Figure 4.1. There was a concentration of unimproved grassland sites along the valley of the River Sense. This valley has also been used as a transportation corridor for the Grand Union Canal and the former Midland Railway, with a consequent subdivision of fields and disruption of drainage patterns. The low-lying nature of some of the fields had previously been a barrier to improvement. On either side of the river, ridge and furrow was absent. This strongly suggests that there has been a long period of non-arable land use. The nature of the riverside grasslands is likely to have changed markedly over the centuries; though individual species associated with traditional alluvial grassland have survived in the valley, e.g. *Sanguisorba officinalis* at three field sites and two canalside banks, *Silaum silaus* at two fields and one bank and *Thalictrum flavum* in emergent vegetation in the

Table 4.3   Land use in the 5 × 5 km square, SP69NW, in 1966 and 1982

	1966			1982		
	Area (ha)	Coeff. of variation*	%	Area (ha)	Coeff. of variation*	%
All 1 × 1 km squares (25)						
Grassland						
• agricultural	1421.8	33.8	56.9	1015.8	56.9	40.6
• amenity	84.3	110.6	3.4	108.6	154.5	4.3
• unmanaged	32.3	106.1	1.3	15.8	105.4	0.6
Arable (excluding reseeded grass)	490.5	64.4	19.6	763.9	69.5	30.6
Woodland	36.3	120.5	1.4	35.0	124.3	1.4
Marsh	1.4	–	0.1	1.4	–	0.1
Residue	433.4	135.9	17.3	599.5	129.2	22.4
Total	2500.0		100.0	2500.0		100.0
Squares remaining rural (18)						
Grassland						
• agricultural	1205.6	21.1	67.0	918.9	45.0	51.0
• amenity	36.0	177.7	2.0	40.2	169.8	2.2
• unmanaged	1.1	73.3	0.1	0.1	–	0.01
Arable (excluding reseeded grass)	421.0	57.1	23.4	703.2	57.2	39.1
Woodland	34.4	107.5	1.9	33.4	108.3	1.9
Marsh	–			–		
Residue	101.9	58.2	5.6	104.2	56.6	5.8
Total	1800.0		100.0	1800.0		100.01
Squares partly or wholly built-up (7)						
Grassland						
• amenity	48.2	46.9	6.9	68.4	38.0	9.8
Residue	331.6	57.0	47.4	455.3	29.8	65.0
Amenity grass as % of residue			14.6			15.0

* Using the component 1 × 1 km squares as replicates; derived from sample (not population) variance.

canal at two locations. (These records date from 1980 to the present.) Away from the River Sense, there was a more or less random pattern of SRPGs throughout the 5 × 5 km square. The mean size of SRPG fields was 2.24 ha.

Grassland Species Richness

The number of A species increased significantly with the vascular plant species richness (S) of grassland fields, each A species corresponding to an average rise of S of 5.25, above a threshold value of $S = 23$ (see Figure 4.2). However, the scatter of points around the regression line, due to variation in the ecological attributes (e.g. grazing/defoliation regime and soil characteristics), should be noted.

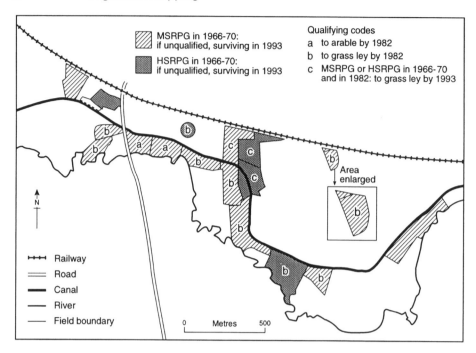

Figure 4.1 Moderately Species Rich Permanent Grassland – MSRPG – and Highly Species Rich Permanent Grassland – HSRPG – in part of the Sense Valley in SP69NW in 1966–70, plus their status in 1982 and 1993

Intensification

Intensification of grassland management was a major factor causing the loss of species from unimproved grassland. In SP69NW this frequently consisted of ploughing up the existing sward and reseeding as shown in some of the entries on Figure 4.1. In addition, permanent grassland swards were sometimes drained, fertilised and treated with herbicide but not ploughed up. Some of the existing MSRPGs had probably once been HSRPGs, and had experienced a loss of species consequent upon intensification. Similarly some PGs may once have been MSRPGs or OSRPGs. Examples of these processes were found on the farmed canal banks (described subsequently).

For much of its length the canal in SP69NW is mounted on a shallow embankment which lies within the farmed land parcel. The slope (25% – mean of five measurements at each of three sites: but extremely variable in profile and overall slope), together with the immediate proximity of the field boundary, makes it difficult for the bank to be ploughed out. Herbicide and fertiliser drift can affect the banks. Nevertheless, the bank represents an essentially unimproved strip (corridor *sensu* Forman and Godron, 1986) which is grazed by the same stock as the body of the field. It therefore gives some indication of the nature of the sward of the entire field pre-improvement, and by extrapolation an indication of the nature of some of the original unimproved swards throughout SP69NW. The banks with their sloping profile and clay soils (with little humus visible) may always have had a somewhat lower nutrient status than the body

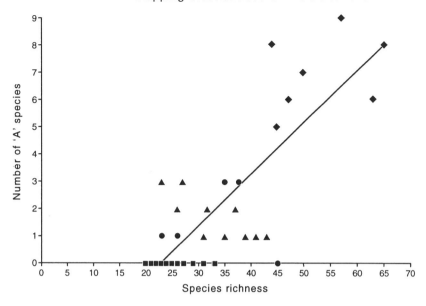

Figure 4.2   The number of A species versus vascular plant species richness for Highly Species Rich Permanent Grassland (HSRPG) – ♦; Moderately Species Rich Permanent Grassland (MRSPG) – ▲; Other Species Rich Permanent Grassland (OSRPG) – ●; and Permanent Grassland (PG) – ■ fields in SP69NW. $y = 0.19x - 4.34$; $p < 0.001$. For definitions see text

of the field. However, some fertiliser granules applied by a spinner to the body of the field will have landed on the banks, and so they are probably becoming progressively nearer to the nutrient status of their associated fields. Around 1984 two of the shallower banks to the east of the A50 road were ploughed out and largely eliminated, while a farm road has been created over another two in the same area, though some original sections do remain intact.

In SP69NW in general some SRPGs and OSRPGs were lost by conversion to arable (e.g. see Figure 4.1), while others were used as house building sites.

Vignette Area

An area of 37.34 ha containing nine surviving banks was selected for further study of the consequences of intensification. This was mainly in SP69NW, but the western part was in SP59NE. Total species lists were made for all banks and the associated body of the field on two occasions in 1991 (in June and July). The anthills of the yellow meadow ant (*Lasius flavus*) were recorded on a per site basis. These anthills are destroyed by mechanical mowing or ploughing, and therefore provide a valuable indication of unimprovement (King, 1981). Ideally, a good number should be present to enable their use in countryside interpretation, as they can also be severely damaged by cattle. A National Vegetation Classification (NVC) community (Rodwell, 1992) and a phytosociological association (Page, 1980) (see list in Table 4.4 and the Discussion section) were applied at the whole site level, with the proviso that areas consisting of

Table 4.4 The phytosociological associations (Page, 1980) and National Vegetation Classification (NVC) (Rodwell, 1992) of grassland communities occurring in the vignette area of SP69NW and SP59NE

Code	Association name	NVC code	Sub-association(s-a)	NVC Sub-community
Ae	Arrhenatheretum elatioris	MG1	Typical s-a	1e
A–Fr	Anthoxantho–Festucetum rubrae	See discussion section	*Lolium perenne* s-a	–
C–C	Centaureo–Cynosuretum cristati	MG5	*Galium verum* s-a	5b
L–C	Lolio–Cynosuretum cristati	MG6	*Anthoxanthum odoratum* s-a*	6b
			Typical s-a	6a
L–P	Lolio–Plantaginetum	MG7a†	–	7a
		MG7e‡	–	7e
P–L	Poo–Lolietum perennis	MG7b†	–	7b
		MG7f‡	–	7f

Notes:
Intermediate stands: Stands which are intermediate between two associations are shown in Table 4.6 thus: A–Fr/L–C.
* Occurred at W533/5F, W536F, W538F and W539F; all other L–C stands were the typical s-a.
† For farmed grasslands.
‡ For regularly mown amenity grasslands; such stands were absent from farmland in the vignette area, but are included in the table to demonstrate the NVC treatment of the L–P and the P–L.

clearly different plant communities (e.g. gateways, *sub*-hedge strips, *Juncus inflexus* stands) were excluded, so that the recording unit was a more or less homogeneous stand. Some stands were intermediate between two communities (associations) in their complement of constant, character and differential species. The term 'site' is used throughout in this account as a topographically neutral phrase; in the vignette area it refers either to a bank site or to the body of the field.

There was no overlap in species richness between the grassland fields and the 'unimproved' banks (Table 4.5) – a clear demonstration of the power of the process of improvement (intensification) in reducing species richness. Furthermore, it was possible to recognise part-improved banks (species richness [S] 27–38) as well as the more or less unimproved banks. The data in Tables 4.5 and 4.6 on number of A species (non-improvement indicators) and number of differentials of the *Centaurea nigra–Cynosurus cristatus* grassland (Mesotrophic Grassland 5 – MG5 – of the NVC) i.e. the Centaureo-Cynosuretum cristati, confirm this basic picture. By contrast, there was no difference between these two categories in number of anthills per site, which did, however, vary greatly from one bank to the next. All eight banks had at least one anthill, which differentiated them from their associated fields. Two of the river valley fields (W536 and W538) (Figure 4.3) had one A species, and W536 (no associated bank) had two anthills near the canal. These features are probably indicative of a former, more species-rich, state.

Whereas the trends are well demonstrated in Table 4.5, they are particularly strongly emphasised in map form (see Figure 4.3), linked to tabular data for individual fields and banks (Table 4.6) as the pattern of the field and bank boundaries helps to

Table 4.5 Characteristics of 'unimproved' and part-improved farmed canal banks in SP69NW and SP59NE compared with the body of the associated field, in 1991

	Plant species richness	Non-improvement indicators	No. of differentials of C-C* and MG5	No. of anthills	Length (m)	Area (ha)
	(S)	(A species)	(D)		(L)	
'Unimproved' banks (n = 4†)						
Bank: Mean	44	4.75	5.0	4.75	113	–
Bank: Range	38–52	3–7	1–8	2–9	33–187	–
Associated fields (n = 4)						
Body of field:						
Mean	23	0.25	0	0	–	4.26
Body of field:						
Range	15–27	0–1	0	0	–	3.59–5.18
Part-improved banks (n = 4)						
Bank: Mean	30	1.0	1.25	5.25	78	–
Bank: Range	27–38	0–2	0–5	1–13	32–117	–
Associated fields (n = 4)						
Body of field:						
Mean	23	0	0	0	–	1.24
Body of field:						
Range	21–24	0	0	0	–	0.39–3.37

Notes:
* Differentials of the Centaureo–Cynosuretum cristati (C-C) are *Briza media, Carex flacca, Centaurea nigra, Lathyrus pratensis, Leontodon hispidus, Leucanthemum vulgare, Lotus corniculatus, Primula veris* and *Rhinanthus minor*. To these Page (1980) adds *Vicia cracca*, absent from the MG5 list.
† Bank W539W on Figure 4.3 was in 1991 part of a British Waterways dredging disposal site, and has been excluded from these figures as it was not subject to agricultural management.

form a mental picture of the ecological relationships involved.

Five of the banks in the vignette area had been previously studied (together with two others elsewhere in SP69NW) between 20 and 24 June 1983 (Table 4.7). The part-improved status of W543 is clearly indicated, with a decline in A species from 9 to 1, and of S from 43 to 27 between 1983 and 1991. There are also indications of a general decline in species richness at the 'unimproved' (1991 categorisation) banks, especially in the case of W542, where the number of A species fell from 6 to 3, though S was hardly affected (40 in 1983, 38 in 1991). The position at W539 is complicated by the elimination of part of the bank *post* 1983 by civil engineering work, carried out by British Waterways.

Quadrat and Whole Site Data

Presence/absence data were gathered at the $1\,\text{m}^2$, $4\,\text{m}^2$, $25\,\text{m}^2$ (all nested) and whole site level at the seven banks studied in 1983 (Table 4.8). Data at the whole site level were also gathered on other occasions (in addition to 1991) with an overall mean of five visits per site. The average number of species rose markedly with increasing sample size; hence for rapid site assessment the use of total species lists is to be preferred.

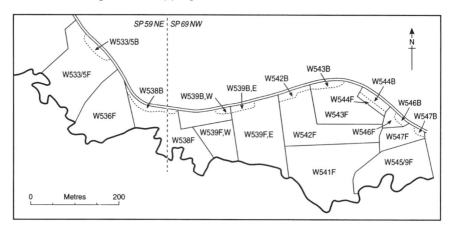

Figure 4.3 Location of canal banks and associated grassland fields in parts of SP69NW and SP59NE. For details of vegetation and other characteristics of each site see Table 4.6

DISCUSSION

Lowland Grassland Plant Communities

In Page's (1980) classification lowland *Agrostis/Festuca* grasslands are recognised as an entity – the Anthoxantho–Festucetum (rubrae), whereas in the NVC they are either considered as Upland Grassland (U4) communities occurring in the lowlands (specifically U4b), or are subsumed in either the Centaureo–Cynosuretum cristati (MG5) or the Lolio–Cynosuretum cristati (MG6). In SP69NW unimproved grasslands frequently contained a matrix of *Festuca rubra* and *Agrostis stolonifera* (e.g. 20–30% combined cover), plus some *Agrostis capillaris* together with small amounts of *Lolium perenne* and *Cynosurus cristatus*. The differentials of the Centaureo–Cynosuretum cristati were wholly or partly absent, and the low abundance levels of *Lolium perenne*, *Cynosurus cristatus* and *Phleum pratense* or *P. bertolonii* distinguished them from pure Lolio-Cynosuretum cristati. These swards are best considered as intermediate between the Anthoxantho–Festucetum rubrae, and the Lolio–Cynosuretum cristati (or occasionally the Centaureo–Cynosuretum cristati); hence both Page's (1980) classification and the NVC have been used in parallel in this account.

It is interesting to note that Centaureo–Cynosuretum cristati (MG5) grasslands are considered to be associated with hay management (Rodwell, 1992), but in the vignette area they were normally grazed continuously either by sheep or by a mixture of sheep and cattle. *Arrhenatherum elatius*, *Anthriscus sylvestris* and *Festuca pratensis* were present on many of the banks, suggesting that at times the level of grazing was light, and/or intermittent.

Mapping techniques

In the initial phase of the study, a total survey of SP69NW was undertaken. This allowed the key features/topics to be identified and investigated further. The data on

Table 4.6 Vegetation and other characteristics of (a) canal banks and (b) associated grassland fields in parts of SP69NW and SP59NE in 1991

(a) Banks

Code	W533/5B	W538B	W539B,W	W539B,E	W542B	W543B	W544B	W546B	W547B
S	40	52	39	44	38	27	27	28	38
A species	6	7	2	3	3	1	0	1	2
Association	C–C	C–C	Ae	C–C	C–C/L–C	A–Fr/L–C	A–Fr/L–C	L–C	C–C
D	8	7	4	4	1	0	0	0	5
Anthills	9	4	0	2	4	13	3	1	4
NVC Code	MG5b	MG5b	MG1e	MG5b	MG6	MG6	MG6	MG6	MG5b
Length (m)	107	187	25	33	125	103	117	32	58
Width (m)	6*	9	10	13	12 and 7	10	7	8	12
Gradient	St	St/Md	St	St	St	St	Sh	Sh	St
Status	U	U	–	U	U	P	P	P	P

(b) Fields

Code	W533/5F	W536F	W538F	W539F,W	W539F,E	W541F	W542F	W543F	W544F	W546F	W547F	W545/9F
S	25	29	27	29	15	22	23	21	24	21	24	16
A species	0	1	1†	0	0	0	0	0	0	0	0	0
Association	L–C	A–Fr/L–C	L–C	L–C	P–L/L–C	P–L	L–C	L–C	L–C	L–C	P–L&L–C(1)	L–P
D	0	0	0	0	0	0	0	0	0	0	0	0
Anthills	0	2	0	0	0	0	0	0	0	0	0	0
NVC Code	MG6	MG6	MG6	MG6	MG7b	MG7b	MG6	MG6	MG6	MG6	MG7b and 6	MG7a
Area (ha)	5.18	3.79	3.59	3	4.04	5.08	4.23	3.37	0.18	0.39	1.03	3.46

Notes:

Code: from the first edition 1:2500 map; W = Wigston (parish); E = East, W = West.

S = Species richness.

A = No. of non-improvement indicator species (Page, 1980): Ae: Arrhenatheretum elatioris, A–Fr: Anthoxantho–Festucetum rubrae, C–C: Centaureo–Cynosuretum cristati, L–C: Lolio–Cynosuretum cristati, L–P: Lolio–Plantaginetum, P–L: Poo–Lolietum perennis. (1) = localised.

D = Number of differentials of the Centaureo–Cynosuretum cristati.

Anthills = number of anthills of *Lasius flavus*.

NVC = National Vegetation Classification community code (Rodwell, 1992), see Table 4.4.

Width: most relevant single figure, width often varies along length (* some sections 1.5 m, some 2 m wide).

Gradient: St = steep, Md = medium, Sh = shallow.

Status: U = unimproved, P = part improved.

the 1:10 560 map for the entire 5×5 km square (a section showing a selection of these findings is presented as Figure 4.1) indicated the following features which could be amplified in the ways described:

(a) A concentration of unimproved sites along the river valley – which could be emphasised by the use of a selective map of the river valley.
(b) A random scatter of other non-improved sites throughout SP69NW – this information could alternatively have been presented as numerical data in a grid, each cell of which represented a 1×1 km square.
(c) Change in SRPG survival – this feature was well shown for part of the area in Figure 4.1 but could be presented in tabular form (e.g. as Table 4.2, supplemented by details of the fate of former SRPGs).
(d) Changes in land use – for site-specific information two full maps are required (or a single map with data from the second survey date coloured or coded differently) – general trends can be shown in tabular form, e.g. Table 4.3.
(e) A major variation in site size – this could be overcome by using enlarged maps of sample or vignette areas (as in Figures 4.1 and 4.3).

The follow-up phase involved the investigation of the intensification process using the canal banks in the vignette area. The resultant map, with its pattern of field boundaries, helps to create a mental picture of the area and its associated ecological relationships which is easily retained in the memory (Figure 4.3). The data in Table 4.6 allow the overall trends to be viewed in association with the map and these trends could be reinforced by the incorporation of both map and tabular data within a geographical information system, thus facilitating display of the spatial distribution of the tabulated features.

Timing of Survey Work

Ground-based *land use* survey can be carried out at almost any time of year, although it may not be possible to predict whether recently ploughed land (especially when visited in autumn or winter) is going to be arable or grass ley in the following season.

For medium-scale (Gulliver, 1994) grassland surveys for *conservation assessment* purposes, species lists consisting of moderately comprehensive data can be obtained from late April to September (but see Smith, 1994, for a demonstration of the effects of recording date on species richness in $0.0625 \, m^2$ quadrats); two (or more) site visits are preferable to one. Many indicator species can be detected most effectively from May to July. The presence of anthills can be observed at any time of year. If all the latter three aspects are studied together, any conclusions will be reinforced by the existence of complementary evidence (as in the vignette area). Conversely, if, for example, anthills alone are recorded, only a sample of the total stock of unimproved sites will be detected. Nevertheless this will be sufficient to show differences between parishes and areas of, for example, 5×5 km or 10×10 km in the level of survival of unimproved grassland. Furthermore, it will provide an adequate baseline for monitoring the subsequent loss of those unimproved grassland sites originally containing anthills.

Table 4.7 Species richness and number of non-improvement indicator plants (A species) at five banks surveyed in 1983 and 1991. Banks are grouped as 'unimproved' or part-improved based on the status allocated in 1991

Site reference number	1983 Whole site data and quadrat data (presented in Table 4.8) gathered at one visit		1991 Two whole site visits	
	A species	Total species richness (S)	A species	Total species richness (S)
'Unimproved' bank				
W533/5*	6	44	6	40
W538†	11	49	7	52
W539‡	9	55	3	44
W542	6	40	3	38
Mean	8	47	4.75	44
Part-improved bank				
W543	9	43	1	27

* In SP59NE.
† Part in SP59NE and part in SP69NW. The other three banks are in SP69NW.
‡ The area of this bank had been reduced by 1991 by civil engineering work (dredging disposal site construction) on the western part of the land parcel.

Table 4.8 The relationship between sample size and number of plant species (S) at seven unimproved, farmed, canal banks in SP69NW (5) and SP59NE (2), in June 1983. From Gulliver (1994), reproduced by permission of the British Grassland Society

	Quadrat			Entire site 1983	Entire site all dates	$4\,m^2$ value as % of 1983 entire site value
	$1\,m^2$	$4\,m^2$	$25\,m^2$			
Mean	27	31	41	45	57	69
Range	21–32	25–34	36–51	40–55	52–58	62–76

Species Richness and Yield Relationships

The Park Grass Plots at Rothamsted (Brenchley and Warington, 1958; Silvertown, 1980; Dodd *et al.*, 1994) were established to demonstrate the effect on hay yield of applying fertilisers and organic matter to grassland. Nitrogen, potassium and phosphates (plus other plant nutrients) and lime were applied in different combinations and at different rates. The net result has been to produce a wide range of soil fertilities. The size of the whole plots (i.e. both half plots or all sub-plots) varies from *c.* 0.2 ha to *c.* 0.05 ha. There is a significant negative relationship between hay yield and species richness of the whole plots (see Figure 4.4), showing the sorts of processes by which the variety of plant species is reduced when fertilisers are applied to farmland swards.

If data are gathered from large plots (e.g. 33 × 33 m) during *field surveys*, the ranking of such sample plots by species richness would probably provide a strong indication of the relative fertility of the swards, and thereby aid the interpretation of the ecology of the sites. The scatter of points around the regression line in Figure 4.4 should be noted:

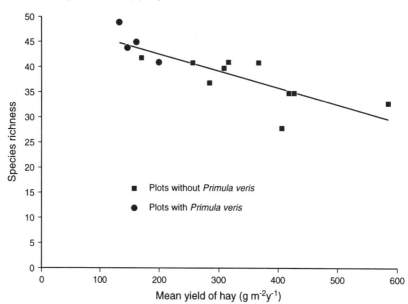

Figure 4.4   Vascular plant species richness versus mean hay yield (1926–49, first cut only) for those Park Grass Plots at Rothamsted with a pH greater than 4.5. $y = -0.033x + 49.3$; $p < 0.001$. Sources: Brenchley and Warington (1958), manuscript records at Rothamsted, author's site records. Yields averaged over limed and unlimed half plots

inherent variability may well be even more evident for survey data from swards subject to a range of different management regimes, and occurring on a variety of soils.

Indicator Species Characteristics

The association between the A species and SRPG is shown in Figure 4.2. The relationship for *Primula veris* (an A1 species) is further amplified in Figure 4.4. *Primula veris* at Rothamsted is restricted absolutely to low fertility, high species-richness plots, and as such is in the 'detector' class of indicator species (Spellerberg, 1992).

ACKNOWLEDGEMENTS

The numerous people who have assisted in the various stages of the study are thanked most sincerely. I would especially like to express my gratitude to my wife.

REFERENCES

Barr, C. (1994) Countryside survey 1990 – an overview. *Ecos*, **15**, 9–18.
Barr, C., Benefield, C., Bunce, B., Ridsdale, H. and Whittaker, M. (1986) *Landscape Changes in*

Britain. Institute of Terrestial Ecology, Huntingdon, UK.

Beresford, J.E. and Wade, P.M. (1982) Field ponds in north Leicestershire. *Transactions of the Leicester Literary and Philosophical Society,* **76,** 25–34.

Bowers, J.K. and Cheshire, P. (1983) *Agriculture, Countryside and Land Use: an Economic Critique.* Methuen, London.

Brenchley, W.E. and Warington, K. (1958) *The Park Grass Plots at Rothamsted 1856–1949.* Rothamsted Experimental Station, Harpenden, UK.

Coleman, A. and Maggs, K.R.A. (1968) *Land Use Survey Handbook.* King's College Department of Geography, London.

Dodd, M.E., Silvertown, J., McConway, K., Potts, J. and Crawley, M. (1994) Stability in the plant communities of the Park Grass Experiment: the relationships between species richness, soil pH and biomass variability. *Philosophical Transactions of the Royal Society London* B, **346,** 185–193.

Forman, R.T.T. and Godron, M. (1986) *Landscape Ecology.* Wiley, New York.

Fuller, R.M. (1987) The changing extent and conservation interest of lowland grasslands in England and Wales: a review of grassland surveys 1930–1984. *Biological Conservation,* **40,** 281–300.

Green, B. H. (1989) Agricultural impacts on the rural environment. *Journal of Applied Ecology,* **26,** 793–802.

Gulliver, R.L. (1986) Land use changes in a study area in central England (SP69NW), together with the results of field boundary and grassland wildlife surveys: an example of providing the scientific basis for landscape planning and management purposes. Manuscript deposited in the Leicester Museum, Leicester, UK.

Gulliver, R.L. (1994) Medium scale surveys of grassland for landscape and conservation purposes. In: R.J. Haggar and S. Peel (eds) *Grassland Management and Nature Conservation.* Occasional Symposium No. 28, British Grassland Society, 1993, Aberystwyth, UK, pp. 281–283.

Harvey, G. (1997) *The Killing of the Countryside.* Jonathan Cape, London.

King, T.J. (1981) Ant-hills and grassland history. *Journal of Biogeography,* **8,** 329–334.

Nature Conservancy Council (1984) *Nature Conservation in Great Britain.* Nature Conservancy Council Interpretative Branch, Shrewsbury, UK.

Page, M.L. (1980) A phytosociological classification of British neutral grasslands. PhD thesis, University of Exeter, UK.

Pollard, E., Hooper, M.D. and Moore, N.W. (1974) *Hedges.* New Naturalist No. 58, Collins, London.

Pye-Smith, C. and Rose, C. (1984) *Crisis and Conservation; Conflict in the British Countryside.* Penguin Books, Harmondsworth, UK.

Rackham, O. (1986) *The History of the Countryside.* Dent, London.

Relton, J. (1972) Disappearance of farm ponds. *Monks Wood Experimental Station Report 1969–1971,* 32.

Rodwell, J.S. (ed.) (1992) *British Plant Communities: 3: Grassland and Montane Communities.* Cambridge University Press, Cambridge.

Shoard, M. (1980) *The Theft of the Countryside.* Maurice Temple Smith, London.

Silvertown, J. (1980) The dynamics of a grassland ecosystem; botanical equilibrium in the Park Grass Experiment. *Journal of Applied Ecology,* **17,** 491–504.

Smith, R.S. (1994) Effects of fertilisers on plant species composition and conservation interest of UK grassland. In: R.J. Haggar and S. Peel (eds), *Grassland Management and Nature Conservation.* Occasional Symposium No. 28, British Grassland Society, 1993, Aberystwyth, UK, pp. 64–73.

Spellerberg, I.F. (1992) *Evaluation and Assessment for Conservation.* Chapman & Hall, London.

Spencer, J.W. and Kirby, L.J. (1992) An inventory of ancient woodland for England and Wales. *Biological Conservation,* **62,** 77–93.

Stace, C.A. (1991) *New Flora of the British Isles.* Cambridge University Press, Cambridge.

5 Non-destructive Sampling of Cretan *Garigue* Biomass as Ground Truth for Remote Sensing

E.W.G. JONES
Department of Geography, University of Cambridge, UK

High-resolution visible/near-infrared sensors such as SPOT and Landsat TM provide information on vegetation variables over large areas at scales of 20–30 m. Before vegetation indices can be used to derive quantitative biomass information from satellite imagery, they must first be calibrated through the collection of ground data. Traditionally, ground truth data has been collected by destructive sampling on systematic and random schemes, but this has produced poor accuracies when sampling heterogeneous vegetation. The Cretan *garigue* dwarf shrub communities, in common with much of the semi-natural vegetation found in the Mediterranean region, are highly heterogeneous on both inter-pixel and sub-pixel scales. Sampling of 1 m^2 quadrats was deemed inappropriate for this reason, and the non-destructive technique of dimensional measurements was applied to 5×5 m and 10×10 m quadrats, covering a larger proportion of the pixel area. A Chow equality test was used to compare regression coefficients obtained for the same species at different field sites, and reduced major axis regression was found to give improved results. Results showed that the (5×5 m) quadrat biomass could be determined to a precision of between 1 and 5% in most cases. These initial results suggest that this is a promising technique for the collection of accurate ground data without causing excessive damage to the shrub communities under study.

INTRODUCTION

Before quantitative interpretations about terrestrial variables can be drawn from remotely sensed data the techniques must be validated by the collection of accurate ground data. This is particularly important for vegetation studies because of the large number of factors that can influence the variables of interest (Williamson, 1986) and the spectral signatures of vegetated land surfaces (Myneni *et al.*, 1992). The work described in this chapter was carried out with the aim of obtaining a data set of above-ground biomass measurements for *garigue* dwarf shrub communities on the Mediterranean island of Crete, to facilitate mapping of semi-natural biomass using Landsat TM and SPOT HRV imagery. For this purpose accurate ground biomass data are required to calibrate vegetation indices. Discussion is limited to the problem of evaluating above-ground biomass on the pixel scale and estimating the uncertainty in the measurements, and not extended to the additional uncertainties introduced by registration of the measurements with remotely sensed data.

Vegetation Mapping: From Patch to Planet. Edited by Roy Alexander and Andrew C. Millington.
© 2000 John Wiley & Sons Ltd.

In common with much of the Mediterranean region, the landscape of the study area is heterogeneous at the scale of Landsat TM and SPOT HRV pixels. The semi-natural vegetation communities exhibit sub-pixel variability down to the scale of the individual shrubs against a soil and rock background. The traditional clip and weigh technique of biomass evaluation has been shown to require large samples when applied to semi-natural grasslands in the UK (Curran and Williamson, 1986), and would be expected to produce even poorer results for shrub communities where the fundamental unit of variability (the individual shrub) is larger. Precision can be improved by optimal sampling and kriging which takes account of spatial autocorrelation in the variable of interest (Atkinson, 1991). However, intuition suggests that for heterogeneous surfaces a large proportion of the pixel area must still be sampled to obtain a precise estimate for the biomass, and it is therefore desirable to seek a rapid non-destructive technique. Such methods for the evaluation of shrub biomass are reviewed by Etienne (1989).

The technique of estimating biomass or other tree and shrub variables from dimension measurements is well established in the ecological literature, and is sometimes referred to as 'dimensional analysis' therein (e.g. Whittaker and Woodwell, 1968), although this term has a different meaning in the physical sciences. In this approach regression analysis is used to relate the parameter of interest to easily measured plant dimensions. A certain amount of destructive sampling is necessary to set up the regression relationships for each species, but once this has been done the biomass of large areas can be determined simply by making a few measurements per plant. The shrub dimensions most commonly used for predicting biomass are crown height and the maximum and minimum diameters of the crown when viewed from above. These are often used to calculate crown volume (see Figure 5.1), thus simplifying the regressions.

The literature contains few examples of this technique being applied in Mediterranean shrublands, but in a comprehensive paper Armand *et al.* (1993) showed very good regression results for samples of 16 shrub species at French Mediterranean sites. They reported correlation coefficients of between 0.83 and 0.98 for log–log regressions of biomass against crown volume, from samples of 18–100 individuals. Pereira *et al.* (1994) used much smaller samples to derive regression relationships for fuel mapping in central Portugal. There are a number of examples of the successful application of dimension measurements in American shrub communities. Among these, Ludwig *et al.* (1975) used crown volume to estimate the biomass of desert shrubs in New Mexico, and Hughes *et al.* (1987) concentrated on the relationship between volume and current season production for Texas shrubs.

Three of the geometric models that can be used to calculate crown volume from shrub dimensions are shown in Figure 5.1. The choice of model is dependent on the growth habit of the shrubs being studied, although they only differ by a constant scaling factor so the decision will not affect the fit of the regression line or the accuracy of biomass predictions made from it. The choice of the regression model is more important. Bryant and Kothmann (1979) and Hughes *et al.* (1987) found that log–log ($\log M = \alpha + \beta \log V$) and quadratic models ($M = \alpha + \beta V + \gamma V^2$) yielded the highest correlation coefficients in the majority of cases, but Hughes *et al.* added that the quadratic model occasionally resulted in non-random residuals. Murray and Jacob-

 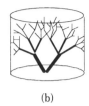  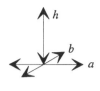

(a)   (b)   (c)

Figure 5.1 Models for calculating crown volume. (a) Inverted cone of elliptical cross-section ($V = \pi/12.abh$), (b) elliptical cylindroid ($V = \pi/4.abh$), (c) upper half ellipsoid ($V = \pi/6.abh$). The perpendicular dimensions *a*, *b* and *h* are defined in the text

son (1982) concluded that simple linear, log–linear, and log–log regressions all provided satisfactory predictors of biomass, but models using height with crown circumference were more reliable. This result is to be expected as a multiple regression should explain more of the variability in biomass than a single independent variable model. The log–log (exponential) model is attractive because it avoids the physically unacceptable prediction of a finite biomass at zero crown volume, and β coefficients in the range 0.67–1.0 can be interpreted as representing a situation somewhere between the two hypothetical extremes of the biomass being distributed entirely over the surface area of the crown and uniformly throughout its volume.

METHOD

Field Sites

Four sites were selected to reflect the variety of *garigue* communities found within the study area. They span the width of the island from north to south, and occupy positions ranging from coastal to an altitude of 650 m in the central mountains (Figure 5.2). The four sites were:

1. Sfakia, on the south coast, east of the village of Chora Sfakion. The site is a large alluvial fan emanating from the mouths of gorges cut deep into the limestone mountains that rise to the north, and descending southwards to the sea. There is a partial cover of thyme (*Thymus capitatus*), thorny burnet (*Sarcopoterium spinosum*), and spiny broom (*Calicotome villosa*), with a large proportion of stones and bare ground visible between the plants.
2. Kournas, a hillside above the road between Lake Kournas and Mathes, a few kilometres from the north coast. The vegetation is the lushest of the four sites, with a more diverse shrub population than Sfakia, and areas of 100% ground cover. Thyme, burnet, *Phlomis fruticosa*, heather (*Erica manipuliflora*), savory (*Satureja thymbra*), and broom (*Genista acanthoclada*) are all evident. Larger shrubs and small trees can be found low on the slopes where there is greater moisture availability.
3. Kumi, inland south of the town of Rethimnon, at the eastern end of the study area. This site is a limestone hillside above the road to the village of Kumi with a partial

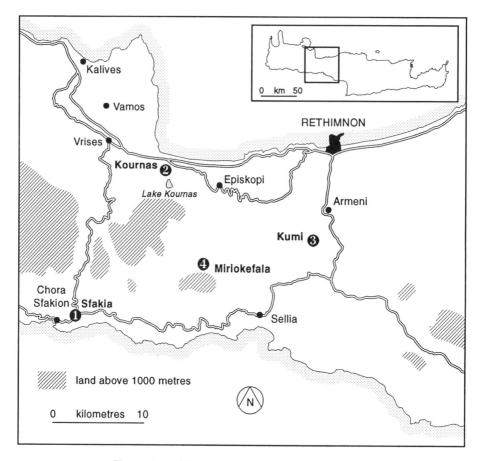

Figure 5.2   Map showing location of field sites

cover of relatively small shrubs. Thyme, burnet, phlomis, heather, savory, broom and *Spartium junceum* are the dominant shrub species.

4. Miriokefala, at an altitude of about 650 m in the central mountains. The site is a steep hillside above the village, and on the northern side of the mountains. Thyme, burnet and gorse are the only *garigue* species evident in significant numbers.

Establishing Size–Biomass Relationships

The dominant shrub species at each field site were identified for destructive sampling to establish the regression relationships. Initially a minimum of five individuals per species per site were measured and harvested. Individuals were selected to reflect the range of sizes evident at each site. An effort was made to select plants with irregular as well as regular crown forms, in order to avoid bias in the fitted size–biomass relationships. The following dimensions were measured for each individual prior to harvesting:

- *a*, the largest diameter of the projection of the crown onto the ground, as determined by eye.
- *b*, the diameter of the projection of the crown onto the ground in the direction perpendicular to *a*.
- *h*, the height of the top of the crown.

For the purposes of the measurements all extremities were defined as the furthest extent of the foliage, excluding flower stalks and stray shoots. Measurements were made with a 2 m rule to an estimated accuracy of $\pm$ 1 cm (although larger error margins were assumed for the dimensions of the larger shrubs). Crown volume was calculated using the elliptical cylindroid model ($V = \pi/4.abh$).

The relative uncertainty in $V = \pi/4.abh$ is given by (Squires, 1985, p. 36)

$$\left(\frac{\Delta V}{V}\right)^2 = \left(\frac{\Delta a}{a}\right)^2 + \left(\frac{\Delta b}{b}\right)^2 + \left(\frac{\Delta h}{h}\right)^2 \qquad (5.1)$$

It can be seen that this is independent of the geometric model.

Once its dimensions had been measured each individual was harvested and weighed. Shrubs were cut to the ground, and weighed in the field using a spring balance with 50 g graduations. Two representative individuals per species per site were retained and sun-dried for 21 days in paper bags, and their mean weight losses used to convert field weights to dry biomass.

When a species occurred at only one field site the sample size was increased to 10 (with the exception of *Rhamnus lycioides* at Kournas, which was found not to be as widespread as at first thought), thus ensuring that a minimum of 10 individuals of any particular species were harvested. The sizes of the calibration samples at each field site are listed in Table 5.1.

Above-ground biomass was then measured non-destructively in 5×5 m square quadrats. The species and dimensions *a*, *b* and *h* of all shrubs within each quadrat were recorded. Individual shrubs were deemed to lie wholly inside a quadrat if the centre of their crowns fell within its boundaries, and wholly outside otherwise. The data were entered into a spreadsheet, and regression relationships were used to estimate the total above-ground (dry) biomass of the quadrat, along with the associated uncertainty. At each location groups of four to six adjacent quadrats were measured, which provided a sample area of at least 100 m^2. The quadrats were orientated at a bearing of 010° so that their sides approximately matched the local along-track and across-track directions of the SPOT HRV and Landsat TM sensors. They were located using a combination of GPS readings and compass bearings to identifiable points.

Regression Analysis

Crown volume was related to above-ground biomass by log–log regression. The function relating the dependent and independent variables is assumed to be of the form

$$M = A \, V^B \qquad (5.2)$$

where biomass, *M*, is the dependent variable, crown volume, *V*, is the independent variable, and *A* and *B* are regression coefficients. Taking logarithms of each side, this function becomes

Table 5.1. Dominant species identified at the four field sites, and numbers of each harvested to set up size–biomass relationships

Species	Sfakia	Kournas	Kumi	Miriokefala
Thymus capitatus	6	9	5	5
Sarcopoterium spinosum	5	5	5	5
Calicotome villosa	10	–	–	–
Phlomis fruticosa	–	10	5	–
Erica manipuliflora	–	5	5	–
Spartium junceum	–	–	10	–
Genista acanthoclada	–	–	5	5
Satureja thymbra	–	6	6	–
Rhamnus lycioides	–	5	–	–

$$\log M = \log A + B \log V \qquad (5.3)$$

Therefore, the coefficients A and B can be determined from a linear regression of $\log M$ against $\log V$.

Because it was found that the relative uncertainties in the volume and biomass measurements were usually of the same order (in both cases the estimated relative errors were larger for smaller shrubs), reduced major axis regression was used to obtain the size–biomass relationships. Till (1973) pointed out that adoption of the classical regression line of Y on X involves an implicit assumption that the measurements of the independent variable, X, are exact (i.e. error free), which is very rarely the case. Curran and Hay (1986) discussed this limitation of classical linear regression in the context of relating remotely sensed variables to ground data, and argued that its application was inappropriate unless the uncertainty in one variable was very much greater than the other. The reduced major axis line (attributed to Jones, 1937) is fitted by minimising the sum of the areas of the right-angled triangles formed by projecting the X and Y coordinates of a set of datapoints onto the line, and hence treats the errors in both variables with equal importance.

The population reduced major axis line, $\hat{Y} = \alpha + \beta X$, can be estimated from a set of measurements, giving

$$\hat{Y} = \hat{\alpha} + \hat{\beta}X \qquad (5.4)$$

where $\hat{\alpha}$ and $\hat{\beta}$ are, respectively, the sample estimates of the slope and intercept parameters α and β.

The slope is given by

$$\hat{\beta} = \sigma_y/\sigma_x \qquad (5.5)$$

and the intercept by

$$\hat{\alpha} = \bar{Y} - \hat{\beta}\bar{X} \qquad (5.6)$$

where σ_x and σ_y are the standard deviations of the X and Y measurements, and $\bar{X}$ and $\bar{Y}$ are their means.

The correlation coefficient, r, is

$$r = \frac{\sum(X_i - \bar{X})(Y_i - \bar{Y})}{[\sum(X_i - \bar{X})^2 \sum(Y_i - \bar{Y})^2]^{0.5}} \tag{5.7}$$

The standard error in the slope estimate is

$$\hat{\sigma}_{\hat{\beta}} = \hat{\beta} \sqrt{\left(\frac{1 - r^2}{n}\right)} \tag{5.8}$$

It can be shown that the mean square deviation perpendicular to the major axis is given by

$$\hat{\sigma}_d^2 = \frac{2\sigma_x^2 \sigma_y^2}{(\sigma_x^2 + \sigma_y^2)}(1 - r) \tag{5.9}$$

And hence, for a given value of X, the mean square prediction error in $\hat{Y}$ due to deviation in the data from the major axis line is given by

$$\hat{\sigma}_{Y.X}^2 = 2\sigma_y^2(1 - r) \tag{5.10}$$

This can be converted to an unbiased estimator by multiplying by $n/(n - 2)$.

Using the approximation $2(1 - r) \approx 1 - r^2$, equations (5.8) and (5.10) can be combined to provide the following (unbiased) estimate for the standard error in $\hat{Y}$ predicted from the major axis line at any value of X

$$\hat{\sigma}_{\hat{Y}} = \sigma_y \sqrt{\frac{1 - r^2}{n - 2}\left[1 + \frac{(X - \bar{X})^2}{\sigma_x^2}\right]} \tag{5.11}$$

However, size–biomass relationships obtained by reduced major axis regression should not be interpreted as functional relationships between biomass and crown volume, but as estimates of an underlying structural relationship obscured by additional random variability (e.g. crown shape, age, local environmental variables). This additional variability is interpreted as *individuality* in the data (Sprent, 1969, pp. 29–46), and is estimated by (5.9). It is treated as an independent error, and hence the estimated standard error in $\hat{Y}$ becomes

$$\hat{\sigma}_{\hat{Y}} = \sigma_y \sqrt{\frac{1 - r^2}{n - 2}\left[1 + n + \frac{(X - \bar{X})^2}{\sigma_x^2}\right]} \tag{5.12}$$

For the purposes of simple spreadsheet calculations this can be rewritten as

$$\hat{\sigma}_{\hat{Y}} = \hat{\sigma}_{\hat{\beta}} \sqrt{\left(\frac{n}{n - 2}\right)\left[(n + 1)\sigma_x^2 + (X - \bar{X})^2\right]} \tag{5.13}$$

Equation (5.13) was used to compute estimates for the standard errors in predicted biomass for all shrubs within each quadrat. The measurement errors in the predictor (volume) were assumed to be included within the regression residuals. The individual errors were then combined in quadrature to provide an estimate for the accuracy of the biomass predicted for the whole quadrat. This requires extensive computation, but is readily achievable with a spreadsheet approach. The predicted biomass values and error estimates are presented in the section on quadrat biomass measurements.

RESULTS

Size–Biomass Relationships

Dry biomass is plotted against crown volume on a log–log scale in Figures 5.3 and 5.4, for thyme and burnet, the two species that were sampled at all four sites. The regression results for the shrub species at each field site are presented in Table 5.2. The correlation coefficients (r^2) are generally very good, and are all higher than 0.9 and significant at the 1% level.

Generality of Size–Biomass Relationships

It is useful to investigate the generality of the size–biomass relationships obtained for a particular species, because if the calibration data from different sites can be treated as being drawn from the same population then the datasets can be combined to form larger, more reliable samples for the derivation of regression relationships. If the same relationship between dimensions and biomass was found to apply over a wide geographical area, it would imply a significant and desirable reduction in the amount of destructive sampling necessary during the collection of ground data.

Chow (1960) presented a series of tests for equality between the sets of coefficients of two linear regressions. He showed that the H_0 null hypothesis, of equality between the p coefficients of two linear regressions estimated from two independent samples of n and m observations (for $n, m > p$), can be tested by the F ratio

$$F(p, n + m - 2p) = \frac{(A - B - C)}{p} \times \frac{(n + m - 2p)}{(B + C)} \tag{5.14}$$

where: A is the residual sum of squares of $n + m$ deviations of the dependent variable from the regression estimated by $n + m$ observations, with $n + m - p$ degrees of freedom; B is the residual sum of squares of n deviations of the dependent variable from the regression estimated by the first n observations, with $n - p$ degrees of freedom; and C is the residual sum of squares of m deviations of the dependent variable from the regression estimated by the second m observations, with $m - p$ degrees of freedom.

Chow derived the above ratio from the matrix formulation of the classical ordinary least squares regression. It is reasonable to assume that the critical arguments in the derivation, regarding the degrees of freedom of $A - B - C$ and $B + C$ and their independence under the null hypothesis, will also hold for the residuals of reduced major axis regressions (i.e. the equivalent sums of the areas of the right-angled triangles formed by projections of the X and Y coordinates of datapoints onto the reduced major axis). Generality of size–biomass relationships was therefore assessed by performing Chow equality tests on reduced major axis residuals. The results for thyme are given as an example in Table 5.3.

Thyme and burnet were found at all four field sites. In both cases the Kournas size–biomass relationships were significantly different from those obtained at all other sites. The interrelationships between the samples from the other three sites were less conclusive, so they were investigated further by grouping pairs of these samples and performing Chow tests against the third. Similar results were again obtained for thyme

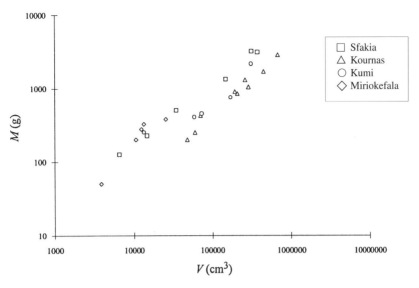

Figure 5.3 Dry biomass versus crown volume for thyme (*Thymus capitatus*) at the four field sites

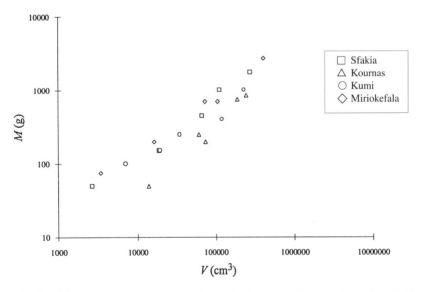

Figure 5.4 Dry biomass versus crown volume for burnet (*Sarcopoterium spinosum*) at the four field sites

Table 5.2 *A, B* regression coefficients and r^2 correlation coefficient for log–log reduced major axis regressions of dry biomass against crown volume for each sampled species at the four field sites

Location	Species	n	A	B	r^2
Sfakia	*T. capitatus*	6	0.1012	0.808	0.998
	S. spinosum	5	0.0769	0.797	0.988
	C. villosa	10	0.0541	0.833	0.939
Kournas	*T. capitatus*	9	0.00797	0.950	0.987
	S. spinosum	5	0.00301	1.014	0.990
	P. fruticosa	10	0.00077	1.085	0.969
	E. manipuliflora	5	0.0153	0.920	0.993
	S. thymbra	6	0.205	0.662	0.986
	R. lycioides	5	0.0653	0.780	0.996
Kumi	*T. capitatus*	5	0.260	0.685	0.923
	S. spinosum	5	0.297	0.641	0.975
	P. fruticosa	5	0.182	0.637	0.954
	E. manipuliflora	5	0.0784	0.820	0.989
	G. acanthoclada	5	0.00054	1.191	0.987
	S. thymbra	6	0.00225	1.066	0.981
	S. junceum	10	0.00044	1.100	0.962
Miriokefala	*T. capitatus*	5	0.00280	1.201	0.953
	S. spinosum	5	0.160	0.744	0.994
	G. acanthoclada	5	0.0824	0.806	0.997

and burnet, with the size–biomass relationships at Kumi being significantly different from those obtained from the combined Sfakia and Miriokefala samples. Hence, only the thyme and burnet samples from the Sfakia and Miriokefala sites were combined. These results are well illustrated by Figures 5.3 and 5.4, where the Kournas samples are seen to have a lower biomass for a given crown volume, and the Sfakia and Miriokefala samples form a line of higher crown density. These observations were interpreted as reflecting a rainfall gradient across the study area, with higher crown densities resulting at the drier sites.

The equality test results for the four species that were found at only two field sites suggested that only heather and savory should be assumed to have the same size–biomass relationship at both of their sites. However, this conclusion for savory was reached despite the Kumi data being distorted by two low biomass datapoints with high relative errors. It was decided to discount these two outliers for the equality test, and this problem illustrates the dangers associated with the use of regression on such small samples. A fit weighted according to the uncertainty in the individual datapoints could improve results.

Quadrat Biomass Measurements

The quadrat biomasses predicted by the reduced major axis regressions are presented in Table 5.4, together with their associated errors. The surprisingly small relative

Table 5.3 Chow equality tests for thyme (*Thymus capitatus*) samples from all four field sites

Site 1	Site 2	n	m	F	5% level	1% level	Significance
Sfakia	Kournas	6	9	78.9	3.98	7.20	1%
Sfakia	Kumi	6	5	3.61	4.74	9.55	No
Sfakia	Miriokefala	6	5	4.83	4.74	9.55	5%
Kournas	Kumi	9	5	5.15	4.10	7.56	5%
Kournas	Miriokefala	9	5	40.4	4.10	7.56	1%
Kumi	Miriokefala	5	5	3.68	5.14	10.92	No

Table 5.4 Predicted dry biomass ($kg\,m^{-2}$) and estimated errors for $25\,m^2$ quadrats using reduced major axis size–biomass relationships

Site	Date	Quadrat	Biomass	Error (%)
Sfakia	5.7.94	A1	0.972	4.49
		A2	0.816	4.31
		B1	0.401	3.55
		B2	0.814	4.51
		C1	0.436	3.93
		C2	0.899	3.95
	Mean = 0.723		S.D. = 0.222	
Sfakia	11.7.94	A1	1.362	4.48
		A2	1.302	4.07
		B1	1.493	4.47
		B2	1.749	4.58
	Mean = 1.476		S.D. = 0.172	
Kournas	12.7.94	A1	0.569	2.48
		A2	0.560	3.64
		B1	0.788	3.90
		B2	0.724	2.43
	Mean = 0.660		S.D. = 0.098	
Kournas	13.7.94	A1	1.838	6.47
		A2	0.970	8.89
		B1	1.758	9.46
		B2	2.089	7.71
	Mean = 1.664		S.D. = 0.418	
Kumi	14.7.94	A1	1.030	6.73
		A2	1.028	6.89
		B1	0.757	5.94
		B2	0.613	3.48
	Mean = 0.857		S.D. = 0.179	

errors can be attributed to a reduction due to summation over the large number of shrubs measured in each quadrat. The reduced major axis technique improved the predicted accuracy by a factor of between 30 and 50% over classical regression for all quadrats (results not presented here). Table 5.4 shows that the variability of mean biomass between adjacent quadrats is generally an order of magnitude higher than the uncertainty in predicted biomass for the individual quadrats. This suggests that the non-destructive method is achieving a more than adequate precision.

Due to the divergent nature of regression errors about the sample mean (see equation (5.11)), the largest errors are associated with individuals of extreme sizes. The poorest results were therefore found when large specimens formed a significant proportion of the quadrat biomass, as their contributions to the total error tended to dominate. A good example of this limitation is the B1 quadrat surveyed at Kumi on 16 July 1994, which contained a specimen of *Spartium junceum* that was much larger than any of the examples used to establish the regression relationship. The predicted error in biomass for this one shrub was so large that it dwarfed the errors of all the other shrubs in the quadrat, and resulted in the unusually large error estimate seen in Table 5.4.

CONCLUSIONS

Log–log regressions of biomass against crown volume showed that the function $M = AV^B$ provides a good model for the size–biomass relationships of several common Cretan *garigue* shrubs. Correlation coefficients of better than 0.9 were obtained for small samples of 5–10 individuals at four field sites. Equality tests on the regression coefficients obtained for the same species at different sites found significant differences in many cases. Samples were combined when found not to differ significantly, but the results suggest that a certain amount of destructive sampling must be performed at each new field site in order to confirm that size–biomass relationships from other sites can be applied.

Reduction of errors due to the large number of shrubs in a 5×5 m quadrat meant that even for such small calibration samples the quadrat biomass could be determined to a precision of better than 10% in most cases. This precision was found to be smaller than the typical variability in biomass between adjacent quadrats, and hence more than adequate. The sampling strategy can be improved for future work by using grids of quadrats so that the biomass of a full remotely sensed pixel can be estimated optimally through kriging.

REFERENCES

Armand, D., Etienne, M., Legrand, C., Marachel, J. and Valette, J.C. (1993) Phytovolume, phytomasse et relations structurales chez quelques arbustes méditerranéens. *Annales des Sciences Forestieres*, **50**(1), 79–89.
Atkinson, P.M. (1991) Optimal ground-based sampling for remote sensing investigations: estimating the regional mean. *International Journal of Remote Sensing*, **12**(3), 559–567.

Bryant, F.C. and Kothmann, M.M. (1979) Variability in predicting edible browse from crown volume. *Journal of Range Management*, **32**(2), 144–146.

Chow, G.C. (1960) Tests of equality between sets of coefficients in two linear regressions. *Econometrica*, **28**(3), 591–605.

Curran, P.J. and Hay, A.M. (1986) The importance of measurement error for certain procedures in remote sensing at optical wavelengths. *Photogrammetric Engineering and Remote Sensing*, **52**(2), 229–241.

Curran, P.J. and Williamson, H.D. (1986) Sample size for ground and remotely sensed data. *Remote Sensing of Environment*, **20**, 31–41.

Etienne, M. (1989) Non destructive methods for evaluating shrub biomass: a review, *Acta Oecologica – Oecol. Applic.*, **10**(2), 115–128.

Hughes, G., Varner, L.W. and Blankenship, L.H. (1987) Estimating shrub production from plant dimensions. *Journal of Range Management*, **40**(4), 367–369.

Jones, H.E. (1937) Some geometrical considerations in the general theory of fitting lines and planes. *Metron*, **13**, 21–30.

Ludwig, J.A., Reynolds, J.F. and Whitson, P.D. (1975) Size–biomass relationships of several Chihuahuan desert shrubs. *American Midland Naturalist*, **94**(2), 451–461.

Murray, R.B. and Jacobson, M.Q. (1982) An evaluation of dimension analysis for predicting shrub biomass. *Journal of Range Management*, **35**(4), 451–454.

Myneni, R.B., Asrar, G., Tanré, D. and Choudhury, B.J. (1992) Remote sensing of solar radiation absorbed and reflected by vegetated land surfaces. *IEEE Transactions on Geoscience and Remote Sensing*, **30**(2), 302–314.

Pereira, J.M.C., Oliveira, T.M. and Paúl, J.C. (1994) Fuel mapping in a Mediterranean shrubland using LANDSAT TM imagery. In: P.J. Kennedy and M. Karteris (eds) *Satellite Technology and GIS for Mediterranean Forest Mapping and Fire Management*. International Workshop, Thessaloniki, 4–6 November 1993, European Commission, Luxembourg, pp. 97–106.

Sprent, P. (1969) *Models in Regression and Related Topics*. Methuen, London.

Squires, G.L. (1985) *Practical Physics*, 3rd edition. Cambridge University Press, Cambridge.

Till, R. (1973) The use of linear regression in geomorphology. *Area*, **5**, 303–308.

Whittaker, R.H. and Woodwell, G.M. (1968) Dimension and production relations of trees and shrubs in the Brookhaven forest, New York. *Journal of Ecology*, **56**(1), 1–25.

Williamson, H.D. (1986) Ground data for remote sensing of vegetation. *Proceedings of a Remote Sensing Workshop on 'Ground Truth'*, 17 April 1986, Department of Geography, University of Nottingham, pp. 41–57.

6 Remote Sensing of Semi-natural Upland Vegetation: the Relationship between Species Composition and Spectral Response

R.P. ARMITAGE[1], R.E. WEAVER[2] and M. KENT[2]
[1] *School of Geography, Kingston University, UK*
[2] *Department of Geographical Sciences, University of Plymouth, UK*

The potential of remote sensing in assisting with the survey and mapping of semi-natural plant communities in the uplands of Britain is assessed. Using two sites on Dartmoor, south-west England, paired quadrat data on floristics and their spectral characteristics derived from ground spectrometry were collected in a series of carefully selected rectangular transects. The spectral data were then processed to simulate the bandsets of both the Airborne Thematic Mapper (ATM) and the Compact Airborne Spectral Imager (CASI). Analysis of the resulting data, using Canonical Correspondence Analysis (CCA), showed a strong relationship between the first axis of the floristic data and near-infrared reflectance (NIR) in both the ATM and CASI bandsets.

Major species could be ordered along the first axis but reflectance was shown to be primarily related to physiognomic properties of dominant species, rather than to detailed floristic composition. Leaf area index was also related to the first axis and to NIR by using the total percentage cover of vegetation in each quadrat as a proxy variable, but only a weakly significant correlation was obtained. Although the results indicate that the pattern of intergrading patches that form semi-natural upland vegetation in Britain should be spectrally identifiable, the complexity of the relationship between spectral response and vegetation composition means that detailed floristic descriptions of communities, like those collected for the National Vegetation Classification (NVC), are difficult to identify from remotely sensed data. The research also demonstrates the potential value of CCA as a means of relating spectral data to floristic composition in environments dominated by semi-natural vegetation such as the uplands of Britain. Further work is required on the application of CCA and related techniques in this area.

INTRODUCTION

Most upland areas in the United Kingdom are covered in semi-natural vegetation, which is distributed as a series of plant communities with differing species compositions (Grant and Armstrong, 1993). The present pattern of grasslands, dwarf shrub communities and, to a lesser extent, mires, is a product of human intervention, which began with the clearance of the native woodlands and has continued with the develop-

Vegetation Mapping: From Patch to Planet. Edited by Roy Alexander and Andrew C. Millington.
© 2000 John Wiley & Sons Ltd.

ment of managed moorlands over a period of 5000 years (Birks, 1988; Ratcliffe and Thompson, 1988). A decline in some traditional management practices, such as controlled burning of mature heather, and greater economic pressures on hill farmers, has led to increased stress on the delicate equilibrium which maintains the present diversity of vegetation (Anderson and Yalden, 1981; Barnes, 1987). The desirability of maintaining the present biodiversity of the uplands is important for both economic and conservation reasons (Grant and Armstrong, 1993; Usher and Thompson, 1993). However, in order to form a viable management strategy, which takes account of economic, recreational and preservation demands, information on the ecology of the uplands, particularly the variation in plant species distributions, is required (Sydes and Miller, 1988; Grant and Armstrong, 1993). The collection of such information using traditional survey techniques is impractical, but multispectral remote sensing may provide a suitable alternative.

A limited amount of research has been carried out into the use of remote sensing for mapping upland vegetation in the United Kingdom (Hume *et al.*, 1986; McMorrow and Hume, 1986; Morton, 1986; Weaver, 1986, 1987; Alam and Southgate, 1987; Wardley *et al.*, 1987). The majority of this work has concentrated on identifying the presence or distribution of single species, particularly *Pteridium aquilinum* and the growth stages of *Calluna vulgaris*, or the differentiation between a number of general vegetation types or key species. Some work has also been done on the problems associated with the boundaries between different plant communities or cover types (Trodd, 1992, 1993).

More general research has indicated that variations in reflected energy from vegetation are not necessarily linked to different species compositions, but relate more to the combined effects of individual leaf biophysical properties, the morphology of individual plants and the structure of the canopy as a whole (Curran, 1994a; Lewis, 1994). Other influences on the spectral response collected by the sensor are time of the year, the health of the vegetation and the nature of the soil beneath the canopy (Curran, 1980, 1983, 1985). In terms of remote sensing of upland areas, this means that the pattern of different spectral responses across an image may not relate, directly, to what is present on the ground in terms of species and plant community variation. The implication of this is that it may not be possible to derive the types of data on plant community distributions from remotely sensed imagery that are required for the sort of management applications discussed above.

The aim of the research discussed in this chapter is to investigate the degree to which variation in spectral response is linked to variation in species composition for a range of typical upland semi-natural plant communities. The approach adopted here involved the collection of spectral response data at ground level and the analysis of their relationship to floristic compositions using multivariate techniques. This chapter will concentrate on the results of analyses carried out using canonical ordination methods.

METHOD

The basic approach adopted in this research was to record the spectral response of a representative range of upland semi-natural vegetation types in visible and near-infrared wavelengths using a handheld spectroradiometer. The collection of ground-level radiance data allows the spectral response of specific objects to be characterised (Alam and Southgate, 1987; Treitz *et al.*, 1992; Taylor, 1993). This characterisation allows a more detailed analysis of the relationship between the properties of the target object and the pattern of recorded reflectance. Steven (1986, 1987) notes that, by gathering data at the ground level, the exact nature of the object and its context can be identified. In the case of vegetation, this means that factors such as species type, structure of the canopy and soil properties can be identified and taken into account.

Study Sites

Two study sites were used for the data collection, both located on the granite massif of Dartmoor, in the south-west of England. The first study site, Homerton Hill (Ordnance Survey grid reference SX 560900), consists of a plateau bounded by two valleys and is located to the south of Meldon Reservoir, near Okehampton (Figure 6.1). The second study area was located in the valley between West Mill Tor and Rowtor (Ordnance Survey grid reference SX 580910), immediately south of Okehampton (Figure 6.2). Both study sites contain examples of the most common vegetation types found in the region covered by the North Dartmoor Site of Special Scientific Interest (SSSI) and Environmentally Sensitive Area (ESA) designations, including unimproved grasslands, heather moorland, bogs and valley mires. The data collection was carried out between 22 July and 7 August 1994.

Sampling Strategy

The sampling strategy was designed to provide a data set containing a wide range of floristic compositions. The first stage in the strategy involved identifying the different community types present in each of the study areas. This was done with the aid of aerial photographs and field reconnaissance. The differentiation between different areas was purely subjective and based on the identification of a change in the general species composition of a stand. Having derived a general pattern of vegetation for both study sites, a series of transects was then identified, each of which ran from one community to another (Figures 6.1 and 6.2). Along each of these transects, a rectangular sampling framework was then established. The dimensions of the sampling rectangles ranged from 10 to 20 m in width and 15 to 40 m in length. The locations of the quadrats from which the floristic and spectral data were to be collected were then selected within the rectangular frameworks. The advantage of selecting sampling points from within the rectangle was that it allowed more of the variation within communities to be sampled than would have been possible with a simple linear transect (Figure 6.3). The actual location of the quadrats was determined subjectively, but was intended to represent the main floristic variation present in each community and its boundary with an adjacent community. In total, 74 quadrats were positioned across the nine sampling rectangles identified in the field.

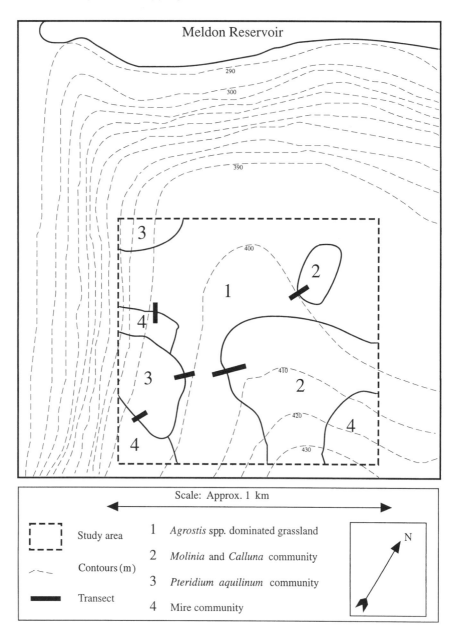

Figure 6.1 Plan showing the study site at Homerton Hill. The dashed rectangle represents the study area and the numbered polygons are the different communities identified. The communities are simply named based on the dominant species present and do not follow a recognised classification

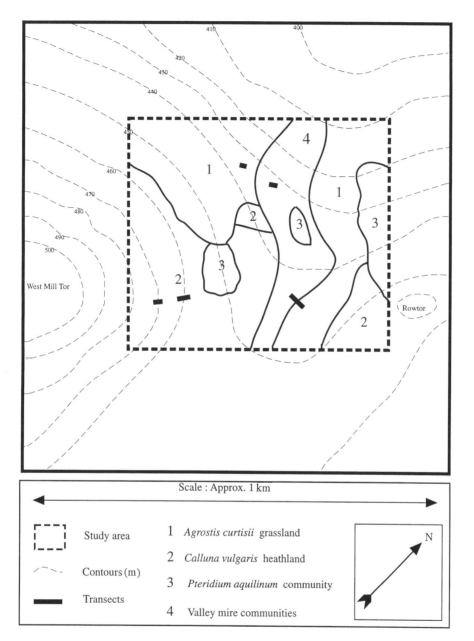

Figure 6.2   Plan showing the study site at Okehampton. The dashed rectangle represents the study area and the numbered polygons are the different communities identified. The communities are simply named based on the dominant species present and do not follow a recognised classification

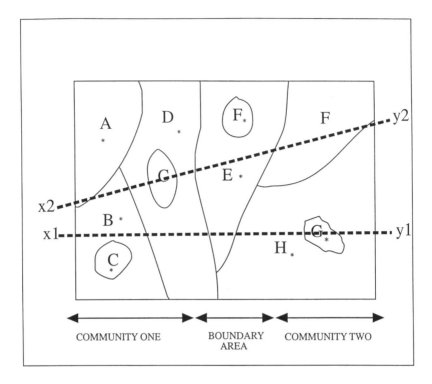

Figure 6.3 The diagram shows the advantage of using the biased sampling approach adopted in this research. The rectangle in the diagram represents the sampling frame placed along each transect. The letters represent areas of different floristic composition within two adjacent communities and their boundary. Sampling along transects x1 y1 or x2 y2 clearly does not allow all the variation present to be sampled. However, by subjectively placing quadrats, indicated by * in the diagram, it is possible to sample the within-community variation more efficiently

Data Collection

For each of the quadrats identified, floristic composition and canopy structure were noted and a measure of spectral response collected. The quadrats used were 0.5 m by 0.5 m. Four sets of vegetation data were collected at each quadrat:

(a) The percentage cover of each species by eye.
(b) The height of each layer in the canopy. This information was collected in order to infer structural properties of the canopy.
(c) A plan of the distribution of the species within the quadrat. This information was collected so that the spectral response could be precisely related to the nature of the vegetation.
(d) An overhead (nadir) photograph of each quadrat was taken as a permanent record.

The spectral data were collected using a field spectrometer on loan from the Natural Environment Research Council (NERC). The instrument, a Spectron SE590,

Table 6.1 Table showing the wavelength range of the spectrometer channels which represent the CASI Default Vegetation bandset

CASI channel	Wavelength (nm)
1	441–461
2	548–557
3	666–674
4	694–703
5	705–711
6	736–744
7	746–753
8	775–784
9	815–824
10	860–870

Table 6.2 Table showing the wavelength range of the spectrometer channels which represent each of the first eight ATM bands

ATM Channel	Wavelength (nm)
1	420–450
2	450–520
3	520–600
4	605–625
5	630–690
6	695–750
7	760–900
8	910–1050

records the levels of radiant energy from a target in 252 contiguous bands covering the electromagnetic spectrum from 400 nm to 1100 nm (Milton *et al.*, 1995). Four measurements were taken from a nadir position for each quadrat. The data were processed in the laboratory, using software provided by NERC, to produce reflectance values (Rollin, 1993a, 1993b). Reflectance is a measure of the incident radiation, at a particular wavelength, that is returned and measured by the sensor, after interaction with the surface (Rees, 1990). The four reflectance values for each quadrat were then averaged.

The spectral data set was processed to represent the bandsets of two airborne sensors (the Airborne Thematic Mapper (ATM) and the Compact Airborne Spectral Imager (CASI)). The rationale for simulating these sensors, using the data from the field spectrometer, was that it allowed assessment of the ability of operational sensors to detect variations in floristic compositions. The simulated data sets produced were a simplification of the bandsets of the two instruments, as no account of the possible variation in sensitivity across individual band widths was made. The bands simulated are shown in Tables 6.1 and 6.2.

DATA ANALYSIS

The aim of the data analysis was to characterise and examine the relationship between floristic composition and spectral response. Two different approaches were adopted. The first involved separate ordination of the floristic and spectral data sets. The floristic data were analysed using Detrended Correspondence Analysis (DCA) (Hill, 1979), while Principal Component Analysis (PCA) was used on the two spectral data sets. The resulting axes of variation and the position of individual quadrats in the DCA and PCA analyses were then compared. This approach was similar to that adopted by Trodd (1992, 1993), which was adapted from work by Gibson and Greig-Smith (1986). The second method, discussed in this chapter, uses Canonical Correspondence

Analysis (CCA), which produces a canonical ordination, in which both floristic and spectral data sets are analysed in a single procedure.

Canonical Correspondence Analysis

Variations in one set of variables can often be explained by differences in a second, related, set of variables. For example, variations in floristic composition can often be explained by differences in environmental variables. Canonical Correspondence Analysis (CCA) provides a method for analysing the relationship between two sets of variables, and is one of a range of techniques called canonical ordination (ter Braak, 1987a, 1987b). The key to CCA and the other canonical methods is the simultaneous analysis of both sets of variables. The ordination axes derived in the analysis are constrained such that the distribution of one set of variables is explained by a linear combination of the other set of variables (ter Braak, 1986, 1987a, 1987b). When used on vegetation and environmental factors, the variation in species composition is related to a gradient formed by a linear combination of environmental factors (Kent and Coker, 1992). In CCA, the plant species are assumed to follow a unimodal response model (ter Braak, 1987b). Further discussion of the theory behind CCA is given in ter Braak (1986, 1987a, 1987b) and ter Braak and Prentice (1988).

The practical application of Canonical Correspondence Analysis has been greatly enhanced by the development of the CANOCO computer package by ter Braak (1987c, 1990). CCA in CANOCO is calculated using a method of ordination and multiple regression of the two sets of variables, in order to derive the relationship between them. Various types of ordination plots, to illustrate the results, can be created using the output from CANOCO in a package called CDLITE (Smilauer, 1992). A detailed description of the CANOCO package is given in ter Braak (1987c, 1990).

Application of CCA to Floristic Composition and Spectral Response

The approach adopted to carry out the CCA involved entering the floristic data as the 'species data'; that is the set of variables whose distribution is to be explained, and the spectral data sets as the 'environmental data'; the set of variables that will constrain the ordination axes and define the gradient along which the 'species' data will be distributed. This approach is unusual, as it would seem more appropriate to look at the variation of spectral response due to changes in floristic composition. However, in this analysis, the aim was to see if a pattern of variation in species compositions, in relation to spectral response, could be determined. In order to do this, the floristic data must be entered as the data set whose distribution is to be explained. In this application, CCA is thus being used as a descriptive tool rather than its more normal role as an analytical one.

Two separate analyses were performed with the CCA, one for each of the simulated spectral data sets with the floristic data. Several runs were carried out in order to refine the identification of trends and patterns present in the data. Three quadrats were identified as outliers, following the initial analyses of both the CASI and ATM data sets. These quadrats were found to be composed entirely of rock and were excluded from further analysis. Following a second run, using the reduced data sets, a further three quadrats were identified as outliers. The presence of rock as part of the quadrat composition accounted for two of these quadrats, the third was found to contain the only occurrence of *Trichophorum cespitosum* in the entire data set. A final run was then carried out on the data sets minus the six outlying quadrats.

The results of each of the final runs are displayed as site-spectral biplots in Figures 6.4 and 6.5. The numbered points on the ordination plots represent the ordering of individual quadrats along each of the first axes of the CCA. The arrows on the plots represent the relationship between individual bands in the spectral data and the ordination axes. Longer arrows, which make smaller angles with a particular axis, are highly correlated to that axis (Kent and Coker, 1992). This means that the distribution of quadrats along an axis can be related directly to their reflectance in particular bands or combinations of bands.

RESULTS

Spectral Data

The relationship between floristic composition and spectral response is shown graphically in Figures 6.4 and 6.5. The first CCA derived axis is strongly related to near-infrared (NIR) reflectance for both CASI and ATM bandsets, with eigenvalues of 0.355 and 0.379, respectively. This relationship is clearly shown by the length of the arrows representing CASI bands six to ten and the angle they make with axis one in Figure 6.4, and by those representing ATM bands six to eight in Figure 6.5. The correlation coefficients between CASI bands six to ten and CCA axis one are all greater than 0.600, and are significant at the 0.05 (95%) level (Table 6.3). The correlation coefficients between the ATM NIR bands were greater than 0.500 and were again significant at the 0.05 (95%) level (Table 6.4).

The variation along the second CCA axis appears not to be related to any part of the sampled electromagnetic spectrum. This conclusion is based on the size and angular relationship of the arrows, representing the individual spectral bands, with the second axis in both Figures 6.4 and 6.5. The visual evidence in the ordination plots is confirmed by the correlation coefficients for spectral bands and axis two, shown in Tables 6.3 and 6.4, none of which rise above 0.200 or are significant at the 0.05 (95%) level. The eigenvalues for axis two are 0.243 and 0.278 for the CASI and ATM analyses, respectively. The second axis is considered in more detail in the discussion.

The indication from the above result is that the order of samples along the first CCA axis is related to some aspect of the near-infrared reflectance from each quadrat. Before considering this relationship in more detail, it is worth examining the ordering of the individual quadrats along the axis in floristic terms and identifying any patterns or trends that may be present.

Ordering of Floristic Compositions along Axis One

The ordering of quadrats along CCA axis one was related to the vegetation composition of each sample. Moving from right to left across the CASI ordination plot, the

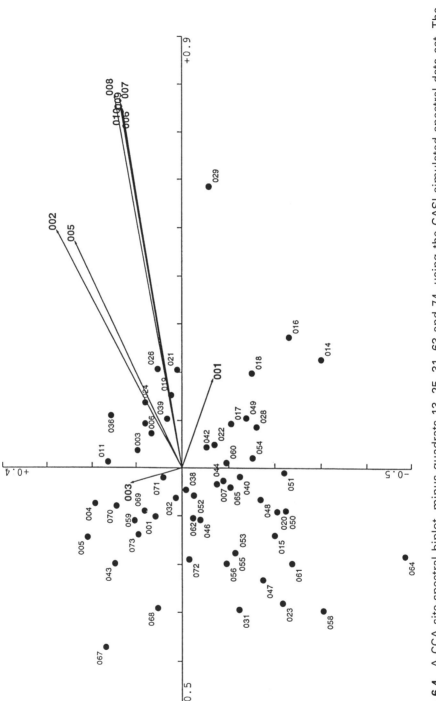

Figure 6.4 A CCA site-spectral biplot, minus quadrats 13, 25, 31, 63 and 74, using the CASI simulated spectral data set. The numbered points represent the sites and the arrows the CASI bands

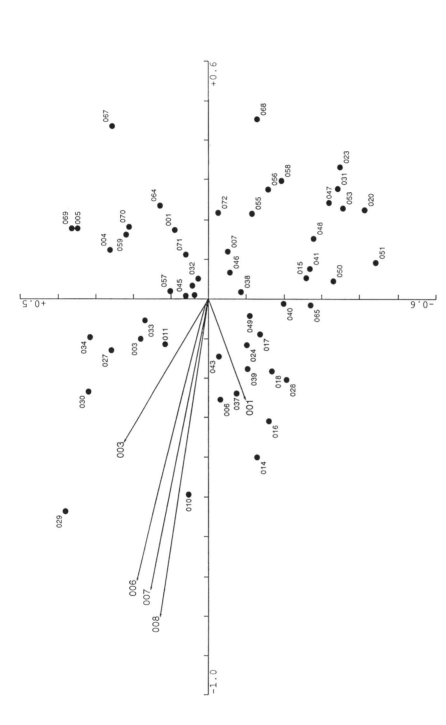

Figure 6.5 A CCA site-spectral biplot, minus quadrats 13, 25, 63 and 74, using the ATM simulated spectral data set. The numbered points represent the sites and the arrows the ATM bands

Table 6.3 Table showing the correlation between CASI bands and the axes derived by CANOCO. The correlation values have been ranked in order of significance for each of the first two axes. Only the bands marked with a * have correlation coefficients significant at the 0.05 (95%) level

Axis one		Axis two	
CASI band	Correlation	CASI band	Correlation
8	0.632*	2	0.194
7	0.627*	5	0.165
10	0.616*	4	0.151
6	0.612*	8	0.105
9	0.609*	7	0.096
2	0.401*	6	0.095
5	0.382*	10	0.093
4	0.195	9	0.091
1	0.100	3	0.075
3	−0.025	1	0.055

Table 6.4 Table showing the correlation between ATM bands and the axes produced by CANOCO. The correlation values have been ranked in order of significance for each of the first two axes. Only the bands marked with a * have correlation coefficients significant at the 0.05 (95%) level

Axis one		Axis two	
Bands	Correlation	Bands	Correlation
5	−0.014	3	0.157
4	−0.090	6	0.117
2	−0.113	7	0.108
1	−0.203	4	0.098
3	−0.292	8	0.090
6*	−0.518	5	0.047
7*	−0.594	2	−0.060
8*	−0.649	1	−0.068

significant changes in vegetation composition identified are shown in Figure 6.6. A clear pattern of change in floristic composition can be identified. *Pteridium aquilinum*-dominated quadrats, located on the far right of the ordination plots, give way to those dominated by *Agrostis capillaris* and *A. curtisii*. Co-dominance between the *Agrostis* species and *Molinia caerulea* then changes to dominance of the latter. The lower strata in the canopy, beneath the *Molinia caerulea*, changes from grass and moss species to *Calluna vulgaris* and *Vaccinium myrtillus*. The dominance of *Molinia caerulea* gives way to a mixture of *Festuca ovina*, *Nardus stricta*, *Vaccinium myrtillus* and *Calluna vulgaris*, the latter two becoming dominant. The final quadrats on axis one, those on the far left of the ordination plots, are dominated by mire species, in particular

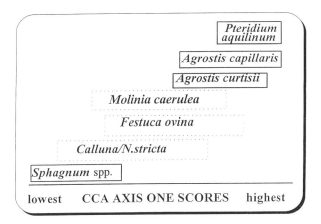

Figure 6.6 Diagrammatic representation of the distribution of species abundance and dominance related to position along the first CCA axis for the CASI and floristic data sets. The species in boxes with a solid line are those found to be related to one part of the axis, or seen to increase or decrease with position along the axis. Those species in dotted boxes appear with a random pattern along the axis. The pattern produced by the ATM and floristic data is similar

Sphagnum spp. A number of quadrats were found not to fit in with this overall pattern of change in floristic composition and these are examined in more detail later. The pattern of results for the ATM data was similar to that of the CASI-based analysis, except it was reversed along the first axis due to the inverse relationship between the axis and the spectral bands.

The strength of the pattern of species distribution along the first CCA axis was examined by correlating the presence or absence of species to quadrat CCA scores. The correlation was carried out using a method called Point Biserial Correlation (PBC). PBC allows a variable which has been measured on a continuous scale, such as the CCA scores, to be correlated with one that has been measured on a binary scale, such as the presence/absence of species (Kent and Coker, 1992). The presence/absence of the key species, identified in Figure 6.6, were correlated with the first axis CCA scores for each quadrat for both CASI and ATM analyses. The results are shown in Table 6.5.

The higher correlation coefficients in Table 6.5 relate to those species which were observed to occur at fixed points along CCA axis one, or were found to increase or decrease in abundance along the axis. For example, *Sphagnum* spp. has a correlation coefficient greater than 0.500 and is found in quadrats occupying one small area of axis one. *Agrostis capillaris* and *Agrostis curtisii* both have high correlation coefficients and are seen to increase in abundance along axis one. Species such as *Calluna vulgaris*, *Vaccinium myrtillus* and *Festuca ovina*, observed to occur without any particular pattern of abundance along axis one, were found to have low correlation coefficients.

The PBC analysis indicates that there is some linkage between species composition and the spectral response gradient derived by the CCA. One end of the axis is dominated by bracken and grass species, the other by mire, with a change from grass to heath domination in the centre. The pattern of change in individual species along the

Table 6.5 Table showing the point biserial correlations between the scores along the first CCA axis and the presence/absence of species for each quadrat. The correlation coefficients are given for both the ATM and CASI spectral data-set-based analyses. Those coefficients marked with a * are significant at the 0.05 (95%) level

Species name	Correlation coefficient for CASI/Floristic CCA	Correlation coefficient for ATM/Floristic CCA
Agrostis capillaris	0.528*	0.528*
Agrostis curtisii	0.520*	0.525*
Calluna vulgaris	0.155	0.080
Carex binervis	0.418*	0.424*
Festuca ovina	0.028	0.014
Galium saxatile	0.411*	0.466*
Hypnum cupressiforme	0.452*	0.464*
Molinia caerulea	0.317*	0.329*
Potentilla erecta	0.435*	0.470*
Sphagnum spp.	0.544*	0.682*
Vaccinium myrtillus	0.033	0.068

first axis was reinforced by the correlation analysis. Some quadrats did not follow the general pattern identified, and these will be considered in more detail in the discussion.

DISCUSSION

The spectral response of vegetation is characterised by low reflectance in the visible part of the spectrum and high in the near-infrared (NIR) (Lillesand and Kiefer, 1994). The reflectance in the visible part of the spectrum is controlled by the pigment in the leaves, specifically the absorption of blue and red light by chlorophyll (Curran, 1980, 1983, 1985, 1994a, 1994b). The NIR reflectance is controlled by the internal structure of the leaves, principally the changes in the refractive index within the mesophyll tissue of the leaves (Colwell, 1974; Tucker *et al.*, 1979; Tucker and Sellers, 1986; Verdebout *et al.*, 1994). This pattern of reflectance, low visible and high NIR, is based on observations of single leaves. However, when measuring the spectral response of canopies, other factors need consideration. The main factors controlling canopy reflectance include the size and orientation of leaves, the vertical structure of the canopy, the physiology of the component species, the nature of the underlying soil and the amount of vegetation cover, which controls the contribution of the background to the spectral response (Curran, 1980, 1983).

Research has indicated that vegetation biomass can be estimated using the levels of visible and NIR reflectance (Tucker and Maxwell, 1976; Filella and Penuelas, 1994). The strong correlation between NIR reflectance and the first CCA axis could indicate that the variation in floristic composition along this axis may be related to differences in biomass between samples. This relationship could not be tested in this case because destructive sampling, to obtain biomass data, was not possible due to the protected status of the study sites. However, the leaf area index (LAI), a biophysical parameter of the canopy that has been shown to be important in remote sensing, can be considered.

LAI is a ratio measure of the unit leaf area in a canopy per unit ground area (Curran, 1985; Lo, 1986). Canopies with a high LAI, such as grasslands, exhibit spectral characteristics similar to those of single leaves (Curran, 1985). In the context of this work, where the gradient along which the samples are arranged is related to NIR reflectance, that could mean that quadrats with high axis one CCA scores would be expected to have high LAI values also.

In order to consider the possible relationship between LAI and NIR reflectance, further analysis was carried out. The total percentage cover of each quadrat was used as a surrogate measure of LAI. The CCA axis scores for both ATM and CASI data sets and the corresponding total percentage cover of each quadrat were plotted against each other. Figure 6.7 shows the results for the CASI data set. There does not seem to be a strong relationship. This observation was confirmed when axis one scores were correlated against total percentage cover. The correlation coefficients were 0.402 and 0.443, significant at the 0.05 level, for the CASI and ATM analyses, respectively. Thus, the indication from these results is that LAI is only one factor influencing spectral response. However, the lack of suitability of percentage cover as a measure of LAI could also be responsible for the results. LAI is a measure of green leaves, whereas the cover figures for each quadrat measure all canopy components.

A second comparison of percentage cover and CCA axis one scores was carried out with a reduced number of the quadrats. The quadrats removed from this analysis were those identified as having either significant amounts of non-leaf plant components in their canopies, or having background factors that may influence their spectral response. First, all the quadrats which were collected in mire-type communities were eliminated from the analysis, due to the likely influence of increased moisture in the background soil, lowering the NIR reflectance. The second group of quadrats removed were those that contained significant amounts of *Calluna vulgaris* or *Vaccinium myrtillus*. The reason for excluding these quadrats was that their percentage cover figures include woody and non-leaf plant components, which strictly fall outside the definition of LAI. The correlation coefficients from this second analysis were 0.546 and 0.589, both significant at the 0.05 (95%) level, for CASI and ATM, respectively. These results show an obvious increase in the strength of the relationship between the first axis site scores and LAI, as represented by percentage cover.

The above discussion is based on the assumption that spectral response in the NIR is related most strongly to variation in LAI, but this may not be the case. As noted earlier, the reflectance in the NIR part of the spectrum is controlled by the internal structure of leaves (Colwell, 1974; Verdebout *et al.*, 1994). Therefore, results of the CCA may be related, to some degree, to variations in the internal structure, particularly the mesophyll tissue, of leaves of different species and combinations of species (Atkinson and Plummer, 1993). The investigation of this as a possible additional factor explaining the variations observed in the analysis requires further research.

Another factor that could influence the pattern identified in the CCA results is the structure of individual plants. Atkinson and Plummer (1993) found that planophile plants, as opposed to erectophiles, are likely to be more effective at photosynthesis, as more of the leaf interacts with the incoming radiation. If planophile plants generally interact more with incident radiation than erectophiles, this may explain some of the results noted in this research. Bracken, for example, is a planophile and has a higher

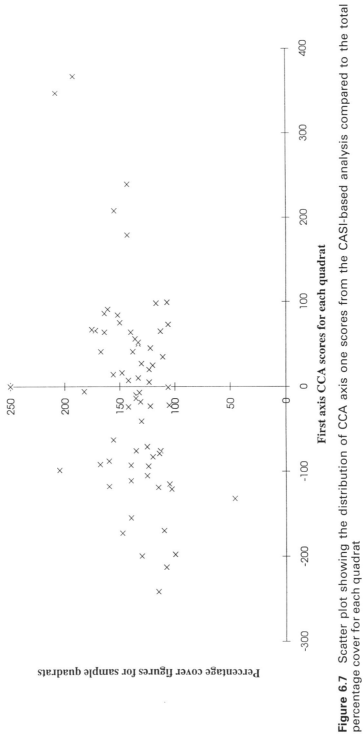

Figure 6.7 Scatter plot showing the distribution of CCA axis one scores from the CASI-based analysis compared to the total percentage cover for each quadrat

spectral response than erectophile grass-dominated communities. The allocation of the highest first axis CCA scores, indicating high NIR reflectance, to full canopy bracken-dominated communities would seem to confirm this pattern. The exceptions to the pattern are heather and *Vaccinium* communities, which tend to have lower NIR reflectance than grasses but look to be predominantly planophile in leaf arrangement. This situation can probably be explained by the presence of non-green plant parts, such as woody stems and flowers, in the canopy, which have different spectral responses to those of leaves (Milton and Rollin, 1988).

The relationship between spectral response and semi-natural vegetation canopies is a complex one. Clearly a number of factors are involved: the structure of the canopy, the nature of the background soil or lower vegetation layers, and the structure of individual plants, both in the distribution of leaves and in the spectral properties of those individual leaves. Some of the parameters relate specifically to particular species, while others do not, meaning a direct correspondence between spectral response and species composition is difficult to identify. Further research is required to understand exactly how all these factors combine to influence the final spectral response collected.

Another area which requires further investigation is the suitability of the different canonical ordination techniques for use with remotely sensed data. CCA was used in this research, as much of the literature (ter Braak, 1987c; Odeh *et al.*, 1991; Goovaerts, 1994) suggests that data should be initially analysed using an approach which assumes a unimodal response model, in order for the nature of the relationship between the data sets to be identified. Further work is also required to explain the variation identified along the second axis of the CCA. One possible explanation, which was investigated, was that the variation identified along axis two was due to the 'arch effect' (Kent and Coker, 1992). The arch effect, a problem which occurs in Correspondence Analysis (CA) based ordination, is where the second axis of variation identified is actually a quadratic distortion of the first axis (ter Braak, 1987b). This effect can be reduced by using Detrended Canonical Correspondence Analysis (DCCA). However, in this case the eigenvalues for the second axis DCCA were similar to those from the CCA, 0.218 compared to 0.243 for the CASI analysis. These results would seem to indicate that there is some form of variation along the second axis, and more work is required.

Although there are a number of problems in determining the exact nature of the relationship interpreted from the CCA, there is, nevertheless, a definite pattern of vegetation composition and corresponding spectral responses. The results discussed in the previous section show a change from high NIR reflectance for bracken, reducing to a low one for heather and mire-dominated communities. This pattern is similar to those derived by other researchers, such as Morton (1986), Weaver (1986, 1987) and Taylor (1993), using different methods.

CONCLUSIONS

The aim of the analysis described in this chapter was to identify whether differences in floristic composition of semi-natural plant communities could be identified using remote sensing techniques. The pattern of results generated by the Canonical Correspondence Analysis (CCA) of floristic and spectral data indicates that different combinations and abundance of species produce differing patterns of reflectance. However, the actual nature of the relationship between vegetation and spectral response is a complex one. The results generated by the analysis indicate that, in principle, the pattern of intergrading patches which form upland vegetation should be spectrally identifiable. However, the complexity of the relationship between spectral response and vegetation composition means that detailed information on plant communities, in terms of floristic descriptions, like those of the National Vegetation Classification (Rodwell, 1991, 1992), are difficult to identify from remotely sensed data. Finally, CCA would seem to be a useful technique for identifying an overall direction of variation between spectral response and vegetation composition.

ACKNOWLEDGEMENTS

The authors would like to acknowledge the assistance of the NERC Equipment Pool for Field Spectroscopy (EPFS) for the loan of the Spectron SE590 field spectroradiometer, loan 176.0194, and thank Dr Liz Rollin and Mr David Emery for their technical assistance. The authors would also like to thank the members of the Department of Geographical Sciences for their assistance in collecting the field data. Finally, the authors would like to thank the reviewer and the editors for their constructive comments on the early draft of this paper.

REFERENCES

Alam, M.S. and Southgate, A.C. (1987) Spectral discrimination of moorland surfaces using ground radiometry in the North York Moors. In: *Advances in Digital Image Processing*, Proceedings of the Annual Conference of the Remote Sensing Society, Nottingham, pp. 321–333.

Anderson, P. and Yalden, D.W. (1981) Increased sheep numbers and the loss of heather moorland in the Peak District, England. *Biological Conservation*, **20**, 195–213.

Atkinson, P.M. and Plummer, S.E. (1993) The influence of the percentage cover and biomass of clover on the reflectance of mixed pasture. *International Journal of Remote Sensing*, **14**, 1439–1444.

Barnes, R.F.W. (1987) Long-term decline of red grouse in Scotland. *Journal of Applied Ecology*, **24**, 735–741.

Birks, H.J.B. (1988) Long-term ecological change in the British uplands. In: M.B. Usher and D.B.A. Thompson (eds) *Ecological Change in the Uplands*. Blackwell Scientific Publications, Oxford, pp. 37–56.

Colwell, J.E. (1974) Vegetation canopy reflectance. *Remote Sensing of Environment*, **3**, 175–183.

Curran, P.J. (1980) Multispectral remote sensing of vegetation amount. *Progress in Physical Geography*, **4**, 315–340.

Curran, P.J. (1983) Problems of remote sensing of vegetation canopies for biomass estimates. In: R.M. Fuller (ed.) *Ecological Mapping from Ground, Air and Space*. ITE, pp. 84–100.

Curran, P.J. (1985) *Principles of Remote Sensing*. Longman Scientific Publications, London.

Curran, P.J. (1994a) Imaging spectrometry. *Progress in Physical Geography*, **18**, 247–266.

Curran, P.J. (1994b) Imaging spectrometry – its present and future role in environmental research. In: J. Hill and J. Megier (eds) *Imaging Spectrometry – a Tool for Environmental*

Observation. Kluwer Academic Publishers, Dordrecht, pp. 1–24.

Filella, I. and Penuelas, J. (1994) The red edge position and shape as indicators of plant chlorophyll content, biomass and hydric status. *International Journal of Remote Sensing*, **15**, 1459–1470.

Gibson, D.J. and Greig-Smith, P. (1986) Community pattern analysis: a method for quantifying community mosaic structure. *Vegetatio*, **60**, 41–77.

Goovaerts, P. (1994) Study of spatial relationships between two sets of variables using multivariate geostatistics. *Geoderma*, **62**, 93–107.

Grant, S.A. and Armstrong, H.M. (1993) Grazing ecology and the conservation of heather moorland: the development of models as aid to management. *Biodiversity and Conservation*, **2**, 79–94.

Hill, M.O. (1979) *DECORANA – a FORTRAN Program for Detrended Correspondence Analysis and Reciprocal Averaging*. Cornell University, Department of Ecology, Ithaca, New York.

Hume, E., McMorrow, J. and Southey, J. (1986) Mapping semi-natural grasslands from panchromatic aerial photographs and digital images at SPOT wavelengths. In: *Mapping from Modern Imagery*, International Archives of ISPRS, ISPRS and Remote Sensing Society, Edinburgh 1986, pp. 386–395.

Kent, M. and Coker, P. (1992) *Vegetation Description and Analysis: A Practical Approach*. Belhaven Press, London.

Lewis, M.M. (1994) Species composition related to spectral classification in an Australian spiniflex hummock grassland. *International Journal of Remote Sensing*, **15**, 3223–3239.

Lillesand, T.M. and Kiefer, R.W. (1994) *Remote Sensing and Image Processing*. Wiley, Chichester.

Lo, C.P. (1986) *Applied Remote Sensing*. Longman Scientific and Technical, London.

McMorrow, J. and Hume, E. (1986) Problems of applying multi-spectral classification to upland vegetation. In: *Mapping from Modern Imagery*, International Archives of ISPRS, ISPRS and Remote Sensing Society, Edinburgh 1987, pp. 610–620.

Milton, E.J. and Rollin, E.M. (1988) The directional reflectance of heather canopies: Towards a descriptive model. In: *Proceedings of IGARSS '88 Symposium*, Edinburgh, Scotland, 13–16 September 1988, pp. 829–832.

Milton, E.J., Rollin, E.M. and Emery, D.R. (1995) Advances in field spectroscopy. In: F.M. Danson and S.E. Plummer (eds) *Advances in Environmental Remote Sensing*. Wiley, Chichester, pp. 9–32.

Morton, A.J. (1986) Moorland plant community recognition using Landsat MSS data. *Remote Sensing of Environment*, **20**, 291–298.

Odeh, I.O.A., Chittleborough, D.J. and McBratney, A.B. (1991) Elucidation of soil–landform interrelationships by canonical ordination analysis. *Geoderma*, **49**, 1–32.

Ratcliffe, D.A. and Thompson, D.B.A. (1988) The British uplands: their ecological character and international significance. In: M.B. Usher and D.B.A. Thompson (eds) *Ecological Change in the Uplands*. Blackwell Scientific Publications, Oxford, pp. 9–36.

Rees, W.G. (1990) *Physical Principles of Remote Sensing*. Cambridge University Press, Cambridge.

Rodwell, J. (ed.) (1991) *British Plant Communities: Volume 2: Mires and Heaths*. Cambridge University Press, Cambridge.

Rodwell, J. (ed.) (1992) *British Plant Communities: Volume 3: Grasslands and Montane Communities*. Cambridge University Press, Cambridge.

Rollin, E.M. (1993a) Aperture–cosine intercalibration routines. *Spectron SE590 Software COSMENU and COSBATCH*, NERC-EPFS Software Manual 2.

Rollin, E.M. (1993b) Primary processing routines. *Spectron SE590 Software SMENU and SBATCH*, NERC-EPFS Software Manual 1.

Smilauer, P. (1992) *CanoDraw: User Guide*. Microcomputer Power, USA.

Steven, M.D. (1986) Ground truth: an underview. In: M.D. Stevens and E. Rollin (eds) *Ground Truth for Remote Sensing*. Remote Sensing Society, Nottingham, pp. 1–13.

Steven, M.D. (1987) Ground truth: an underview. *International Journal of Remote Sensing*, **8**, 1033–1038.

Sydes, C. and Miller, G.R. (1988) Range management and nature conservation in the British uplands. In: M.B. Usher and D.B.A. Thompson (eds) *Ecological Change in the Uplands*. Blackwell Scientific Publications, Oxford, pp. 323–338.

Taylor, J.E. (1993) Factors causing variation in the reflectance measurements from bracken in Eastern Australia. *Remote Sensing of Environment*, **43**, 217–229.

ter Braak, C.J.F. (1986) Canonical correspondence analysis: A new eigenvector technique for multivariate direct gradient analysis. *Ecology*, **67**(5), 1167–1179.

ter Braak, C.J.F. (1987a) The analysis of vegetation–environmental relationships by canonical correspondence analysis. *Vegetatio*, **69**, 69–77.

ter Braak, C.J.F. (1987b) Ordination. In: R.H.G. Jongman, C.J.F. ter Braak and F.R. van Tongeren (eds) *Data Analysis in Community and Landscape Ecology*. PUDOC, Wageningen, pp. 91–173.

ter Braak, C.J.F. (1987c) *CANOCO – A FORTRAN Program for Canonical Community Ordination by [Partial] [Detrended] [Canonical] Correspondence Analysis, Principal Component Analysis and Redundancy Analysis (Version 2.1)*. ITI-TNO, Wageningen.

ter Braak, C.J.F. (1990) *Update Notes: CANOCO Version 3.10*. Agricultural Mathematics Group, Wageningen.

ter Braak, C.J.F. and Prentice, I.C. (1988) A theory of gradient analysis. *Advances in Ecological Research*, **18**, 271–313.

Treitz, P.M., Howarth, P.J., Suffling, R.C. and Smith, P. (1992) Application of detailed ground information to vegetation mapping with high spatial resolution digital imagery. *Remote Sensing of Environment*, **42**, 65–82.

Trodd, N.M. (1992) Gradient analysis of heathland vegetation from near ground level remotely sensed data. *Remote Sensing and Spatial Information*, Proceedings of the Sixth Australasian Remote Sensing Conference, pp. 406–409.

Trodd, N.M. (1993) Characterising semi-natural vegetation: continua and ecotones. In: K. Hilton (ed.) *Towards Operational Applications*, Proceedings of the 19th Annual Conference of the Remote Sensing Society, pp. 191–198.

Tucker, C.J. and Maxwell (1976) Sensor design for monitoring vegetation canopies. *Photogrammetric Engineering and Remote Sensing*, **42**, 1399–1410.

Tucker, C.J. and Sellers, P.J. (1986) Satellite remote sensing of primary production. *International Journal of Remote Sensing*, **7**, 1395–1411.

Tucker, C.J., Elgin, J.H., McMurtrey, J.E. and Fan, C.J. (1979) Monitoring corn and soybean crop development with hand-held radiometer spectral data. *Remote Sensing of Environment*, **8**, 237–248.

Usher, M.B. and Thompson, D.B.A. (1993) Variation in the upland heathlands of Great Britain: conservation importance. *Biological Conservation*, **66**, 69–81.

Verdebout, J., Jacquemound, S. and Schmuck, G. (1994) Optical properties of leaves: modelling and experimental studies. In: J. Hill and J. Megier (eds) *Imaging Spectrometry – a Tool for Environmental Observation*. Kluwer Academic Publishers, Dordrecht, pp. 169–192.

Wardley, N.W., Milton, E.J. and Hill, C.T. (1987) Remote sensing of structurally complex semi-natural vegetation – an example from heathland. *International Journal of Remote Sensing*, **8**, 31–42.

Weaver, R.E. (1986) Use of remote sensing to monitor bracken encroachment in North York Moors. In: R.T. Smith and J.A. Taylor (eds) *Bracken, Ecology, Land Use and Control*. Parthenon, London, pp. 65–76.

Weaver, R.E. (1987) Spectral separation of moorland vegetation in airborne Thematic Mapper data. *International Journal of Remote Sensing*, **8**, 43–55.

7   Integrating Photointerpretation and GIS for Vegetation Mapping: Some Issues of Error

D.R. GREEN[1] **and S. HARTLEY**[2]
[1]*Centre for Remote Sensing and Mapping Science, Department of Geography, University of Aberdeen, UK*
[2]*Institute of Terrestrial Ecology, Hill of Brathens, Banchory, UK (Present address: School of Biology, University of Leeds, UK)*

At the landscape scale, temporal and spatial studies of semi-natural vegetation change, successional pathways and pattern, frequently make use of both up-to-date and archival aerial photography as a data source. However, the relatively straightforward task of mapping vegetation patch boundaries from aerial photography can be affected by multiple sources of error associated with the processes of georeferencing, digitising and subjective photointerpretation. Different measurements of landscape pattern and change will vary in their sensitivity to the different types of error. To date, relatively little attention has been focused on developing an integrated approach to the identification and estimation of error in vegetation mapping tasks where the data subsequently forms the basis for the quantitative measurements of vegetation change. In this chapter a simple empirical method is proposed for estimating positional boundary error due to the separate and combined effects of georeferencing, digitising and subjective photointerpretation. The findings of this study, which reveal that up to one-third of the total area may be in error, lead to a strong recommendation that researchers make use of such empirical methods to assess the constraints inherent in their data prior to undertaking complex or time-consuming analyses.

INTRODUCTION

In recent years a growing body of interest has been expressed in the study of landscape ecology (see e.g. Forman and Godron, 1986; Haines-Young *et al.*, 1993), defined by Vink (1983, p. 2) 'as the study of the relationships between phenomena and processes in the landscape or geosphere including the communities of plants, animals and man'. In part, this interest stems from an increasing desire to bridge the gap between traditional small-scale ecological experiments (of a necessarily limited scope) and the larger-scale patterns and processes of ecosystems that are of relevance to environmental scientists, resource managers and politicians (May, 1994; Pimm, 1994; Brown, 1995; Hornung and Reynolds, 1995). In addition, significant technical advances in data acquisition and data handling have made large-scale studies more accessible and feasible. In

Vegetation Mapping: From Patch to Planet. Edited by Roy Alexander and Andrew C. Millington.
© 2000 John Wiley & Sons Ltd.

particular, many of these studies have begun to apply the two technologies of remote sensing and Geographical Information Systems (GIS) as an aid to the study of the processes and dynamics of landscape change.

One such area of interest has been vegetation mapping and monitoring over both space and time (e.g. Green *et al.*, 1993a; Reid *et al.*, 1994). Aerial photography, satellite imagery, airborne video and, increasingly, radar data have been used to map a range of different vegetation types (natural, semi-natural and anthropogenic) at a variety of different scales. For reasons of spatial resolution and historical availability, aerial photography has proved to be particularly useful for change detection studies (e.g. Lo and Shipman, 1990; Miller *et al.*, 1991), and for the investigation of successional pathways (e.g. Lowell and Astroth, 1989; Lowell, 1991; Callaway and Davis, 1993; Green *et al.*, 1993a).

In studies which seek to analyse changing landscapes three considerations are important:

- the choice of an appropriate practical method,
- the choice of measurement for quantifying change, and
- estimating the magnitude of error and uncertainty in the results.

From a purely practical perspective a methodology can easily be developed to permit the spatial and temporal mapping of vegetation from remotely sensed data, and the subsequent handling and analysis of this data within a GIS. However, to go beyond simple map-making towards quantification, modelling and ultimately prediction, requires that serious consideration be given to the problems of error and the provision of confidence intervals. It is the sources and estimation of error, the effect they can have on subsequent analyses and the choice of appropriate solutions which this chapter seeks to address.

QUANTIFYING LANDSCAPE CHANGE

Categorical Maps as Single-date Abstractions

One of the first steps towards a scientific approach to any problem is quantification of the property of concern, for if one cannot measure something it is impossible to know whether it is changing or responding to other variables. In most conservation or environmental programmes the units of measurement for vegetation are based on taxonomic and/or structural assemblages of plants, that is to say a higher level of organisation than plant species *per se*.

In order to record spatial information about these vegetation units their positions and extent are mapped by drawing a line or so-called 'hard' boundary around what is perceived to be a uniform or acceptably homogeneous area, namely the vegetation patch. In many respects the inadequacy of this method for representing the continuous and heterogeneous nature of semi-natural vegetation is at the heart of the problem of trying to achieve a consistent and meaningful quantification of the landscape. However, in practice, delineating polygons is easy to apply and provides one way in which it is possible to represent the landscape. As such it is a perfectly acceptable abstraction provided that the following points are borne in mind:

- In photointerpretation the definition of a vegetation patch is primarily based upon the criteria of tone, texture and the spatial relationships between other recognisable features (Avery, 1970).
- The interpretation based upon tone, texture and relational information reflects the spatial and spectral resolution of the sensor used to record the information, the quality of the interpreter's eyesight, and his or her experience of photointerpretation.
- The boundaries drawn do not necessarily reflect the existence of sharp transitions between vegetation patches at the ground level.

Undoubtedly some 'hard' boundaries do exist in reality for features such as roads, rivers, forest plantations and field boundaries. However, semi-natural vegetation is character-ised by gradients (ecotones) and mosaics, for which the use of fuzzy or 'soft' boundaries is far more appropriate. By logical extension some boundaries will be 'softer' than others, depending upon the classes being divided (Aspinall and Pearson, 1995).

Nonetheless, if polygon data are to be manipulated in a standard GIS, hard boundaries with precise locations are still required, if only to maintain topological consistency. One must simply acknowledge that these only represent the best estimate of location within a zone of uncertainty. The crucial point is how to quantify this variance or uncertainty, i.e. *error* in the statistical sense.

Measurements of Landscape Pattern

Once a categorical map of vegetation polygons has been produced it is possible to describe the landscape in many more ways other than simply summating the area of each land cover class. The manner in which the units are topologically distributed and interrelated can be of crucial importance to many ecological and hydrogeological processes. Turner (1989) and Levin (1992) discuss the interrelationship between pat-tern and process in population dynamics, the aggregation of prey species or the spread of organisms, disease and genes (see also With and Crist, 1995). The spatial measures of habitat patch size and patch *isolation* are fundamental parameters in island biogeogra-phy theory (MacArthur and Wilson, 1967) and its subsequent development towards metapopulation studies (e.g. Hanski and Gilpin, 1991; Rolstad, 1991). For some species or communities, topological features of the landscape are a resource in themselves, e.g. edge habitats, river corridors, mosaics and habitat interiors (Burgess and Sharpe, 1981; Bridgewater, 1993). Thus the description of a landscape pattern can be an important component of vegetation mapping.

Quantification of landscape pattern is achieved through the use of various math-ematical indices. Each index provides, in a single figure, a description of a certain aspect of the landscape pattern. Some commonly used indices are diversity (or domi-nance), the fractal dimension, mean patch size (or number of patches per unit area), contagion, length of boundaries between different classes and spatial autocorrelation (e.g. Burrough, 1987; O'Neill *et al.*, 1988; Turner and Gardner, 1991).

Quantifying Change for Vegetation Monitoring

The key objective of any monitoring programme is to detect change. In some cases change may be expected and considered desirable (e.g. the response to a new conservation strategy) or in other situations change, if it occurs, is indicative of a problem that may need addressing elsewhere (e.g. air pollution monitoring). In either case, considerable thought must be given to what is being measured, how accurately and precisely it can be measured, and what information this can yield (Goldsmith, 1991). Unfortunately, many monitoring schemes are poorly designed and so by default they become a series of repeated surveys unable to address the original aims of the project. Likewise, repeating a mapping exercise for two or more points in time is not always the best strategy for vegetation monitoring. However, even if multitemporal mapping cannot always tell us exactly what we want to know, it may be the only option available for obtaining any information at all, and therefore it is important to know what it can tell us.

Quantification of landscape change can take several forms. Beyond a purely verbal description, based upon visual examination, the simplest approach comprises an areal measurement of each land cover type at a number of time intervals (t_1, t_2, ... t_n). Change over time is measured as a net increase or decrease for each cover type considered. In the case of pattern indices, their value, too, can be followed over time.

On its own, however, such an approach often fails to reveal some of the more important changes occurring in the landscape. If the landscape is in 'dynamic equilibrium' or a 'steady-state shifting mosaic' (Shugart, 1984), then comparing measurements of areal cover at t_1, t_2, ... t_n, will show no apparent change. In contrast, overlaying vegetation cover maps for two or more dates will reveal the magnitude and direction of the transitions that are occurring. Overlay is a basic operation within GIS and as such lends itself well to quantifying and modelling landscape change through the application of Markovian-style transition matrices. Soil type, contagion, distance from seed source and vegetation cover one or more time periods ago are possible polygon map layers that may be used to estimate transition probabilities (e.g. Turner, 1987; Lowell and Astroth, 1989; Hartley, 1994), and in each case a polygon overlay operation will be required.

In summary, single date measurements can include numerous pattern indices as well as simple area estimates. All of these measurements can be repeated for images of different dates to reveal change over time but, in addition, transition matrices can provide information regarding the pathways of change, revealing dynamic properties not detected by the other 'one-off' measurements.

CONCEPTS OF ERROR

Out of necessity the process of map-making involves generalisation of complex 'real-world' features (Campbell, 1993; Goodchild, 1993). As such no map will be an exact replica of the real world and therefore may be considered to be in error. Irrespective of the level of generalisation chosen, however, there will almost always be additional errors (unwanted deviations) introduced by the instrument, the methodology of data capture, and the processing. Since abstracting 'real-world' features onto a map (be it a paper or digital map) involves many separate processes (see Thapa and Bossler, 1992), different types of error may be introduced at each stage and/or the same type of error may be repeatedly introduced and magnified (Chrisman, 1989).

Traditionally, end-users of map products have judged the quality of a map's content by its appearance (Burrough, 1994; von Rimscha, 1996), which is tantamount to assuming that all errors are within the limits of acceptable generalisation. Obviously this is neither a satisfactory nor a positive response to the problem of error, especially when planning and decision-making may rest upon the outcome of data analysis. However, practitioners have often been reluctant or unable to supply quantified estimates of error, which is hardly surprising given that virtually none of the commercial GIS products currently offer any error-handling facilities (Brunsdon and Openshaw, 1993). Consequently, error estimation has remained largely the province of theoreticians rather than GIS practitioners and end-users. Burrough (1986), Walsh *et al.* (1987) and Lunetta *et al.* (1991) have all provided comprehensive reviews on the issue of error in GIS, while Veregin (1989b) has compiled an extensive bibliography of literature dealing with error.

Many authors distinguish between positional errors (i.e. class boundary location) and attribute errors (i.e. the correct classification or labelling of polygons (e.g. Chrisman, 1989; Veregin, 1989a; Goodchild *et al.*, 1992)). Others (e.g. Thapa and Bossler, 1992) sometimes refer to these as quantitative and non-quantitative errors, respectively. In practice, however, these two broad types of error are interrelated (Chrisman, 1989), and both can affect the results of quantitative analyses. Aspinall and Pearson (1995) identify three components of uncertainty (potential error) in thematic maps:

- class identity
- class heterogeneity within a polygon
- class boundary location.

In this chapter discussion will mainly be confined to the errors associated with boundary placement in the delineation of vegetation patches (polygons) and the subsequent use of this data for the derivation of quantitative measures and overlay analysis. In addition, two broad types of boundary positional error are distinguished. One involves delineating two different-shaped polygons for the same area, whilst the other is more akin to a translational error involving the position of the whole polygon (see Figure 7.1). These two levels of positional error correspond to Burrough's local and global positional errors respectively (Burrough, 1986).

Other schemes of categorising errors have also been developed, often based upon the conceptual source of the error: for example, *primary* errors accumulated during data collection/measurement as opposed to *secondary* errors arising through subsequent data handling and processing (Thapa and Bossler, 1992); or, errors arising due to the *inherent properties of nature* (fuzzy boundaries); the *nature of the measurement* and the *model* of abstraction used to communicate the measurement (Maffini *et al.*, 1989). These classifications are less useful in leading us to a quantitative assessment of the magnitude of error, although they may help to focus our ideas as regards selecting an appropriate methodology for data capture and handling.

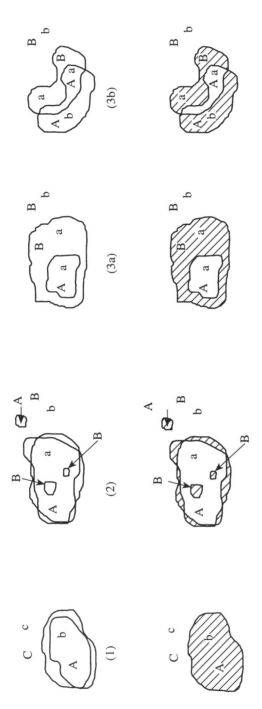

Figure 7.1 Four types of error which may manifest themselves as differences between a map and 'reality' or as inconsistency between two maps of the same reality. (1) Misclassification of the entire polygon. (2) Within-polygon heterogeneity due to mapping a greater or lesser level of detail. (3a) Positional boundary error due to different interpretation of polygon shape. (3b) Positional boundary error due to translational errors. Letters a,b,c, and A,B,C represent the respective labelling of land cover classes by the two interpreters (or the difference between map and 'reality'). The lower set of diagrams illustrate these differences as areas of 'class mismatch' between the two versions

SOURCES OF ERROR

Thapa and Bossler (1992) identify up to 14 separate processes that may contribute to positional and categorical errors in the final copy of a map. When mapping vegetation from aerial photographs these processes can be grouped under three broader headings:

- inaccuracies during the georeferencing and rectification procedure
- inaccuracies in the transcription and digitising phase
- the subjectivity involved in photointerpretation

The subjectivity of photointerpretation may introduce all three of the error types described previously – misclassification, depiction of different levels of detail (i.e. within-polygon heterogeneity) and positional errors of the boundary. The digitising and georeferencing of mapped lines can add further positional errors to boundary locations (Table 7.1). During the digitising process polygons may also become erroneously labelled, though with adequate proofing such gross errors or blunders should be avoidable.

In the sections that follow we discuss traditional methods of visualising and modelling positional boundary error followed by an analysis of how the error introduced by each of the major map-making processes can be quantified and combined to give a measure of the total error present.

QUANTIFICATION OF ERROR

The Epsilon-band Model of Positional Boundary Error

The idea that there is an error associated with the position of a boundary is not new and the use of either constant width epsilon bands (Figure 7.2) or variable width probability bands has been proposed elsewhere (e.g. Perkal, 1966; Chrisman, 1983; Blakemore, 1984; Burrough, 1986; Caspary and Scheuring, 1992; Wenzhong et al., 1994; Aspinall and Pearson, 1995). Yet, nearly always these remain as a concept that is rarely, if ever, put into practice with 'real-world' applications, Aspinall and Pearson (1995) being a notable exception.

Assuming a correct classification of polygons at a gross scale one could argue that a point precisely on the boundary could equally well belong to either class (Blakemore, 1984). Thus, the probability of a correct classification is 50% and the boundary is simply the best estimate for dividing the two classes. Moving away from the boundary towards the centre of the polygon the probability of a correct classification increases and conversely the probability that this is where the boundary should lie decreases. The manner in which this uncertainty drops off depends upon the form of uncertainty or error associated with that particular boundary.

Often, epsilon bands are illustrated simply as a token, visual acknowledgement to the fact that there is positional error or uncertainty associated with the location of boundaries. However, in their more sophisticated form epsilon bands enclose a zone which has a specific probability of including the 'true' location of the boundary (Dunn et al., 1990; Goodchild, 1993). One definition applied by Chrisman (1983) is that

Table 7.1 Types of error and their possible sources. The number of asterisks indicates an increasing likelihood that a particular type of error originates from a given source

	Type of error		
Source of error	Misclassification	Within-polygon heterogeneity	Positional boundary error
Subjectivity of interpretation	***	***	***
Georeferencing			***
Digitising	*		***

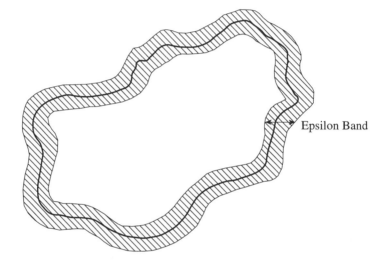

Figure 7.2 Example of an epsilon band

'deviations outside are exactly balanced by those inside', in other words, the epsilon band encompasses the 'true' location of the boundary 50% of the time (Dunn *et al.*, 1990).

In order to calculate how wide a statistically defined epsilon band should be, it is necessary to know the magnitude and distribution of errors (i.e. deviations) associated with the boundaries, in other words their probability distribution function (p.d.f.). Bolstad *et al.* (1990), Walsby (1995) and Dunn *et al.* (1990) found bell-shaped distributions of point, vertex and line displacement, respectively, although often the observed deviations were slightly more peaked than a normal distribution of equivalent mean and standard deviation. Part of the reason for this must be that a theoretical normal distribution is unbounded at the tails, whereas in practice there must be an upper limit to the magnitude of positional errors that will go undetected and unedited (Dunn *et al.*, 1990). Nonetheless, a normal or Gaussian distribution of errors remains a reasonable approximation so that an epsilon band extended out by one standard deviation in either direction would be expected to include the 'true' location of the boundary approximately 68% of the time, whilst a band extended out 1.96 standard deviations will include 95% of the 'true' locations (Drummond, 1990). Similarly, an epsilon band of half-width equal to the mean deviation would fulfil Chrisman's (1983) definition as it would enclose the central 50%, or interquartile range, of boundary errors (Dunn *et al.*, 1990). Unfortunately, many of the studies which do apply epsilon bands fail to state why a certain width was chosen, what probability zone it encloses, or which probability distribution function applies. One way to calculate the width of probabilistic epsilon bands is to measure the positional error being introduced by the three processes: georeferencing, digitising and subjective photointerpretation. The next three sections describe how this may be achieved.

Positional Errors 1 – Geo-referencing and Co-registration

Georeferencing refers to the geometric correction of an image or interpretation to a planimetrically correct coordinate system. It involves identifying points on the image (Ground Control Points or GCPs) that can be referenced to known map locations so that the image as a whole can be transformed to map coordinates. The accuracy of georeferencing is dependent upon the number of GCPs, the accuracy with which their image and map positions can be determined, and the nature of the transformation equation employed (e.g. 1st order, 2nd order, or polynomial trend surfaces versus localised rubber sheeting). Whichever transformation is selected to rectify the image there will always be some areas whose position has not been exactly matched to their corresponding map coordinates. This is particularly likely when georeferencing aerial photographs of hilly terrain in which relief displacement creates complex warping in the image.

Problems can arise in the selection of the GCPs, particularly in areas of semi-natural vegetation where there are few recognisable mapped features. To this end Global Positioning Systems (GPS) technology can be a valuable aid in fixing the ground coordinates of features obvious in the imagery and on the ground, but not necessarily depicted on reference maps.

For simple estimates of vegetation change, e.g. net change, it is necessary only to ensure the accurate scaling of the maps so that changes in area are measured in common units for the two or more dates used. However, where quantitative estimates of dynamic change are required, e.g. transition frequencies, then accurate georeferencing of the imagery is imperative to avoid the appearance of erroneous changes caused by one vegetation map being offset relative to the other, i.e. the boundary error of translation described by Figures 7.1 (3b). This is a recurrent problem in any investigation attempting to calculate transition matrices for land cover change (see Dunn *et al.*, 1990).

If accurate geo-referencing is difficult to achieve then an alternative approach, but one which still allows a dynamic description of vegetation change, is image to image co-registration. This involves 'fitting' one image to another, without reference to ground coordinates.

In either case the 'goodness of fit', either between the two sets of points or coordinates can be assessed using the RMS (Root Mean Square) value. Essentially this is the equivalent of a two-dimensional standard deviation (Figure 7.3) and is expressed mathematically by equation 7.1:

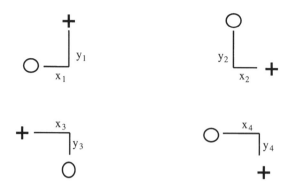

Figure 7.3 Two-dimensional displacement between the recorded positions of GCPs (crosses) and their map position (open circles)

$$\text{RMS} = (\Sigma\,(x^2 + y^2)\,/\,n)^{\frac{1}{2}} \tag{7.1}$$

where RMS = Root Mean Square deviation or error,
    x = difference along the x-axis,
    y = difference along the y-axis, and
    n = number of ground control points (Erdas, 1990; Caspary and Scheuring, 1992).

The positional errors introduced by poor geo-referencing will generally be directed in the same direction across large areas of the photograph. Thus, boundary positions will display a systematic bias, or in Burrough's terminology a 'global' positional error, rather than random variation or 'local' positional error in their mapped position (Burrough, 1986). This has important consequences for assessing the impact of geo-referencing errors when overlaying two maps of the same area produced from photographs taken from the same position. If their geo-referencing errors are systematically biased in the same directions across the map, then the overlay error will be much less than predicted by the combination of independent geo-referencing errors.

Positional Errors 2 – Digitising

The process of manual digitising is another source of error. However, in the context of vegetation mapping it will normally be much less significant than the error associated with subjective photointerpretation (Goodchild, 1993; Aspinall and Pearson, 1995).

The most common causes of digitising errors are (Bolstad *et al.*, 1990; Giovachino, 1993):

- unsteadiness of hand
- parallax between the cross-hairs and map
- the thickness of the line as depicted on the base map
- generalisation of curved lines into a series of short straight line segments
- the digitiser puck recording off-centre

- warping/shrinkage of base maps
- distortion of base maps through photocopying

The accuracy of digitising can also be measured using an RMS error. It is obtained by digitising a set of reference points a number of times, and calculating the RMS displacement between subsequent attempts. This is often a requirement of any digitising session and the RMS error is reported in digitiser units. Once scaled up to the ground units of the map being digitised, this can be added to the geo-referencing error as a further source of positional inaccuracy. Assuming that the two sources of error are random and independent then the combined positional RMS error of entering data from the map into the computer is given according to equation 7.2 (Thapa and Bossler, 1992).

$$E = (e^2_{\text{dig}} + e^2_{\text{geo-ref}})^{\frac{1}{2}} \qquad (7.2)$$

where E = standard deviation (or RMS) of the positional error, e_{dig} = standard deviation (or RMS) of the digitising error, and $e_{\text{geo-ref}}$ = standard deviation (or RMS) of the geo-referencing error.

Equation 7.2 is derived from the Law of Propagation of Errors for the addition of variances, rearranged to give the result as a standard deviation or RMS error (Chrisman, 1983).

The magnitude of digitising errors in ground units will obviously depend upon the scale of the imagery being used. Careful digitising by itself typically introduces an average error in line placement of less than 0.5 mm (Goodchild, 1993). Operating on 1:10 000 scale base maps this corresponds to a ground error of 5 m. Most of the error will be random positional variation about the depicted line due to the accuracy limits of manual tracing, although there may be systematic biases in the sense of which side of the line is digitised, parallax and off-centre pucks (see Walsby, 1995). Table 7.2 illustrates a range of digitising error values reported in the literature. Note that the lowest values are recorded for studies of point-on-point error, whereas the higher values are average line displacements calculated after completion of a much larger study. With respect to the values reported it is also interesting to note that Li and Openshaw (1992) consider 0.5–0.6 mm (at contact scale) as an appropriate lower limit for the diameter of 'smallest visible objects' to be retained during the process of line generalisation.

Positional Errors 3 – Subjectivity

Despite the fact that polygon maps are an abstraction from the reality of vegetation gradients and heterogeneous patches, this would not be so much of a problem in change detection studies if the same observer could be entirely consistent in his or her interpretations, since the same 'abstraction' would be applied in each case. However, photointerpretation is exactly what it says it is, an *interpretation* of features visible on an aerial *photograph* followed by judgement as to how best to represent the feature.

The two main problems encountered in actually mapping boundaries, independent of the problem of classification, are where to place the dividing line along a continuous vegetational gradient and how to cope with complex mosaics and edges. The level of detail mapped will depend upon the minimum mappable unit chosen at the outset of

Table 7.2 A range of published RMS errors due to the digitising process. Where values were reported in map units they have been converted back to digitiser units for ease of comparison. Likewise, where a measure of the distribution of individual deviations was reported (other than RMS) these values have been converted to an RMS value by assuming a normal distribution of deviations

Published error due to digitising (deviations)	RMS equivalent (mm)	Measurement used	Source
0.001″ (RMS)	0.025	Point-on-point	Giovachino, 1993
0.054 mm (mean) 0.261 mm (max)	0.08	Point-on-point	Bolstad *et al.*, 1990
90% < 0.4 mm 33% < 0.08 mm	0.24	Vertex displacement	Maffini *et al.*, 1989
0.25 mm	0.25	Not reported	Petrie, 1990 (quoted in Thapa and Bossler, 1992)
Majority < 0.15 mm	< 0.22	Vertex displacement	Walsby, 1995
0.25 ± 0.17 mm perkal width (mean, ± S.D.)	0.25 (if perkal width = RMS)	Not reported	Aspinall and Pearson, 1995
0.33 mm (mean RMS)	0.33	Point-on-point	Green and Hartley (this study)

the project. Nevertheless, deciding where to put a boundary line around, for example, a scattered group of trees, can remain problematic. The problem lies not only with where to draw the boundary, but also the shape of the boundary, and how much detail and precision the boundary should reflect. It is difficult enough to be consistent across a single image, let alone across different photographs taken at different times and scales. Individual trees, which at one scale make up separate units, at another scale may be mapped as part of a larger stand of trees (Figure 7.4).

In actual fact, the placement of vegetation boundaries on an aerial photograph is also inextricably linked to the classification and identification of vegetation types. So, although most photointerpretation exercises begin with the drawing of boundaries around areas of apparent uniformity, the fact that the interpreter is setting out to map vegetation types means that, consciously or unconsciously, the boundaries being sought and delineated are only those that differentiate the vegetation classes in the chosen classification scheme (Bie and Beckett, 1973).

In effect the classification scheme acts as a type of mask or filter, thereby making the delineation of the vegetation boundaries of interest that much easier.

As well as the psychological aspects of subjectivity there are also a number of reasons why images from different photographs will appear different, and therefore may be interpreted differently. This problem is particularly relevant to multitemporal studies using aerial photographs taken under different conditions:

- season has a major effect upon the appearance of vegetation etc.
- lighting conditions
- effects of the atmosphere, e.g. clouds
- scale
- resolution and contrast of photographs
- processing

Figure 7.4 Five interpretations of the same clump of trees

Using aerial photographs taken at different times of the year may mean that the apparent boundary location and detail mapped will change even though there is no real change. For example, trees may lose leaves, bracken may have died back etc. Even the time of day and weather will affect the appearance of vegetation due to changes in illumination and shadow. In some instances the interpreter may end up mapping the shadow edge rather than the vegetation edge, especially where the vegetation has a dark tone or the vegetation stand contains a lot of shadow.

All of the above sources of error are important in the context of vegetation mapping and although experience of photointerpretation, familiarity with the area and photo-interpretation keys may help to improve the consistency of results they will never entirely eradicate variation in the interpretation. However, combined with a measure of the other sources of error, it is the very existence of differences between interpretations that enables one to quantify the positional error due to subjectivity.

Total Positional Error

Given a measurement of positional error due to geo-referencing and digitising, what is required is an equivalent standard deviation to describe the error introduced by subjective boundary placement. If one existed it could then be combined with the previous two sources of positional error to give a final estimate, E_{total} as expressed by equation 7.3:

$$E_{\text{total}} = (e^2{}_{\text{dig}} + e^2{}_{\text{geo-ref}} + e^2{}_{\text{subj}})^{\frac{1}{2}} \qquad (7.3)$$

where E_{total} = standard deviation (or RMS) of the total positional error, e_{dig} = standard deviation (or RMS) of the digitising error, $e_{\text{geo-ref}}$ = standard deviation (or RMS) of the geo-referencing error, and e_{subj} = standard deviation (or RMS) of the subjectivity error.

It is this need for a standard measure of the variation introduced by subjectivity which we seek to address in a later section.

Misclassification and Within-polygon Heterogeneity

The errors associated with misclassification may be quantified through the use of a confusion matrix (Green *et al.*, 1993b). These are constructed by comparing the classification of a stratified random sample of points or areal units against the 'true' class. Typically the 'true' class is determined by ground-truthing, but interpretations could also be validated against larger-scale (more accurate) aerial photography (Drummond, 1990; Chrisman and Lester, 1991). A problem with multitemporal studies is that there is often no way of validating classifications of historical aerial photography where these are the only data source available. All that can be done is to assume similar rates of confusion as with contemporary ones, though this may be an unsatisfactory assumption if the photographs are of different scales and/or clarity.

Since the validation of polygon class typically involves determining the true class from larger-scale, more detailed information (e.g. ground-truthing), a point sampling strategy will record instances of misclassification in proportion to the level of within-polygon heterogeneity that exists as well as the cases of gross error involving the whole polygon. Although these two attribute errors are interrelated, it would often be very helpful to know to what extent each is contributing to the reported classification accuracy statistics. One possible method for decomposing attribute error would be through the use of two separate confusion matrices. The first, obtained from point-in-polygon validation, would assess misclassification and heterogeneity errors combined, whilst a second confusion matrix, where whole polygons are taken at face value and validated for majority class accuracy, would reveal misclassification errors *per se*.

Combining Errors in Map Overlay

As was made clear in the introduction to quantification of landscape change, it may often be desirable to overlay vegetation maps from two or more time periods in order to quantify the frequencies of transition from one vegetation class to another (or in other situations to intersect a vegetation map with a soil map). Naturally, the boundaries of both maps will have their own positional error, E, and the same logic for combining independent, random errors suggests that when two maps are combined the average positional discrepancy between them will be expressed by equation 7.4:

$$E_{\text{overlay}} = (E_1{}^2 + E_2{}^2)^{\frac{1}{2}} \qquad (7.4)$$

where E_{overlay} = standard deviation (or RMS) of positional mismatch between maps 1 and 2, E_1 = standard deviation (or RMS) of total positional error in map 1, and E_2 = standard deviation (or RMS) of total positional error in map 2.

This is the standard deviation of spurious boundary shift to expect when overlaying two maps. It is not to be confused with the joint probability of a correct classification occurring at a specific point for both maps. This is modelled as $P(A) \times P(B)$ where $P(A)$ and $P(B)$ are the independent probabilities of a correct classification in each of the parent maps A and B (Burrough, 1986; Drummond, 1990; Lanter and Veregin, 1992).

AN EMPIRICAL APPROACH TO QUANTIFYING POSITIONAL BOUNDARY ERRORS

To estimate the positional error of boundaries that is due to the subjectivity of photointerpretation consider interpreting the same image twice. This could either be the same person making two interpretations or two different people outlining their own individual interpretations. Inevitably some boundaries will be definite and very consistent, whilst others will vary widely (e.g. Drummond, 1990; Walsby, 1995), and when overlaid there will probably be many sliver polygons whose class differs between the individual interpretations (e.g. Chrisman and Lester, 1991). Figure 7.5 shows two parent boundary lines with the sliver polygons that result from their intersection.

It is reasonable to assume that the interpreted lines are two equally valid, independent estimates for the position of the true boundary, and that the observed deviations between them are therefore equivalent to the overlay of two parent lines each with a total positional error of E (RMS or standard deviation). The magnitude of deviation could be measured empirically at a number of points as the width of the sliver polygon. If repeated a sufficient number of times this would produce a probability distribution function for the positional errors of the overlay. From the distribution one could then calculate the standard deviation (RMS) of the overlay and, if the errors in map 1 are assumed equal to those in map 2, the positional error of the component maps (see equation 7.4).

However, if one assumes a normal distribution of the deviations (or any other parameterised distribution) then the standard deviation can be calculated directly from the mean deviation. The utility of this relationship lies in the fact that the mean displacement or deviation between two lines is essentially the mean width, w, of the sliver polygons created by the overlay of the two parent lines (Dunn *et al.*, 1990; Goodchild, 1993). The value w can be approximated by dividing the cumulative area of the sliver polygons, A, by the mean length, l, of the two parent boundary lines (two figures that can be easily extracted from a GIS). Although, strictly speaking, this is the division of area by half the perimeter, rather than the division of area by length; for long, thin sliver polygons the two values will be sufficiently similar to justify the use of this much simpler and more practical method (but see Lester and Chrisman, 1991). In the case of a normal distribution the standard deviation equals 1.48 mean deviations (Fisher and Yates, 1957).

If w is calculated for the overlay of the two 'hard' lines then the subjectivity component of positional error will be negligible and so the deviation is due only to geo-referencing and digitising errors. Converting the empirically derived mean deviation to a standard deviation (RMS error) for the combined effect of geo-referencing and digitising, it can then be compared with the appropriate summation (equation 7.2) of the separate point-based estimates of these two errors. Likewise, by comparing the standard deviation of 'soft' boundaries with that of 'hard' boundaries, one can calculate the amount of positional error due solely to the subjectivity of photointerpretation. In this example, we have only distinguished between 'hard' and 'soft' boundaries, but in principle the same method could be applied to obtain boundary-specific standard deviations of subjectivity that depend upon precisely which two classes are being separated (see Aspinall and Pearson, 1995).

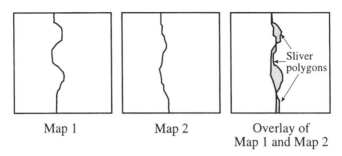

Map 1   Map 2   Overlay of
Map 1 and Map 2

Figure 7.5   Spurious sliver polygons

Thus, simply by re-interpreting a few lines or areas of a map and overlaying them with the original interpretation it is possible to estimate appropriate epsilon bands of boundary uncertainty. By assuming a certain distribution of errors, or by measuring the actual distribution, one can generate buffer zones of different widths that encompass the 'true' position of a boundary to different levels of probability. Within a map, probability zones based upon boundary errors could be combined with probability zones based upon classification accuracy *per se* to generate a visualisation of the distribution of errors (Drummond, 1990; Wenzhong *et al.*, 1994). Subsequently, the certainty of an overlay analysis between two maps could also be estimated. This so-called 'ideal situation' for error reporting and handling has been suggested many times in the past (Goodchild, 1993), and yet it remains almost unknown in everyday applications of GIS.

By taking an internal and external buffer of polygons, upper and lower bounds for the area covered by each class can also be generated (e.g. Chrisman, 1983), but it must be realised that these will not represent an equivalent probabilistic confidence interval for measurements of area. This is because errors in boundary position are not 100% autocorrelated in respect of which side of the polygon perimeter they fall, so deviations in boundary position tend to cancel themselves out to some extent, thereby producing a lower variance in area measurements (Goodchild, 1993). The precise relationship between boundary variance and area variance will depend upon the geometry of the polygons and the degree of autocorrelation in positional error along a polygon's outline (see Prisley *et al.*, 1989, for a mathematical treatment of the subject).

The only way to obtain error estimates for all quantifiable map traits would be through stochastic simulation of the observed map geometry incorporating appropriate degrees of positional variation and misclassification (with an element of spatial autocorrelation in both). Just such an approach is encouraged by Heuvelink *et al.* (1989), Goodchild *et al.* (1992) and Brunsdon and Openshaw (1993). Inevitably though, there must be a trade-off between the ease of applying a model of error (using current GIS software), the number of parameters the model requires and ultimately how well it can inform one in the use of available data. In many situations, simpler models are more likely to be applied by GIS practitioners than complex ones, although the danger remains that simple models may also yield misleading estimates about the degree of error present (Brunsdon and Openshaw, 1993).

AN EXAMPLE OF ERROR ASSESSMENT WHEN MAPPING VEGETATION CHANGE

To illustrate the potential magnitude of error involved in studies of vegetation change and to demonstrate ways by which it can be minimised the following investigation was conducted.

Method

Three aerial photographs of the Dinnet National Nature Reserve, Grampian Region, north-east Scotland, were selected for interpretation (see Plate 1 (Figure 7.6) for a map of the study area). Dinnet National Nature Reserve is of particular conservation interest because of its diverse range of semi-natural habitats and the extensive natural regeneration of birch woodland that has occurred over the past 50 years (Marren, 1980). In this sense it represents a very real situation in which the assessment of landscape change and successional processes are of interest to ecologists and reserve managers, and where investigations must make the best possible use of the available data sources. The first aerial photograph chosen for interpretation was flown in late May 1947, at a scale of 1:10 700 (photo 1947a). Despite the large scale of this photograph it suffered from poor contrast and resolution across much of the image. The second was flown in August 1986 at a scale of 1:15 000 (photo 1986a), and the third was flown six weeks later in September 1986, also at a scale of 1:15 000 (photo 1986b). If there has been significant vegetation change since the 1940s it should be apparent by comparison of the 1947 photograph with either of the 1986 photographs, whereas comparison of the two 1986 photographs should reveal negligible change. A visual comparison of the photographs confirms this opinion, but can the change be quantified with confidence?

Photographs 1947a and 1986b were interpreted once each by interpreter 1, whilst photograph 1986a was interpreted twice, once each by two different interpreters, 2 and 3. Interpreter 1 originally used a 25 class classification scheme during interpretation, and maps were traced onto transparent acetate before enlargement to a 1:7500 paper map ready for subsequent digitising into the ARC/INFO GIS. Interpreters 2 and 3 used a 16-class classification scheme and their interpretations were digitised directly from the transparent acetate maps. All digitising was performed by the same person and for the purposes of this project the 25 class scheme and the 16 class scheme were reduced to a common 13 class scheme. Again this represents a realistic project situation in which digital maps are available from different sources collected using slightly different methodologies.

Pairs of digital maps were then superimposed and the differences between them were highlighted as potential vegetation 'change', the amount of change being measured as a percentage of the total area. As it was assumed that there had been negligible real change in vegetation cover between the 1986a and 1986b photographs, all recorded differences were regarded as 'false change' caused by the types of error already discussed in this paper. In contrast, it was expected that an element of actual change would have occurred between 1947 and 1986, the challenge being to try and detect this with confidence over and above the 'spurious changes' introduced as a result of error.

Since multitemporal studies must necessarily utilise different photographs, of potentially differing appearance, this element of variation should be present in all of the pairwise comparisons. Therefore, all overlays were made between maps derived from different photographs, and in four out of five cases interpreted by independent interpreters.

The amount of 'false change' contributed by the different types and sources of errors was estimated by the methods for quantifying error already discussed in this chapter (i.e. confusion matrices, point-on-point RMS measures, and empirical measurements of boundary shift). Based upon an understanding of the sources and causes of error, two 'treatments' were applied to the map pairs in order to try to reduce the contribution of errors to the measurement of vegetation change.

The first treatment aimed to reduce classification errors by amalgamating easily confused and/or rare classes into fewer, simpler aggregate classes. The second treatment was to generate a buffer zone around all of the boundaries that encompassed approximately 50% of the positional boundary error introduced by digitising, geo-referencing and, where appropriate, subjectivity. The area within the buffer zone was then excluded from subsequent change detection analysis since one could not have a high degree of confidence in any of the 'changes' recorded within this zone. The widths of the buffer zones were determined from empirical measurement of the mean boundary shift observed when overlaying two interpretations of the same boundary. Boundaries involving water or agricultural land were considered to be 'hard' (non-subjective) boundaries whilst the transitions from 'heath' to 'establishing birch' were the 'soft' (subjective) boundaries.

Results

Misclassification Errors

Table 7.3 lists the five pairwise overlays that were generated, and the percentage of the total area classified differently between the two maps. In the case of two maps for 1986 this is almost entirely 'spurious' change caused by a combination of misclassification, different levels of delineation and positional boundary errors discussed earlier. The disturbing point from a practitioner's point of view is that the errors can cause as much as one-third of the land cover to be recorded differently in two maps of the same area for the same time.

Figure 7.7 illustrates one such overlay, that between 1986a–interpreter 1 and 1986b–interpreter 2, in which all areas of land that were classified differently have been shaded black. The single largest source of error appears to be in the inconsistent classification of entire polygons, but it is also possible to see that to the south one of the interpreters has mapped the heath/birch mosaic to a much greater level of detail and there are also some spurious boundary shifts due to digitising and geo-referencing error. Construction of a confusion matrix (Table 7.4) helps to reveal the combined effect of these errors on a class by class basis.

Note that the amount of 'difference' recorded between these two interpretations covers 32% of the total area and of the 13 categories only five coincide in their mapped extent more than 50% of the time. There is very little reciprocal confusion

Table 7.3 Table summarising the amount of vegetation 'change' detected as a percentage of the total area. 1986a and 1986b are two different photographs taken of the same area, at the same scale, approximately six weeks apart

Photograph, ID Interpreter:	1986a Interpreter 1	1986b Interpreter 2	1986b Interpreter 3	
1947a Interpreter 1	41%	54%	50%	⇐ 'real' and 'false' change
1986a Interpreter 1	–	32%	33%	⇐ 'false' change only
	–	22%	17%	⇐ assumed amount of 'real' change

which suggests systematic differences in the way in which each interpreter operated (Chrisman and Lester, 1991) and some of the categories are completely unused by one or other of the interpreters. The most understandable classification confusion would seem to be between 'wet heath' and 'dry heath', two vegetational cover types that are extremely difficult to distinguish from each other, whilst disagreements over whether an area is 'birch' or 'heath' probably stem from different decisions over whether or not to delineate small clumps of trees or small patches of heath in the heath/birch mosaic.

The first step towards reducing the incidence of 'false' change is to ensure a consistent classification at the level of whole polygons. Guided by the confusion matrix in Table 7.4 a new classification scheme was devised in which 'wet heath' and 'dry heath' were amalgamated into a single class, as were the various categories of tree, including bracken. Anthropogenic and/or barren habitats formed another obvious grouping and finally, reeds and rough grassland were also combined. The new confusion matrix is shown in Table 7.5.

Having amalgamated some classes there is now much better agreement between the two interpreters, although the total mismatch between maps is still 26% of the total area. Some residual confusion remains between 'heath' and 'grass/reeds', and reclassifying has not removed the differences between 'tree' and 'heath' caused by the birch/heath mosaic; in fact, these are the only two categories that display any sign of reciprocal confusion, which is the sort of confusion that would be expected in a mosaic situation. In addition, the large percentage of land classed as 'developed or agricultural' by interpreter 1 and 'trees' by interpreter 2, is actually a road passing through the woodland. Interpreter 1 delineated this as a separate polygon whilst interpreter 2 subsumed it into the background vegetation. This discrepancy emphasises the need for explicit rules regarding the treatment of minimal mappable areas and features.

Positional Boundary Errors

The digitising error as determined by redigitising a set of reference points was 0.33 mm RMS (which is equivalent to 4.6 m 'on the ground' for both of the 1986 photographs), whilst the RMS values for the linear geo-referencing transformations were 11.5 m for photograph 1986a and 4.6 m for photograph 1986b. Discounting any subjectivity errors

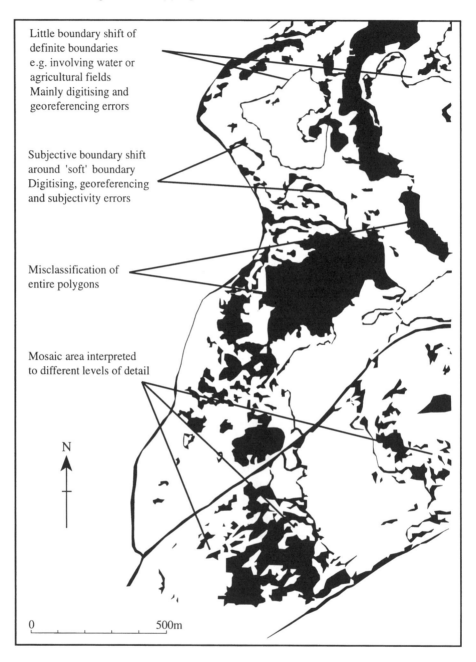

Little boundary shift of
definite boundaries
e.g. involving water or
agricultural fields
Mainly digitising and
georeferencing errors

Subjective boundary shift
around 'soft' boundary
Digitising, georeferencing
and subjectivity errors

Misclassification of
entire polygons

Mosaic area interpreted
to different levels of detail

N

0 _____ 500m

Figure 7.7 Map highlighting the areas of inconsistent classification between two independent interpretations of the 1986 vegetation cover. Note the four main types of error

Table 7.4 Confusion matrix or 'apparent transitions' resulting from two independent interpretations of the same area from different photographs taken six weeks apart. Values are standardised to a total area of 1000

Interpreter 1 \ Interpreter 2	Dry Heath	Wet Heath	Birch	Pine	Willow	Bracken	Reeds	Mineral	Water	Agric.	Rough Grass	Buildings	Other
Dry Heath	111.8	0.3	72.6	0	0	0	0	0	0	0	0	0	0
Wet Heath	56.1	7.5	40.6	0	0	0	0	0	0	0	0	0	0
Birch	38.6	6.1	455.9	0	0.3	0	0	0	0.3	0	4.5	0	3.0
Pine	0.1	0	0.4	0	0	0	0	0	0	0	0	0	0
Willow	0	0.3	0	0	0	0	0	0	0	0	0	0	0
Bracken	0	0	1.9	0	0	0	0	0	0	0	0	0	0
Reeds	0	32.5	5.0	0	0	0	3.6	0.7	0	4.9	0	0	0
Mineral	0	0	0.2	0	0	0	0	0	0	0	0	0	0
Water	0	2.9	0.8	0	0.3	0	2.9	0	67.7	0	0	0	0
Agric.	0	0	0.7	0	0	0	0	0	0	12.3	0	0	0
Rough Grass	0	0	5.5	0	0	0	0	0	0	0	36.5	0	0
Buildings	3.0	0.3	16.4	0	0	0	0	0	0	0	0	0	2.5
Other	0	0	0	0	0	0	0	0	0	0	0	0	0

Table 7.5 Confusion matrix or 'apparent transitions' using a simplified classification scheme. Both interpretations are from 1986 photographs. Values are standardised to a total area of 1000 units

Interpreter 2 / Interpreter 1	Trees	Heath	Develop/ Agric.	Reeds/ Grass	Water	Other/ Missing
Trees	405.9	40.4	0.9	4.0	0.4	2.7
Heath	144.9	222.2	0	0.7	0	0
Develop/Agric.	15.2	3.0	11.0	0	0	2.3
Reeds/Grass	9.4	28.8	0	40.1	0.7	0
Water	1.0	2.6	0	2.6	60.3	0
Other/Missing	0.2	0	0	0	0	0

these two sources alone predict a cumulative positional error $(e^2_{dig} + e^2_{geo\text{-}ref})^{1/2}$, of 12.4 m RMS for photograph 1986a and 6.5 m RMS for photograph 1986b, which combined in an overlay would yield an average boundary displacement of 14 m RMS.

By comparison, the mean width of sliver polygons (w) created by an overlay of 'hard' (non-subjective) boundaries, such as those bounding water bodies or agricultural land was measured as 6.9 m. Assuming a normal distribution of errors this translates into a standard deviation (RMS) for the overlay of 10.2 m. This is less than the value predicted above by the combination of independent digitising and geo-referencing errors. The most likely explanation for the difference is that because the two photographs are taken from similar positions, they probably share similar topographic distortions and relief displacement so that their residual geo-referencing 'errors' are in phase across most of the study area. Consequently, the mapped lines are registered to each other better than they are referenced to planimetric ground coordinates, and the geo-referencing errors do not combine as independent random errors. This enables one to distinguish between the geo-referencing error of a particular map and the co-registration error that exists between specific pairs of maps. In this example, the overlay error expected from digitising errors alone would be 6.5 m, which implies that the actual co-registration error between the two maps is 7.9 m.

Finally, the average width of 'soft' boundary shifts, i.e. the heath/tree boundaries, was measured as 10.2 m (15.1 m RMS). Decomposing this error according to equations 7.3 and 7.4, the error due to subjectivity alone is found to be approximately 7.9 m for each map. See Table 7.6 for a full listing of the component errors.

Taking 6.9 m as the mean displacement, w, expected in an overlay of 'hard' boundaries and 10.2 m as the equivalent measure for 'soft' boundaries, these values were used to generate epsilon bands of total width $2w$. These zones were then excluded from subsequent analysis, so that sliver polygons of 'false' change would only be formed when the two parent boundary lines deviated by an amount greater than two mean deviations. Assuming a normal distribution of errors, only 18% of deviations will exceed twice the mean deviation so, conversely, 82% of positional boundary errors should have been removed from the overlay analysis. In addition, an unknown proportion of the

Table 7.6 Results of decomposing positional error. Values in bold are the empirical measurements used to calculate all other values, assuming a normal distribution of errors and according to the equations presented earlier in this paper. Since the georeferencing errors do not combine as independent variables a co-registration error has been calculated for the overlay using the relationship $E_{total} = (e^2_{dig} + e^2_{co-reg})^{1/2}$

	Positional error in one map (m)	Discrepancy in an overlay (m)		
	RMS	Mean deviation (50th percentile)	RMS (68th percentile)	2* mean devn. (82nd percentile)
e_{dig}	**4.6**		6.5	
$e_{geo-ref}$	**4.6,11.5**		6.5,16.3	
E_{total} 'hard' boundaries	7.2	**6.9**	10.2	13.8
e_{co-reg}	N/A		7.9	
E_{total} 'soft' boundaries	10.7	**10.2**	15.1	20.4
e_{subj}	7.9		11.1	

'small island' errors caused by interpretations to different levels of detail will also have been removed by reclassification and buffering. Figure 7.8 displays the areas of 'false' change remaining in an overlay of 1986a and 1986b after application of the new classification scheme and exclusion of the boundary buffer zones, whilst the particular transitions are detailed in Table 7.7.

Through the application of these two treatments, the total amount of 'false' change recorded has been reduced from 32% to 9% of the area available for analysis (i.e. a new classification consistency of 91%). By way of contrast, when these same two treatments were applied to a 1947/1986b pair of photographs the area of 'change' dropped from 54% to 46% (or from 50% to 36% in the case of interpreters 1 and 3) (see Figure 7.9). The values for a between-date comparison are a mixture of 'real' and 'false' change, whereas the overlay of 1986a/1986b is entirely 'false' change. If we assume that equal amounts of 'false' change are present in both sets of comparison we can calculate the residual 'real' change that occurred between 1947 and 1986 (cross-reference to Table 7.3). Carrying out this calculation both before and after the treatments we find that the ratio of 'real' change to 'false' change has been altered from approximately 2:3 to 3:1. In other words the ratio of signal to noise has been much improved. Consequently, one can interpret the final transition matrix with much greater confidence.

Table 7.8 presents a matrix of the transitions occurring between 1947 and 1986 after re-classification and exclusion of boundary buffer zones. Note from this table that the most striking transition is from heath to trees, which agrees well with what is already known of the area (Marren, 1980). In addition, by cross-referencing to Table 7.7, the appropriate confusion matrix for this overlay, we can see that the transition from reeds/grass to heath is probably a systematic inconsistency between interpreters 1 and 2, whereas the transition from reeds/grass to trees is more likely to be a real effect. Also, many of the small, insubstantial fluxes recorded in previous matrices (e.g. Table 7.5) are now filtered out and replaced by zeroes.

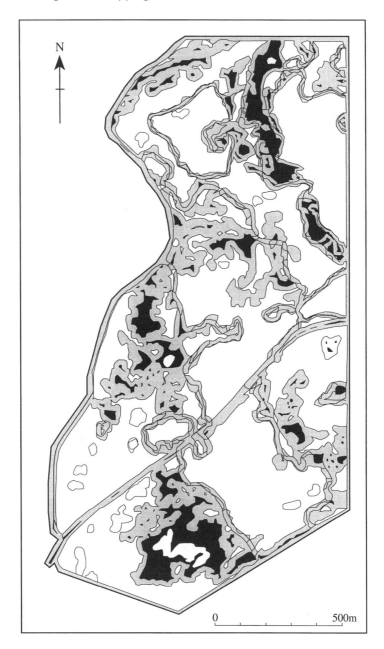

Figure 7.8 Map highlighting the areas of inconsistent classification remaining after simplification of the classification scheme and masking out of a 82% probabilistic epsilon band. 'Soft' boundaries have a broader epsilon band than do 'hard' boundaries

Table 7.7 Confusion matrix or 'apparent transitions' using a simplified classification scheme and excluding areas inside the boundary epsilon bands. Both interpretations are from 1986 photographs. Values are standardised to a total area of 1000 units

Interpreter 2 / Interpreter 1	Trees	Heath	Develop/ Agric.	Reeds/ Grass	Water	Other/ Missing
Trees	435.3	10.0	0	0.3	0	0.1
Heath	80.1	279.4	0	0.3	0	0
Develop/Agric.	1.4	0	14.9	0	0	0
Reeds/Grass	3.3	31.8	0	50.5	0	0
Water	0	0.6	0	0.9	90.0	0
Other/Missing	0	0	0	0	0	0

DISCUSSION

In this investigation the entire study area was reinterpreted and redigitised a number of times to illustrate the potential for error and ways in which it can be assessed. In practice though, the same assessments can be made simply by reworking a random sub-set of a larger study area. The empirical estimates and visualisations of error, thus produced, are likely to provide a much better guide to the types and magnitudes of errors occurring than any theoretical discussion. For example, in this study we found that the errors of geo-referencing did not combine as independent variables and were better expressed by a single error of co-registration.

One important reason for wishing to know how much different sources are contributing to positional error is so that the data capture and processing procedures can be reviewed. By knowing which element of the map-making process is contributing the largest errors, effort can be directed towards improving the accuracy of this particular process. Conversely, time may be saved in other areas by avoiding working to unnecessary levels of accuracy and precision. For example, is it worth obtaining more ground control points and applying more complex rectification algorithms if the subjectivity of boundary placement is contributing over 90% of the positional error?

A knowledge of the relative contribution of the different sources of error can also inform one about the different types of error to expect. In the worked example we did not attempt an analytical separation of the total error into different types, as we felt this was reasonably transparent from the overlay presented in Figure 7.7. Nonetheless, quantitative methods are possible. For example, as mentioned in an earlier section the construction of two confusion matrices, one based upon point-in-polygon validation and the other based upon whole polygon validation, might help to separate misclassification *per se* from errors due to within-polygon heterogeneity. Chrisman and Lester (1991) present a complementary diagnostic test for distinguishing between the sliver

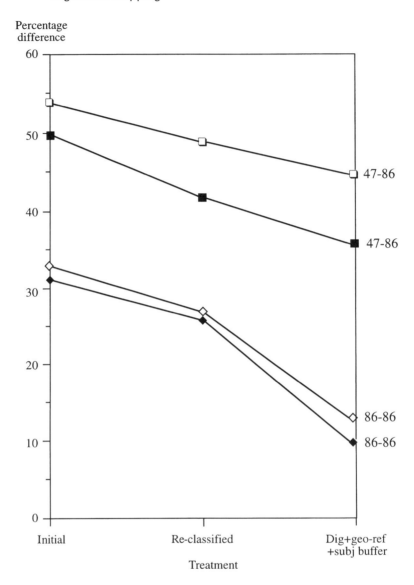

Figure 7.9 Graph depicting the percentage difference between pairs of interpretations before and after re-classification and exclusion of a boundary buffer zone. For pairs comparing same-date interpretations all of the difference is regarded as 'false' change

Table 7.8 Mapped transitions between 1947 and 1986 using a simplified classification scheme and excluding regions close to boundaries. Values are standardised to a total area of 1000 units

1986 Int. 2 / 1947 Int. 1	Trees	Heath	Develop/ Agric.	Reeds/ Grass	Water	Other/ Missing
Trees	111.8	0.3	0	0.4	0	0
Heath	395.1	308.9	0	0	0	0
Develop/Agric.	1.3	0	12.5	0.6	0	0
Reeds/Grass	23.2	25.0	0	38.3	0	0
Water	0.9	2.2	0	2.8	75.0	0
Other/Missing	0.6	0	0	0	0	0

polygons arising from positional boundary error and the inconsistent, small inclusions of within-polygon heterogeneity. Put simply, sliver polygons are characterised by the fact that they are the product of the intersection of two parent boundary lines of approximately equal length.

Having established the nature and magnitude of error that is present, it is then possible to decide whether the data is of sufficient quality for the intended methods of analysis. Throughout this chapter we have stressed how a dynamic description of change based upon transition frequencies requires good co-registration, consistent classification and consistent generalisation. By comparison, estimates of net change will be relatively unaffected by positional boundary errors. Take, for example, the road that was mapped by interpreter 1 in both 1947 and 1986. There was very little subjectivity involved in its mapping and as an estimate of the area of land covered by tarmac it was probably reasonably accurate. However, simply due to geo-referencing and digitising errors, a dynamic description of change would suggest that all the 'road' changed to 'trees' between 1947 and 1986, whilst an equivalent area of 'trees' was transformed into 'road'.

Returning to some of the pattern indices mentioned in an earlier section, each one describes a different aspect of landscape topology. Although the fractal dimension is used to describe the relationship between geometry at different scales, the measure itself is also scale dependent (Krummel *et al.*, 1987; Williamson and Lawton, 1991). Thus, it is wrong to consider the fractal dimension as an intrinsic, unvarying property of a landscape, as the value obtained will be critically dependent upon the scale of the photographs and the detail of mapping. However, as a quantification of landscape pattern it will be little affected by the global positional errors of geo-referencing. Mean patch size is another statistic that will be unreliable if maps are produced to different levels of detail but which will be relatively insensitive to the other types of error. Diversity indices measure the relative proportions of different land cover classes and as such will be highly susceptible to errors in misclassifications or systematic inclusions or exclusions of within-polygon heterogeneity. In addition it almost goes without saying that diversity indices can only be compared between maps employing the same classification scheme (O'Neill *et al.*, 1988). Finally, contagion is mainly concerned with what border is what, rather than precisely where the boundary occurs, so it will be relatively immune to errors in boundary position, as long as the overall relationships between polygons are maintained.

The final incentive for investigating the error present in a map is to enable some form of *post hoc* error management to be undertaken. We have presented two such examples: class amalgamation and buffering, which deal mainly with classification and positional errors, respectively. In addition, other methods, such as the filtering out of small polygons or the application of mathematical shape generalising algorithms (e.g. Li and Openshaw, 1992), can be applied to standardise the level of within-polygon heterogeneity that is expressed.

It should be appreciated though, that none of the error management and standardisation techniques are a panacea, as all will tend to reduce the information content of the map. For example, consider the situation in which class *a* is sometimes confused with class *b*, *b* with *c* but *c* is never confused with *a*. Having amalgamated all three classes to avoid misclassifications one has lost any discrimination between classes *a* and *c* that might usefully have existed. Just as amalgamating reduces the categorical resolution of a map, so buffering reduces the spatial resolution of polygon boundary position. Since most change occurs by expansion or contraction at the edges, by discounting these areas from the analysis one is also ignoring the zone that is most sensitive to change.

Based upon a consideration of the nature and sources of error associated with vegetation mapping, we have derived some general rules that may be applied when deciding whether certain types of analyses or treatment are appropriate. These guidelines are presented in Table 7.9.

SUMMARY AND CONCLUSION

The objective of this chapter has been to examine the sources and nature of error associated with mapping vegetation polygons from aerial photography and to raise awareness of their significance when undertaking different spatial analyses and measurements. Moreover it set out to develop a simple empirical approach to error estimation designed to encourage researchers involved in similar studies to assess the appropriateness of their data and methodology for describing and analysing vegetation change. All maps contain error. In some cases the magnitude of the error will be as great as or greater than the change that one is trying to detect. However, in other cases it will not affect the main conclusions drawn from the data. For this reason it is important to quantify and characterise the error present if data are to be given scientific credibility.

Although the theory and practice of error modelling can become quite complex, simple, empirical methods for quantifying the margin of error are likely to be the most illuminating in the context of project work.

Table 7.9 Types of error, their likely sources, how they affect different analyses, and how the situation may be remedied

Type of error	Many misclassifications	Different levels of delineation	Large hard line boundary shifts, mainly translational	Large subjective boundary shifts
Likely source of error	Subjective photointerpretation	Different scale images, subjective photointerpretation	Digitising errors or georeferencing mismatch	Subjective photointerpretation
Analysis possible with this type of error present				
Descriptive mapping	yes	yes	yes	yes
Quantify net changes	no	no	yes	no
Quantify transitions	no	no	no	no
Fractal analysis	yes	no	yes	no
Diversity indices	no	no	yes	no
Contagion analysis	no	no	yes	yes
Will the following treatment help?				
Reclassifying	yes	possibly	no	possibly
Buffering	no	yes	yes	yes

ACKNOWLEDGEMENTS

The authors would like to thank the students of the 1994/95 Environmental Remote Sensing MSc course for providing some of the photointerpretations used in this study and the University of Aberdeen and the Institute of Terrestrial Ecology (ITE) for providing computing and digitising facilities. We are also grateful to Jack Lennon, Philip Bacon, Giuliana Torta, Bryan Shorrocks and the anonymous referees for their valuable comments on earlier drafts of this manuscript and to Stan Openshaw for discussions on 'error'. Finally, thanks must also go to Alison Sandison, Cartographer in the Department of Geography, at the University of Aberdeen, for drafting and re-drafting the tables and figures included in this chapter.

REFERENCES

Aspinall, R.J. and Pearson, D.M. (1995) Describing and managing uncertainty of categorical maps in GIS. In: P. Fisher (ed.) *Innovations in GIS 2*. Taylor & Francis, London, pp. 71–83.
Avery, T.E. (1970) *Interpretation of Aerial Photographs*. Burgess, Minneapolis.
Bie, S.W. and Beckett, P.H.T. (1973) Comparison of four independent soil surveys by air-photo interpretation, Paphos Area (Cyprus). *Photogrammetria*, **29**, 189–202.
Blakemore, M. (1984) Generalisation and error in spatial databases. *Cartographica*, **21**, 131–139.
Bolstad, P.V., Gessler, P. and Lillesand, T.M. (1990) Positional uncertainty in manually digitised maps data. *International Journal of Geographical Information Systems*, **4**, 399–412.

Bridgewater, P.B. (1993) Landscape ecology, geographic information systems and nature conservation. In: R. Haines-Young, D.R. Green, D.R. and S.H. Cousins (eds) *Landscape Ecology and Geographic Information Systems*. Taylor & Francis, London, pp. 23–36.

Brown, J.H. (1995) *Macroecology*. University of Chicago Press, Chicago.

Brunsdon, C. and Openshaw, S. (1993) Simulating the effects of error in GIS. In: P.M. Mather (ed.) *Geographical Information Handling – Research and Applications*. Wiley, London, pp. 47–61.

Burgess, R.L. and Sharpe, P.M. (1981) *Forest Island Dynamics in Man-dominated Landscapes*. Springer-Verlag, New York.

Burrough, P.A. (1986) Data quality, errors, and natural variation. In: P.A. Burrough (ed.) *Principles of Geographical Information Systems for Land Resources Assessment. Monographs on Soil and Resources Survey No. 12*. Oxford Science Publications, Oxford, pp. 103–135.

Burrough, P.A. (1987) Spatial aspects of ecological data. In: R.H.G. Jongman, C.J.F. ter Braak and O.F.R. van Tongeren (eds) *Data Analysis in Community and Landscape Ecology*. Centre for Agricultural Publishing and Documentation, Wageningen, pp. 213–251.

Burrough, P.A. (1994) Accuracy and error in GIS. In: D.R. Green and D. Rix (eds) *The AGI Sourcebook for Geographic Information Systems 1995*. AGI, London, pp. 87–91.

Callaway, R.M. and Davis, F.W. (1993) Vegetation dynamics, fire, and the physical environment in coastal central California. *Ecology*, **74**, 1567–1578.

Campbell, J. (1993) *Map Use and Analysis*. WCB Publishers, Dubuque, Iowa.

Caspary, W. and Scheuring, R. (1992) Error bands as measures of geometrical accuracy. *Proceedings of the 3rd EGIS Conference*, Munich, Germany, 23–26 March 1992, pp. 226–233.

Chrisman, N.R. (1983) A theory of cartographic error and its measurement in digital databases. *Proceedings of AUTO-CARTO 5*, Environmental Assessment and Resources Management: Proceedings, Fifth International Symposium on Computer-Assisted Cartography, Hyatt Regency Crystal City, Crystal City, Virginia, 22–28 August 1982. ASPRS/ACSM, Falls Church, pp. 159–168.

Chrisman, N.R. (1989) Modeling error in overlaid categorical maps. In: M. Goodchild and S. Gopal (eds) *Accuracy in Spatial Databases*. Taylor & Francis, London, pp. 21–34.

Chrisman, N.R. and Lester, M.K. (1991) A diagnostic test for categorical maps. *Technical Papers*, ACSM–ASPRS Annual Convention, **6**, 330–348.

Drummond, J. (1990) A framework for handling error in geographic data manipulation. In: *Fundamentals of Geographic Information Systems: A Compendium*, ASPRS, pp. 109–118.

Dunn, R., Harrison, A.R. and White, J.C. (1990) Positional accuracy and measurement error in digital databases of land use: an empirical study. *International Journal of Geographical Information Systems*, **4**, 385–398.

Erdas (1990) *Field Guide*. Version 7.4. Erdas Inc.

Fisher, R.A. and Yates, F. (1957) *Statistical Tables for Biological, Agricultural and Medical Research*. Oliver & Boyd, Edinburgh.

Forman, R.T. and Godron, M. (1986) *Landscape Ecology*. Wiley, New York.

Giovachino, D. (1993) How to determine the accuracy of your digitiser. *GeoInfo Systems*, March 1993, pp. 50–53.

Goldsmith, B. (1991) *Monitoring for Conservation and Ecology*. Chapman & Hall, London.

Goodchild, M.F. (1993) Data models and data quality: problems and prospects. In: M.F. Goodchild, B.O. Parks and L.T. Steyaert (eds) *Environmental Modeling with GIS*. Oxford University Press, New York, pp. 94–104.

Goodchild, M.F. (1994) Integrating GIS and remote sensing for vegetation analysis and modelling: methodological issues. *Journal of Vegetation Science*, **5**, 615–626.

Goodchild, M.F., Guoqing, S. and Shiren, Y. (1992) Development and test of an error model for categorical data. *International Journal of Geographical Information Systems*, **6**, 87–104.

Green, D.R., Cummins, R., Wright, R. and Miles, J. (1993a) A methodology for acquiring information on vegetation succession from remotely sensed imagery. In: R. Haines-Young, D.R. Green and S.H. Cousins (eds) *Landscape Ecology and Geographic Information Systems*. Taylor & Francis, London, pp. 111–128.

Green, E.J., Strawderman, W.E. and Airola, T.M. (1993b) Assessing classification probabilities for thematic maps. *Photogrammetric Engineering and Remote Sensing*, **59**, 635–639.

Haines-Young, R., Green, D.R. and Cousins, S.H. (1993) *Landscape Ecology and Geographic Information Systems*. Taylor & Francis, London.

Hanski, I. and Gilpin, M. (1991) Metapopulation dynamics: brief history and conceptual domain. *Biological Journal of the Linnean Society*, **42**, 3–16.

Hartley, S. (1994) Modelling natural succession and landscape pattern at Dinnet National Nature Reserve. Unpublished MSc thesis, University of Aberdeen.

Heuvelink, G.B.M., Burrough, P.A. and Stein, A. (1989) Propagation of errors in spatial modelling with GIS. *International Journal of Geographical Information Systems*, **3**, 303–322.

Hornung, M. and Reynolds, B. (1995) The effects of natural and anthropogenic environmental changes on ecosystem processes at the catchment scale. *Trends in Ecology and Evolution*, **11**, 443–449.

Krummel, J.R., Gardner, R.H., Sugihara, G., O'Neill, R.V. and Coleman, P.R. (1987) Landscape patterns in a disturbed environment. *Oikos*, **48**, 321–324.

Lanter, D.P. and Veregin, H. (1992) A research paradigm for propagating error in layer-based GIS. *Photogrammetric Engineering and Remote Sensing*, **58**, 825–833.

Lester, M. and Chrisman, N.R. (1991) Not all slivers are skinny: a comparison of two methods for detecting positional error in categorical maps. *Proceedings of GIS/LIS 91*, ASPRS/ACSM, pp. 648–658.

Levin, S.A. (1992) The problem of pattern and scale in ecology. *Ecology*, **73**, 1943–1967.

Li, Z. and Openshaw, S. (1992) Algorithms for automated line generalization based on a natural principle of objective generalization. *International Journal of Geographical Information Systems*, **6**, 373–389.

Lo, C.P. and Shipman, R.L. (1990) A GIS approach to land-use change dynamics detection. *Photogrammetric Engineering and Remote Sensing*, **56**, 1483–1491.

Lowell, K. (1991) Utilizing discriminant function analysis with a geographical information system to model ecological succession spatially. *International Journal of Geographical Information Systems*, **5**, 175–191.

Lowell, K.E. and Astroth, J.H. (1989) Vegetative succession and controlled fire in a glades ecosystem: a geographical information system approach. *International Journal of Geographical Information Systems*, **3**, 69–81.

Lunetta, R.S., Congalton, R.G., Fenstermaker, L.K., Jensen, J.R., McGwire, K.C. and Tinney, L.R. (1991) Remote sensing and geographic information system data integration: error sources and research issues. *Photogrammetric Engineering and Remote Sensing*, **57**, 677–687.

MacArthur, R.H. and Wilson, E.O. (1967) *The Theory of Island Biogeography*. Princeton University Press, New Jersey.

Maffini, G., Arno, M. and Bitterlich, W. (1989) Observations and comments on the generation and treatment of error in digital GIS data. In: M. Goodchild and S. Gopal (eds) *Accuracy of Spatial Databases*. Taylor & Francis, London, pp. 55–67.

Marren, P. (1980) *Muir of Dinnet: Portrait of a National Nature Reserve*. Nature Conservancy Council, Aberdeen.

May, R.M. (1994) The effects of spatial scale on ecological questions and answers. In: P.J. Edwards, R.M. May and N.R. Webb (eds) *Large Scale Ecology and Conservation Biology*. Blackwell Scientific Publications, Oxford, pp. 1–17.

Miller, D.R., Gauld, J.H., Bell, J.S. and Towers, W. (1991) Land cover change in the Cairngorms. *Proceedings of the AGI '91*. Association for Geographic Information (AGI) Conference and Exhibition, Birmingham, England, pp. 2.13.1–2.13.5.

O'Neill, R.V., Krummel J.R., Gardner, R.H., Sugihara, G., Jackson, B., DeAngelis, D.L., Milne, B.T., Turner, M.G., Zygmunt, B., Christensen, S.W., Dale, V.H. and Graham, R.L. (1988) Indices of landscape pattern. *Landscape Ecology*, **1**, 153–162.

Perkal, J. (1966) On the length of empirical curves. *Discussion Paper 10*, Ann Arbor, Michigan Inter-University Community of Mathematical Geographers.

Petrie, G. (1990) Digital mapping technology: procedures and applications. In: T.J.M. Kennie and G. Petrie (eds) *Engineering Surveying Technology*. Blackie, Glasgow, pp. 329–389.

Pimm, S.L. (1994) *The Balance of Nature: Ecological Issues in the Conservation of Species and*

Communities. University of Chicago Press, Chicago.

Prisley, S.P., Gregorie, T.G. and Smith, J.L. (1989) The mean and variance of area estimates computed in an arc-node Geographic Information System. *Photogrammetric Engineering and Remote Sensing*, **55**, 1601–1612.

Reid, E., Mortimer, G.N., Lindsay, R.A. and Thompson, D.B.A. (1994) Blanket bogs in Great Britain: an assessment of large-scale pattern and distribution using remote sensing and GIS. In: P.J. Edwards, R.M. May and N.R. Webb (eds) *Large Scale Ecology and Conservation Biology*. Blackwell Scientific Publications, London, pp. 229–246.

Rolstad, J. (1991) Consequences of forest fragmentation for the dynamics of bird populations: conceptual issues and the evidence. *Biological Journal of the Linnean Society*, **42**, 149–163.

Shugart, H.H. (1994) *A Theory of Forest Dynamics: The Ecological Implications of Forest Succession Models*. Springer-Verlag, New York.

Thapa, K. and Bossler, J. (1992) Accuracy of spatial data used in geographic information systems. *Photogrammetric Engineering and Remote Sensing*, **58**, 835–841.

Turner, M.G. (1987) Spatial simulation of landscape change in Georgia: a comparison of three transition models. *Landscape Ecology*, **1**, 29–36.

Turner, M.G. (1989) Landscape ecology: the effect of pattern on process. *Annual Review of Ecological Systems*, **20**, 171–197.

Turner, M.G. and Gardner, R.H. (1991) Quantitative methods in landscape ecology: an introduction. In: M.G. Turner and R.H. Gardner (eds) *Quantitative Methods in Landscape Ecology: The Analysis and Interpretation of Landscape Heterogeneity. Ecological Studies 82*. Springer-Verlag, New York, pp. 3–14.

Veregin, H. (1989a) Accuracy of spatial databases: annotated bibliography. *Technical Paper 89–9*, National Center for Geographic Information and Analysis (NCGIA), Santa Barbara.

Veregin, H. (1989b) Error modeling for the map overlay operation. In: M. Goodchild and S. Gopal (eds) *Accuracy of Spatial Databases*. Taylor & Francis, London, pp. 3–17.

Vink, A.P.A. (1983) *Landscape Ecology and Land Use* (English translation editor D.A. Davidson). Longman, London.

von Rimscha, M. (1996) Seeing is not believing. *GIS Europe*, **5**, 14–15.

Walsby, J.C. (1995) The causes and effects of manual digitising on error creation in data input to GIS. In: P. Fisher (ed.) *Innovations in GIS 2*. Taylor & Francis, London, pp. 113–122.

Walsh, S.J., Lightfoot, D.R. and Butler, D.R. (1987) Recognition and assessment of error in GIS. *Photogrammetric Engineering and Remote Sensing*, **53**, 1423–1430.

Wenzhong, S., Ehlers, M. and Tempfli, K. (1994) Modelling and visualizing uncertainties in multi-data-based spatial analysis. In: *Proceedings of 5th EGIS/MARI*, Paris, France, 29 March–1 April, 1994, pp. 454–464.

Williamson, M.H. and Lawton, J.H. (1991) Fractal geometry of ecological habitats. In: S.S. Bell, E.D. McCoy and H.R. Mushinsky (eds) *Habitat Structure: The Physical Arrangement of Objects in Space*. Chapman & Hall, London, pp. 69–86.

With, K.A. and Crist, T.O. (1995) Critical thresholds in species' responses to landscape structure. *Ecology*, **76**, 2446–2459.

8 Vegetation Analysis, Mapping and Environmental Relationships at a Local Scale, Jotunheimen, Southern Norway

P.D. COKER
Department of Environmental Sciences, University of Plymouth, UK

The research area in Høydal (61° 40′ N, 8° 06′ E) is about 20 km^2 in extent and includes a wide range of habitat types from fjellfeld to anthropogenic grassland. The area is geologically complex with extensive areas of calcareous rocks; limited woodland clearance and summer grazing have taken place over a long period of time, and the area has been under threat for several years in respect of the development of its potential as a source of hydroelectric power.

Satellite and aerial photographic imagery were used to provide a base on which a stratified random sampling design of 500 relevés (sites), each 4 m^2 in extent, was imposed. One hundred sites were selected at random and additionally investigated for a range of environmental factors, including soil chemistry, slope, aspect and duration of snow-lie.

Vegetation data were classified using the TWINSPAN (Two-Way Indicator Species Analysis) package, and the two-way table thus produced for the 500 sites showed evidence of 13 alliances and 25 associations. The sub-sample of 100 sites for which environmental data had also been collected was analysed in a similar fashion and 10 vegetation groups were distinguished and mapped. The distribution of these groups among the sampling points was used to provide ground truth for the aerial photographic cover, and for subsequent GIS work. The 100 site data set and its associated environmental data were ordinated using the CANOCO (Canonical Correspondence Analysis) package. The ordinations showed that duration of snow-lie was strongly correlated with the first vegetational axis and that exchangeable calcium and pH were closely allied to the second axis. Examination of the environmental biplots for both species and sites revealed some species groups that were strongly correlated with snow-patches and others which preferred exposed sites or calcareous soils.

Trend surface analysis (TSA) was applied to the data for exchangeable calcium and this showed a good correspondence between the predicted pattern of the fourth-order surface, and the pattern of regional variability in the geology of the area.

The evidence of the survey work supports the contention that the area is of national importance in terms of its biodiversity and the number of nationally rare species which are found there.

Vegetation Mapping: From Patch to Planet. Edited by Roy Alexander and Andrew C. Millington.
© 2000 John Wiley & Sons Ltd.

PREAMBLE

The work described in this chapter is an attempt to discover and quantify the roles of the major factors underlying the distribution and environmental interactions of plant communities in an extreme montane environment. It clearly shows the importance of snow-cover and soil nutrient status in determining the species content of vegetation mosaics and demonstrates the power and utility of modern analytical packages in elucidating the complex relationships and the underlying structure of montane vegetation at both a meso- and a microscale.

Nomenclature follows the *Flora Europaea* for higher plants, Dahl and Krog (1973) for lichens and Arnell (1956) and Nyholm (1954–1969) for bryophytes.

INTRODUCTION

This long-term investigation was prompted by local environmental anxieties regarding a proposal by the Norwegian electricity generating authorities to raise the level of Høydalsvatnet by about 30 m. The area has been known for a number of years as being of great botanical and geological interest and, unusually for such an important site, it was not included in the Jotunheimen National Park, the north-western boundary of which stops less than 5 km from the proposed dam site. Whatever the reasons for its exclusion, and bearing in mind the fate of some of the richest montane grasslands and wetlands in Upper Teesdale, Northern England, which were flooded as a consequence of a similar project 25 years ago, it was decided to carry out a detailed survey, based upon a statistically sound sampling system which would enable the vegetation and environmental data to be analysed using multivariate methods, including classification and ordination. The area which was surveyed included the mountain Høyrokampen, and the north-eastern shores of Bøvertunvatn and Høydalsvatnet (Figure 8.1). Surveying took place during the period 1976–1980, with re-surveying of selected sites in 1992 to investigate successional changes in selected habitats.

The area is close to a main road and access for tourists is quite easy. Fortunately, numbers are low at present, but established footpaths and tracks are showing signs of damage, particularly in the area around Bøvertun and Vassenden, and campers are causing some damage to communities in the Vassenden area. Most tourists walk through the area in summer, but some cross-country skiing takes place during the winter when access to Høydal is impossible by other means. Some grassland areas, notably near Bøvertun and near Høydalsæter, have been 'improved' for hay production and sheep, cattle and goats have been seen grazing and browsing throughout the area. Goats are currently kept at Høydalsæter during the summer period, but, at present, damage to sensitive communities is very slight.

The area is of outstanding biogeographical importance, partly because it contains isolated and important stands of plant species which are nationally or internationally rare, and partly because it represents one of the largest and most diverse calcareous montane habitats anywhere in Norway. Typical of these rare species are *Braya linearis*, *Rhododendron lapponicum* and the recently discovered bryophytes, *Seligeria oelandica*

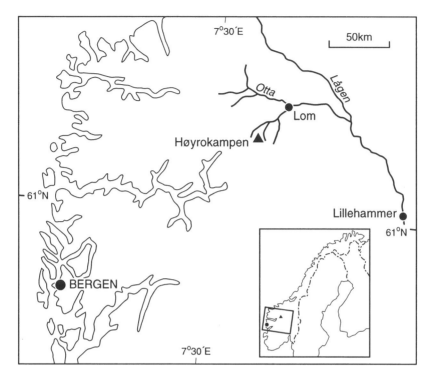

Figure 8.1   Location map of Høydal and Høyrokampen in Norway

and *S. carniolica* (Lid, 1963; Coker, 1983). In phytosociological terms, it is particularly rich in communities within the calcicolous classes SESLERIETEA ALBICANTIS and the CARICI RUPESTRIS–KOBRIESIETEA BELLARDII, while the leached soils are favoured by the LOISELEURIO–VACCINETEA. The SALICETEA HERBA-CEAE are well represented in the wide range of early to late snow-bed communities, particularly on the north-western part of the study area. In wetter sites, a range of communities from the orders MONTIO–CARDAMINETALIA and TOFIEL-DIETALIA are quite widespread, and include some calcareous bryophyte-dominated mires with *Meesia tristichia*, *Helodium blandowii* and *Paludella squarrosa*.

The initial purpose of the investigation was to produce a reasonably complete species list of all plant groups (with the exception of algae and fungi) and their distribution throughout the area, based on a minimum of 500 samples, and to collect environmental data from a minimum of 80 sites. As the project developed, it became clear that a combination of the geology (large amounts of lime-rich rocks at 900–1400 m), the effects of glaciation, changes in land use and the variety of slopes and exposures has resulted in an intricate patchwork of habitats and, in consequence of this, areas of great biodiversity, including some unique botanical sites. The pressing need was to map these and, making use of Landsat MSS imagery (bands 4, 5 and 7) and aerial photography (1:25 000), major vegetation boundaries were transferred onto a base map and verified by ground-truth observation. A final aspect of the work was to develop a vegetation/environment synthesis from this data that would examine the existence of a range of phytosociological syntaxa within the area, and, where possible, their correspondence with vegetation groups produced by multivariate analysis.

GEOLOGY

In common with most of the Jotunheimen, the rocks in the study area date from the Caledonian orogeny (Landmark, 1950), and the lime-rich schists are predominantly of Dalradian age (Bowen, 1981). Banham (1979) is of the opinion that most of the rocks are part of a sedimentary complex of Lower Cambrian age (post 1000 Ma.) and represent a fining-upwards sequence of conglomerates, limestones and shales which were subjected to later greenschist metamorphism.

The majority of the study area is composed of calcareous rocks and this has profound implications for the richness and diversity of the vegetation of the area. A geological sketch map (Figure 8.2) shows the extent of the six major rock types identified on the 1;100 000 map (NGU, 1968). Rocks with moderate to high levels of available calcium occupy about 60% of the study area.

METHODS

Site Selection

A large number of computer-generated random coordinates were used and 500 of these were selected: the criterion being that open water and high cliffs or unstable slopes were eliminated for reasons of safety. These were plotted onto a 1:10 000 base map of the area, kindly provided by Dr P. Banham, and onto an enlarged (1:25 000) aerial photograph which was used to locate the sites in the field. An enlarged Landsat MSS infrared false-colour composite was used to provide basic information on the vegetation of the area (all areas with photosynthetically active vegetation reflect infrared radiation which appears on the false-colour composite as a magenta coloration). The low resolution of this data nevertheless permitted the outlining on the base map of larger (> 30 m) stands of vegetation which could be interpreted by a later site visit. Monochrome aerial photograph stereo pairs were used to help define the extent of major features such as woodland, scree, cliffs and unvegetated areas, which were also transferred onto the base map. Extensive ground-truthing was carried out during the survey work and the data thus obtained is being transferred to a GIS archive. The most important advantage of this approach is that all sample sites are chosen on an objective, rather than a subjective basis, and operator bias which might affect the reliability of the sampling or its analysis is reduced to minimal proportions.

The precise size of the sample area matters considerably and consistency is required if the data which are gathered are to be considered as statistically valid and the results of the multivariate analyses to be acceptable. Minimal area estimations in a range of habitats confirmed that an area of $4 \, m^2$ was appropriate for all except mature woodland where a $100 \, m^2$ area would be more appropriate (cf. McVean and Ratcliffe, 1962).

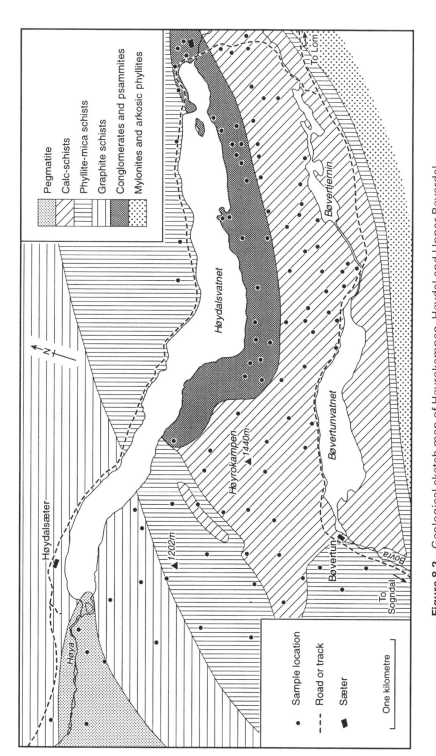

Figure 8.2 Geological sketch map of Høyrokampen, Høydal and Upper Bøverdal

Legend:

Pegmatite
Calc-schists
Phyllite-mica schists
Graphite schists
Conglomerates and psammites
Mylonites and arkosic phyllites

• Sample location
- - - Road or track
▮ Sæter
One kilometre

Høydalsæter
Høya
Høydalsvatnet
Høyrokampen ▲1440m
▲1202m
Bøvertunvatnet
Bøvertjernin
Bøvertun
Bøvra
To Sogndal
To Lom
N

This survey used the $4\,m^2$ option, even for woodland sites. The groups which include these sites are sufficiently distinctive in all the analyses for the sub-optimal size to be ignored (Coker, 1988).

Site Recording

Each site was recorded systematically for all plant groups (apart from algae and fungi) using a percentage cover scale of abundance, rather than the Braun–Blanquet or Domin scales. This was because it was thought best to preserve as much information in the raw data as was possible. At each site, the sample area was marked out with a portable frame and its location fixed with two compass bearings and a marker peg. This gave a reasonable chance of re-locating the sites for future investigations and proved to be moderately efficient during the 1992 re-survey. The altitude and coordinates of each site, together with its aspect and slope, were collected prior to the botanical investigation. At each of the 100 sites in the sub-sample, a soil core was taken, and an assessment of the duration of snow-lie was made *within the sample area* on a five-point scale as follows:

0   No snow lie likely (too exposed or steep)
1   Some snow lie, disappeared early in the summer
2   Moderate snow-lie, usually gone by mid-summer
3   Late snow-lie, occasionally present in late summer
4   Permanent or near-permanent snow patches

Information about likely snow-lie was provided by detailed observation of the dominant species in the vegetation (and reference to Dahl, 1956, and McVean and Ratcliffe, 1962) since many plants have particular affinities with snow patches (e.g. those which occur in the SALICETALIA) while others are chionophobous (such as the Kobresio–Dryadion). Cryptogams also provide useful information of a similar nature – mountain soils dominated by the lichen *Solorina crocea* are rarely snow-covered but are particularly prone to frost-heaving and solifluction – while the snow-bed moss *Polytrichum sexangulare* forms part of a community which is characteristically snow-covered until late in the season (3–4 on the above scale). Very exposed sites where snow rarely, if ever, lies, are frequently dominated by chionophobous lichens such as *Alectoria fuscescens* and *Thamnolia vermicularis* and were regarded as 0 on the snow-lie scale. The transition zone between the 'yellow map lichen' *Rhizocarpon geographicum* agg. and the grey or black 'rock tripe' lichens *Umbilicaria* spp. (which are widely found on rocks) is a useful indication of snow depth (Coker, 1988), corresponding to 2 on the scale.

On return to base, the pH of each soil sample was measured (1:3 soil/water), and the sample was sealed and chilled prior to laboratory analysis (Coker, 1988) for exchangeable calcium, available phosphate, total nitrogen and organic content, making a total of eight environmental variables per site. Moisture content was not included because of unpredictably wet days during the sampling periods.

DATA SET PROPERTIES

The main data set contained 425 species and details of their abundance in each of the 500 sample sites. Some species – mainly bryophytes and lichens, were rare and occurred in very small quantity in two or three sites. In order to reduce the 'noise' in this data set (*sensu* Gauch, 1982), species were excluded which occurred in less than four sites. This reduced the species count to 366, which was felt to be acceptable, without losing too much of the diversity of the flora, compared with the data as originally collected. The reduction in species also enabled the computer to deal with eight levels of division in TWINSPAN without running out of memory, compared with five when the 425 species data set was processed. The analysis divides the data sequentially and in order to obtain a reasonably fine classification, the division process may need to be applied several times. For example, in an area as diverse as this, at least eight levels of division would be required in order to separate snow-meltwater-fed bryophyte flushes from similar flushes in which *Taraxacum* and *Sibbaldia* predominated.

The secondary data set (100 sites with corresponding vegetational and environmental data) contained 171 species which occurred in three or more sites. There is a risk of redundancy in large data sets, and it was felt that a randomised sub-sample of 1 in 5 sites would overcome this problem as well as provide a good-sized data set for further manipulation. It was also realised that it would be totally impractical to attempt to record environmental variables and chemical analyses for 500 sites. The location of the sites in this data set is shown in Figure 8.3.

ANALYSES

In order to investigate the data efficiently, a two-way classification of each of the data sets was carried out using TWINSPAN (Hill, 1979), together with a canonical ordination and plotting (the Canonical Correspondence Analysis (CCA) option from CANOCO (ter Braak, 1987) and CANOPLOT (van Tongeren)) for the 100 site joint data sets and the standard ordination (Detrended Correspondence Analysis (DCA)) option for the 500 site floristic data set.

The 100 site environmental data set was also used as the input for a Trend Surface Analysis (TSA) program. This version used an orthogonal polynomial model since the data points were irregularly spaced (Whitten, 1970). Burrough (1987) has applied TSA to a range of ecological and environmental data and concluded that it represents a valuable tool in vegetation mapping and for relating mathematically extracted components back to field observations.

RESULTS AND DISCUSSION

Vegetation Mapping Using the 100 Site Data Set

Ten clearly distinguished groups of sites were extracted from the two-way table produced by TWINSPAN and a constancy table drawn up (Table 8.1). It was decided,

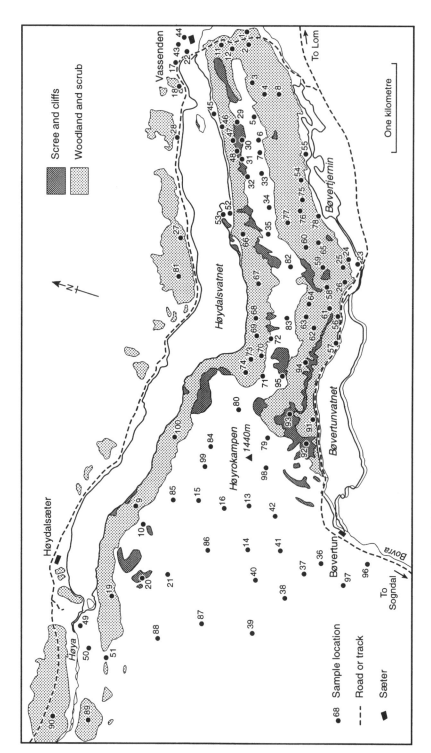

Figure 8.3 Location of sites used for the 100 site sub-sample

Table 8.1 Constancy table for Høydal plant communities, based upon 100 site TWIN-SPAN analysis (see Appendix 2)

Group number	1	2	3	4	5	6	7	8	9	10
Number of relevés	5	15	20	17	13	8	8	4	5	5
Vaccinium myrtillus +		I	V	I	III	I	II			
Deschampsia flexuosa			V		III				1	
Juniperus nana		IV	IV	I	I	I				
Ranunculus acris +		IV	I				I			
Aconitum septentrionale		IV	I			I				
Betula tortuosa		IV	V	I			I			
Geranium sylvaticum		V	IV	I						
Pyrola rotundifolia		III	III	I		I				
Rubus saxatilis		III	III	I						
Pleurozium schreberi		I	III		I					
Melampyrum sylvaticum		II	III		I					
Dicranum scoparium		I	II	I						
Gentiana campestris		I	I	I						
Gymnocarpium dryopteris		I	II							
Luzula sylvatica			II							
Trientalis europaeus			II							
Cirsium heterophyllum		I	I							
Tritomaria quinquedentata		II	I							
Convallaria majalis		I	I							
Succisa pratensis			I							
Prunella vulgaris +		II								
Melica uniflora		I		I						
Viola palustris		V	I		I					II
Parnassia palustris		IV	I	II						
Fragaria vesca		III								
Rhytidiadelphus triquetrus		II	I	I						
Silene vulgaris		II								
Stellaria nemorum		I								
Veronica s. serpyllifolia		I	I							
Rhinathus borealis		I								
Circaea alpina		I								
Viola riviniana		I		I						
Agrostis stolonifera		II								
Geum rivale		I	I							
Valeriana officinalis		I								
Cotoneaster intergerrimum		I								
Crepis mollis		II								
Vaccinium vitis-idaea +		II	V	III	III	I				
Campanula rotundifolia		V	III	III	I					
Selaginella selaginoides		III	I	II						
Astragalus alpinus		IV	I	II						
Gentianella amarella		I			I					
Tortella tortuosa		II		I						
Ctenidium molluscum		I		I						
Atennaria diocia		II	I	II						
Galium boreale		I			I					

Table 8.1   (*cont.*)

Group number	1	2	3	4	5	6	7	8	9	10
Number of relevés	5	15	20	17	13	8	8	4	5	5
Empetrum hermaphroditum +		I	IV	II	V	I				
Vaccinium uliginosum		I	III	III	II	I				
Campylopus flexuosus			III	II	II	I				
Cladonia furcata			I	I	II					
Bartsia alpina		I	I	II	I					
Astragalus norevegicus		I		I						
Calluna vulgaris			I		II					
Cladonia gracilis			I		II					
Carex flacca			I		II					
Polytrichum commune			I		I					
Ptilidium ciliare			I	I	I					
Chamaeorchis alpina				I						
Aulacomnium palustre +	IV									
Homalothecium nitens	V									
Paludella squarrosa	V									
Helodium blandowii	V									
Plagiomnium elatum	V									
Saxifraga aizoides	IV	II	I	I		II				
Salix lanata	I	II	I		I	I				
Carex atrata	II									
Carex capillaris	II									
Pinguicula vulgaris	II	I	I	I		I				
Poa alpina	I		II	I	II	II	I		II	II
Botrychium lunaris +				II						
Racomitrium lanuginosum				I	I					
Thamnolia vermicularis				II	I					
Arctostaphylos uva-ursi				II	I					
Pulsatilla vernalis				II						
Kobresia myosuroides				II						
Arctostaphylos alpina +		II	I	I						
Cornicularia aculeata			I	I						
Alectoria fuscesens			II	I						
Salix mysinites +		II	I	I			I			
Alchemilla glomerulans		II	I	I			I			
Euphrasia frigida		II	I	II		I	I		I	
Equisetum variegatum		I	I	I		I				
Cetraria nivalis +				II	III	I				
Alectoria ochroleuca				II	II	I	I			
Dryas octapetala		I	I	IV		II	I			
Carex rupestris		I		III	I	I	I			
Cladonia alpicola			I	I	II	I				
Betula nana +			II	III	IV	I		I		
Pedicularis lapponica			I	I	I		I			
Hieracium alpinum agg.		I	I		I					
Saussurea alpina +		IV	I	II	I	II				
Thalictrum alpinum		V	I	II	I	III			I	
Potentilla crantzii		IV		II	I	II	I			

Table 8.1 (*cont.*)

Group number	1	2	3	4	5	6	7	8	9	10
Number of relevés	5	15	20	17	13	8	8	4	5	5
Trifolium campestre +		I					I			
Agrostis tenuis		III	II		I			II	II	
Deschampsia caespitosa		V	I						I	
Anthoxanthum odoratum		II	IV	I			II	II	II	
Hylocomium splendens		II	IV	I	I	I	I			
Equisetum silvaticum		I	II				I		II	
Equisetum pratense		I							II	
Salix phylicifolia		II	II						II	
Festuca ovina +		II	III	IV	IV	II	II			
Tofieldia pusilla		I		III	I		I			
Cetraria nivalis				III	II	I				
Equisetum scripoides			I	I		I				
Cetraria delisei			I	I	II					
Phytidium rugosum				I		I				
Veronica fruticans				I		I				
Cladonia coccifera +					II	I				
Lycopodium annotinum					II	I				
Nardus stricta			I		I				II	
Cetraria islandica +			I	II	III	I		II		
Cladonia rangiferina			I	I	III	I				
Cladonia arbuscula			III	I	IV	II		II		
Carex bigelowii				II	IV	II			IV	
Loiseleuria procumbens					I				III	
Juncus trifidus				I	I	I			III	
Peltigera venosa			I	I	I	I				
Cladonia unicalis					II			II		
Stereocaulon arcticum					I					
Ochroleichia frigida				I	II	I		II		
Stereocaulon evolutum				I	I	I		II		
Silene acaulis +		II		III	I	III	I	II		
Polygonum viviparum		III	II	IV	II	IV			II	I
Antennaria alpina		II		II	I	III				
Phyllodoce caerulea			I	I	III	I				
Leontodon autumnalis		II		I		I			III	
Cladonia pyxidata			I		I	I	I			
Saxifraga oppositifolia +				IV		II	I			
Salix reticulata		II		IV		II	I			
Distichium capillaceum			I	I	I		I			
Gentiana nivalis				I			I	II		
Pedicularis oederi		I		II		I	I			
Salix glauca +			I	II		I	IV			
Solidago virgaurea			I	II	I	II	III			
Scapania gracilis			I				I			
Alchemilla alpina			I						II	II
Peltigera spuria			I	I			I			
Philonotis fontana +	II								IV	
Saxifraga stellaris								I	III	

Table 8.1 (*cont.*)

Group number	1	2	3	4	5	6	7	8	9	10
Number of relevés	5	15	20	17	13	8	8	4	5	5
Carex dioica		I	I					III		
Saxifraga rivularis										I
Cladonia squamosa +		I	I			I				
Deschampsia alpina		I				II				
Rumex acetosa		I					I		II	II
Solorina saccata			I				II			
Polytrichum juniperinum +					II	II			I	
Pohlia nutans					II	II	I			
Cladonia cenotea					I	I				
Rhodobryum roseum +						I	II			
Trisetum spicatum						II	I			
Sibbaldia procumbens		I	I	I		IV	II	II		
Taraxacum croceum						IV	I	II		II
Veronica alpina		I		I			I	I	II	
Cerastium alpinum		I		I		II		I		
Polytrichum alpinum		II						I		
Hypnum callichroum								II		II
Oxyria digyna					II	II	II	II		
Saxifraga cernua +						I	I	I		
Cerastium cerastoides						I	III			
Salix herbacea					I	V	III	V		
Cassiope hypnoides			I		I	II		IV	III	II
Anthelia julacea							II	II	II	II
Omalotheca supina						I	I	II	I	II
Gymnomitrion concinnatum							II			
Marsupella ustulata							III		II	
Solorina crocea			I				III	I		
Polytrichum sexangulare				I			IV	III		
Kiaeria falcata				II		I	II	III		
Kiaeria starkei				I			II	III		
Conostomum boreale							I			

for reasons of clarity, to map the 10 site groups only, and to indicate groups of species which occurred with high constancy (III or more) as sub-groups a, b etc. This has provided a total of 25 habitat types which, it is believed, represent the commonest vegetation types in the area (Figure 8.4).

The communities were distinguished as follows:

1. **Mire communities**

   A complex of bryophyte-dominated communities of sites with impeded drainage. Most mire sites in this area are calcareous and *Sphagnum*-dominated sites (which are quite rare in this area) are confined to riverside sites near Høydalsæter. Altogether, five sites were included and formed one extremely well-defined group.

2. **Wood pasture**

   Open *Betula tortuosa* woodland, with significant areas of mesotrophic grassland and tall herb communities. There were 15 sites in this group.

   2a.   Well-defined open grassland with tall herbs – notably *Geranium sylvaticum*, *Ranunculus acris* and *Aconitum septentrionale* – found mainly on south-facing slopes.

   2b/c.   Communities in which low-growing herbs occur widely. Type 2b, dominated by *Viola palustris* and *Parnassia* tends to occur on deeper and moister soils than 2c, where *Campanula rotundifolia* and *Astragalus* occur with high constancy and appear to favour dry, slightly leached soils.

   2d/e.   These communities occur sporadically in the area. Type 2d tends to occur on patches of thin, rocky and calcareous soil where tree cover is low, and type 2e appears most frequently in areas where trees have been felled, thus opening up the site for colonisation.

3. **Woodland**

   Usually dense woodland dominated by *Betula tortuosa* with some regeneration and a well-developed understorey of *Juniperus* and a shrub layer consisting mainly of *Vaccinium* spp. in particular *V. myrtillus*. Bryophytes and lichens are widespread throughout the 19 sites.

   3a.   This community occurs in exposed sites, where tree cover is intermittent. *Vaccinium myrtillus* and *Deschampsia flexuosa* appear with high constancy.

   3b.   A community which is broadly similar to type 2a, but lacking tall herbs and with a well-developed bryophyte layer consisting mainly of *Pleurozium schreberi*.

   3c.   This weakly chionophilous community is found in areas of moderate snow-cover on north-facing slopes near the upper limit of the tree line. *Vaccinium vitis-idaea* occurs with high constancy.

   3d.   A chionophobous community which is dominated by *Empetrum*. *Vaccinium uliginosum* occurs in the most exposed sites where tree cover is least well developed or where it has been removed by human activity.

4. *Dryas* **heath**

   Areas dominated by *Dryas octopetala*, together with calcicole grasses and herbs; confined in this area to soils of high base status. There are 17 sites in this group.

   4a.   *Dryas* and *Carex rupestris* sites in which dry, thin, highly calcareous soils predominate.

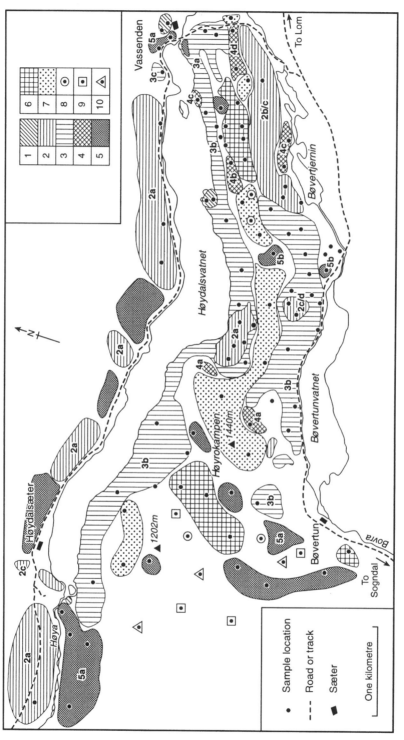

Figure 8.4 Location of the TWINSPAN groups. Major groups 1–10 are denoted by shading or symbols while sub-groups are indicated by the group number and the suffix a, b etc

4b. *Festuca ovina*, together with *Tofieldia*, are the characteristic species of this vegetation type. The soils tend to be calcareous, with some surface leaching and the community tends to have a high lichen and bryophyte cover.

4c/d. Typically, communities of calcareous substrates. Type 4c tends to occur on moist sites and *Polygonum viviparum* and *Silene acaulis* are constant, while type 4d is characteristic of wet rock ledges with *Saxifraga oppositifolia* and *Salix reticulata*.

5. Dwarf-shrub heath

Tree-less areas in which the majority of the cover is provided by shrubs such as *Empetrum* or *Vaccinium* spp. together with bryophytes, lichens and occasional grasses. The group has 13 sites.

5a. Open community in which *Vaccinium myrtillus* and *Deschampsia flexuosa* occur with moderate constancy – characteristically found on exposed sub-alpine areas.

5b. A closed and low-growing community, dominated by *Empetrum* and to a lesser extent, *Betula nana* and widespread, particularly in sites which are unsuitable for the development of birch woodland.

5c. A lichen heath community typically found on exposed sites at all altitudes – *Cetraria* spp. are common and the shrubby *Cladonia* spp. occur with high constancy.

6. Irrigated sites

A group of eight relevés in which species are irrigated predominantly by snow-meltwater. The soil type tends to be neutral.

6a. This is a relatively uncommon type which has some affinities with type 4c – the difference is chiefly in the neutral to acidic soil pH and the greater constancy of *Antennaria alpina* in this site group.

6b. Found only in close proximity to late snow-patches and receiving considerable amounts of meltwater. A community dominated by *Taraxacum* and *Sibbaldia*.

7. Pioneer sites

A small group of eight sites which show many pioneer traits.

7a. A severely stressed community in which solifluction plays a major part; characterised by the presence of bryophyte tufts and crusts and the lichen *Solorina crocea*. Poorly grown *Salix herbacea* and *Cerastium cerastoides* are moderately constant.

7b. Sites in which *Salix glauca* is an early and constant colonist are quite rare except on the moraines near Høydalsæter, where the species occurs with *Solidago*.

8. Snow-beds

A well-defined group of six sites.

8a. *Salix herbacea* and *Cassiope hypnoides* are present with high constancy.

8b. These sites are bryophyte-dominated, with *Kiaeria* spp. and *Polytrichum sexangulare* forming the majority of the ground cover.

9. Fjellfeld

A widespread vegetation type in which *Carex bigelowii*, *Loiseleuria* and *Juncus trifidus* occur in moderate to high constancy on the most exposed sites. Lichens,

such as *Stereocaulon* and *Ochroleichia*, are moderately constant, but bryophytes
are less important in the five sites in this group.

10. Meltwater flushes

A group of four sites in which *Philonotis fontana* and *Saxifraga stellaris* occurred
with high constancy.

Small areas of *anthropogenic* grassland were encountered in which the tree or shrub
cover has been removed and the growth of pasture grasses and herbs such as *Agrostis*,
Phleum, *Festuca*, *Ranunculus acris* and *Cirsium* spp. has been encouraged by mowing
and, occasionally, by the application of fertiliser. The crop is cut and removed and
some incidental grazing by cattle and sheep may occur. These sites (at Høydalsæter
and near Rustadsæter) occurred in less than 1% of the study area and have been
omitted from the map for reasons of clarity.

The vegetation groupings which have been mapped at a scale of 1:30 000 were
compared with those proposed for the Natural Vegetation Map of Europe (at a scale of
1:2.5M) (Bohn, 1994). It proved possible to identify only three of the categories as
definitely occurring in the study area, and the categories are as follows:

A2a Icelandic/Scandinavian sub-nival vegetation of high mountains
Open vegetation of crustose/foliose lichens, *Ranunculus glacialis*, *Koenigia* etc.
B33 Eastern Scottish–Scandinavian dwarf-shrub heath
Empetrum, *Loiseleuria*, *Diapensia* (or *Dryas* on calcareous sites); in areas with longer
duration of snow-lie, *Vaccinium myrtillus* and *Phyllodoce* or *Betula nana* scrub in
combination with *Salix lapponum*, *S. lanata*, *S. glauca* or *S. phylicifolia*.
C5a Fennoscandian moist birch forests
Betula tortuosa with *Chamaepericlymenum suecicum*, *Vaccinium myrtillus*, *Bar-
bilophozia* (and tall herbs and ferns including *Lactuca alpina*, *Geranium sylvaticum*,
Aconitum septentrionale, *Dryopteris* spp. and *Gymnocarpium*).

Type A2a corresponds to groups 7, 8 and 9 above, type B33 to groups 4 and 5 and type
C5a to types 2 and 3. These three types account for 83% of the relevés in the sample;
the remaining 17% consist of sites which are too small for satisfactory mapping on
anything other than a large scale.

The TWINSPAN groups were also compared with traditional phytosociological
syntaxa which had been identified in the area (Coker, 1988). Each group was allocated
to the alliance(s) and associations which most closely reflected the species composition
of the sites. Where present, each sub-group was compared with the closest published
phytosociological association or, in some cases, sub-association.

Group 1 sites fall naturally into the alliance Sphagneto–Tomenthypnion, in particular
the association defined by *Carex dioica* and *Tomenthypnum nitens*.
Group 2 sites are most closely linked with the Lactucion alliance, with the exception of
group 2e which is part of the Phyllodoco–Vaccinion.
2a – Aconitetum septentrionalis
2b – Rumiceto–Salicetum lapponae
2c – Corno–Betuletum
2d – no close correspondence found, but near to the Potentilleto–Polygonion vivipari
    alliance

2e – Phyllodoco–Vaccinetum myrtilli
Group 3 sites share features of the Lactucion, Nardeto–Caricion and Phyllodoco–Vaccinion.
3a – Geranio–Betuletum
3b – no close correspondence, but nearest to the Phyllodoco–Vaccinetum myrtilli dicranetosum sub-association
3c – Hylocomieto–Betuletum nanae salicetosum glaucae
3d – Phyllodoco–Vaccinetum myrtilli empetrosum
Group 4 sites are split between the xeric Kobresio–Dryadion and the increasingly hygrophilous Potentilleto–Polygonion and Cratoneureto–-Saxifragion.
4a – Kobresietum myosuroidis
4b – Dryadetum octopetalae
4c – Potentilleto–Polygonetum vivipari
4d – Cratoneureto–Saxifragetum aizoidis
Group 5 sites most closely correspond with associations found in the Phyllodoco–Vaccinion, the Arctostaphyleto–Cetrarion and the Nardeto–Caricion.
5a – Phyllodoco–Vaccinetum myrtilli lichenetosum
5b – Cladonietum alpestris betuletosum
5c – Cetrarietum delisei caricetosum
Group 6 sites are quite distinctive and dwarf willow (*Salix herbacea*) is constant.
6a – A variant of the Polygoneto–Salicetum (4c) – possibly a sub-association
6b – Lophozieto–Salicetum herbaceae typicum
Group 7 sites are closely related to those in group 6b and belong to the Cassiopeto–Salicion.
7a – Gymnocoleo–Salicetum herbaceae
7b – Lophozieto–Salicetum herbaceae conostometosum
Group 8 sites are part of the same alliance as group 7.
8a – Cassiopeto–Salicetum herbaceae
8b – Dicranetum starkei
The group 9 sites are all part of the Arctostaphyleto–Cetrarion, and seem to accord most closely with the Cetrarietum nivalis trifidetosum, although one of the sites might be intermediate between this sub-association and the Loiseleurio–Diapensietum association.

   The sites in group 10 were all part of the cold (spring) flush alliance (the Mniobryo–Epilobion) and clearly corresponded with the Philonoto–Saxifragetum stellaris.

Careful examination of the groups and their phytosociological affinities showed that the individual TWINSPAN groups did not always correspond closely with any one alliance, particularly in the botanically distinctive woodland and grassland habitats. The closest correcpondence was seen in those cases where a well-defined habitat with character species (*sensu* Whittaker, 1978) of high constancy occurred (such as around snow-patches or in soligenous (calcareous) mires).

   The correspondence issue is one which will be investigated in future work using the full 500 site data set, but at present it appears that the clearest view of the vegetation plexi in the area and their environmental relationships is given by the objective classification of TWINSPAN, rather than the almost 'contrived' comparisons with published phytosociological syntaxa. Dirkse (1995) comes to the same conclusion with respect to the inventory of Dutch forests which was undertaken using a stratified random sampling design. He states that the sampling technique 'violates the Braun–Blanquet rule of homogeneity, therefore the TWINSPAN clusters differ fundamentally from the classical (phytosociological) associations'.

CANOCO Analyses

The extent and direction of each of the eight environmental biplot scores were superimposed upon the plots of sites and species scores, as produced by CANOPLOT.

Species–environment Interrelationships

Figure 8.5 demonstrates clearly that duration of snow-lie and calcium levels are the most important of the environmental factors that were measured, and that pH and nitrogen levels are also important in this area. Soil organic content and angle of slope were both, surprisingly, much less important. It appeared that aspect and phosphate concentrations had little bearing upon the distribution of species.

Examining the species associated most closely with the first four factors demonstrates the importance of each in this area. *Salix herbacea*, *Veronica alpina* and *Saxifraga cernua* occur along the line of the snow-lie trace while well-known snow-bed species such as *Polytrichum sexangulare*, *Cassiope hypnoides*, *Omalia supina* and *Cerastium cerastoides* occur in close proximity to it. This relationship is even more clearly demonstrated in the full data set of 500 sites (Coker, 1988). Similarly, the clear correlation between species with a preference for habitats with high pH soils (due in this case to high levels of calcium) is demonstrated by mire species such as *Paludella*, *Helodium* and *Plagiomnium elatum* or strongly calcicole species such as *Carex capillaris* and *Solorina saccata*. Species such as *Tortella tortuosa* and *Chamaeorchis alpina* occur very close to *Kobresia* and, like it, grow well under highly calcareous conditions. Several woodland species are associated with the nitrogen trace, including *Stellaria nemoreum*, *Melica*, *Valeriana* and *Viola riviniana*, and this is not surprising since all the accessible woodland areas tend to be more heavily trampled by cattle than the open grassland areas of the valley bottoms which tend to provide inadequate grazing due, in large measure, to low species diversity and biomass on podsolised substrates. These would be characterised by species such as lichens including several species of *Cladonia*, *Calluna*, *Rumex acetosella* and *Deschampsia flexuosa*, seen in the south-east quadrant of the plot. It is also significant that species in this quadrant are predominantly calcifuge and characteristically found in habitats in which nitrogen levels are low.

There is a striking group of species which are associated with sites in which snow-lie is negligible (in the south-west quadrant of the plot). This includes *Arctostaphylos* spp., *Juniperus*, *Racomitrium lanuginosum* and a wide range of lichens including *Cetraria cucullata*, *Cornicularia aculeata*, *Thamnolia vermicularis*, *Cetraria nivalis*, *Alectoria* spp. and *Ochroleichia frigida*. The group occurs widely throughout the area, predominantly on fjellfeld sites and also close to the eastern end of the lake, on an exposed site on a conglomerate outcrop.

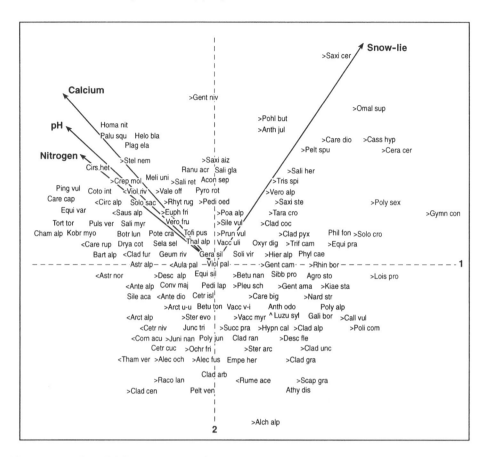

Figure 8.5 CANOCO ordination of the 100 site data set. Superimposed upon the species plot are the four most significant environmental factor biplots (in order, snow-lie, calcium, pH and nitrogen); the direction and length of the arrow indicates the importance of each factor. Four less important factors (slope, aspect, organic content and phosphate) are not shown, because their biplots would obscure the species names

The analysis picked out the extremely good correspondence between slope and *Saxifraga aizoides*. This species is dominant on dripping wet rock faces and steep slopes, particularly on neutral to calcareous soils and rock.

Site–Environment Relationships

Consideration of Figure 8.6 confirms that the overriding factors are, as might be expected, duration of snow-cover, and nitrogen, calcium and pH. This figure has been prepared by overlaying the direction and importance of the environmental factors as shown in the environmental biplot onto the corresponding Axis 1 and 2 plot of the site ordination. The arrowheads represent the maximum extent and the location of the environmental factors and the numbers are site numbers.

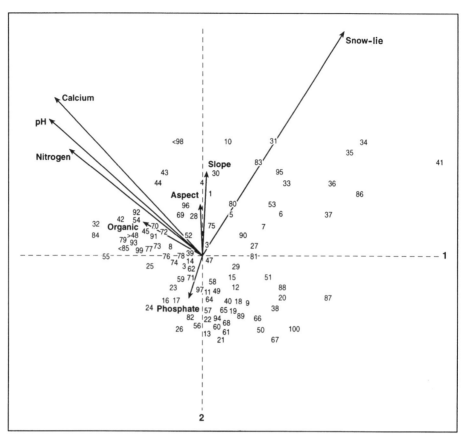

Figure 8.6 A site ordination in which the same environmental biplots are shown superimposed. The site ordination enables those biplots not displayed in Figure 8.5 to be clearly seen

Sites 34–37, 41 and 86 are sites where snow lies for a considerable part of the year, while, for example, sites 3 and 23–26 are rarely snow-covered. Chionophobous communities such as these are quite frequent in the study area and include lichen heaths, dominated by *Alectoria* and *Cetraria* spp. or relevés that appear to belong to the Kobresio–Dryadion; they occur on dry, well-drained sites, often on south-facing slopes. The extent to which snow accumulates in the birch woods is variable, but sites such as 27, 28, 81 and 90 are characterised by late snow-lie and sites 63, 69 and 74–76 lose their snow cover rather earlier in the season.

Sites in the south-east quadrant are characteristically dominated by dwarf-shrub species such as *Empetrum* and *Vaccinium* spp.; sites in which *V. vitis-idaea* occurs tend to be further from the origin of the plot than those in which chionophobous species such as *Empetrum, V. myrtillus* or *V. uliginosum* are present in quantity.

Calcareous mires and flushes are widespread in the area – typically sites such as 43 and 44 are dominated by bryophytes (such as *Helodium* and *Paludella*) and sites 32 and 84 by *Carex* spp. such as *C. capillaris*.

The use of a direct gradient analysis such as CCA demonstrates that the four major environmental factors which were measured directly are not related in any *simple* fashion to either axis 1 or 2 of the ordination (cf. Kent and Coker, 1992).

An inspection of the site ordination reveals that axis 1 is most probably linked with soil moisture; sites on the left-hand side of Figure 8.6 contain vegetation that is either more hygrophytic or has soil that was, when collected, wetter than those at the right-hand of the horizontal axis (1). The vertical axis (2) is, in this case, clearly related to slope and, to a lesser extent, aspect.

Sites such as 80, 30 and, to a greater extent, 10 are snow-bed sites on increasingly steep, north-facing slopes. Sites 21, 26 and 82 are on very gentle slopes facing south or south-west. These conclusions were confirmed by an examination of the output of the 500 site data set. While it is often quite difficult to determine the likely affiliation of a particular axis when using indirect gradient analysis methods, the task can be made a lot easier if results from a direct gradient analysis are available for consultation. Output plots from axes 2/3 and 1/3 yielded no significant information.

Trend Surface Analysis (TSA)

The results of applying TSA to the exchangeable calcium data set are extremely interesting, and the map is shown as Figure 8.7. Given the randomised distribution of the datapoints, as shown in Figure 8.3, the comparison between the solid geology and the calcium residuals is strikingly good, and demonstrates the predictive power of this technique. The program that was used to produce the map (TREND) has produced predictions for areas where no data has been collected and the trend for these areas is negative, becoming more negative the further they are from sites where actual data are available. This is in some respects a fault, but it serves to illustrate very clearly the value of well-spaced sampling points when attempting to determine a trend in spatially distributed data. By way of comparison, TSA was also applied to snow-lie and nitrogen data; the results were of some interest but did not reveal any clear trend.

CONCLUSIONS

The application of multivariate analyses and TSA to floristic and environmental data has provided a great deal of information on the distribution of vegetation types within a geologically diverse montane region. In particular, it has demonstrated the value of programs such as TWINSPAN and CANOCO for handling complex data sets and revealing and displaying the underlying structure of the data and its relationship to environmental factors.

The relationship of the TWINSPAN groups delineated in this research to those mapping units used for the current version of the map of the natural vegetation of Europe (Bohn, 1994) is straightforward. Three vegetation types from the 1:2.5M map are thought to be present in the area, but owing to the significant difference in scales, it is quite impossible at the smaller scale of the European map to distinguish more than

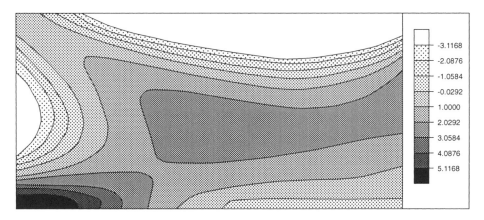

Figure 8.7 Trend surface (fourth order) of calcium concentration

the broadest of categories corresponding to types A2a, B33 and C5a which account for 83% of all the sites in the 100 site vegetation map.

The compartments shown on the vegetation map are a fair representation of the existing plant cover in the study area. It is intended to produce a further map in which the full TWINSPAN analysis (500 site) is used, and to investigate the improvement in information content that this provides. A long-term aim is to collect environmental data from a much larger sample of sites – at least 200 – and to investigate the correlations which may exist between the most important factors by the use of TSA, which has already shown its value as a tool for reconnaissance studies (cf. Mucina *et al.*, 1991).

ACKNOWLEDGEMENTS

This work was carried out in conjunction with the Jotunheimen Research Expeditions, led by John Matthews, and I am grateful to him both for the initial opportunity to work in the Jotunheimen in 1974, and for many of the logistical arrangements since then. Martin Kent helped clarify some of my ideas on ordination and classification and Peter Banham did the same for the geology of the area. G.M. Dirkse's refreshingly candid and enlightening views on phytosociology were of great assistance in the development and production of this chapter. I am particularly grateful to Rosemary Coker who has shown great patience both as a field assistant and also as a classificatory and computational widow over the past few years.

This paper is Jotunheimen Research Expeditions Contribution No. 121.

REFERENCES

Arnell, S. (1956) *Illustrated Moss Flora of Fennoscandia. I. Hepaticae.* Gleerups, Lund.
Banham, P.H. (1979) Geological evidence in favour of a Jotunheimen Caledonian suture. *Nature (London)*, **277**, 289–291.

Bohn, U. (1994) International project for the construction of a map of the natural vegetation of Europe at a scale of 1:2.5M. Its concept, problems of harmonization and application for nature protection. *37th IAVS Symposium*, Bailleul, France.

Bowen, G.S. (1981) The geology of the Vassenden area, Bøverdalen, Norway. BSc dissertation, Bedford College, University of London.

Burrough, P. (1987) Spatial variation in vegetation, soil and precipitation across the Hawkesbury Sandstone Plateau in New South Wales, Australia: an analysis using ordination and trend surfaces. In: R.H.G. Jongman, C.J.F ter Braak and O.F.R. van Tongeren (eds) *Data Analysis in Community and Landscape Ecology. PUDOC, Wageningen*, pp. 252–257.

Coker, P.D. (1983) *Seligeria carniolica* and *Seligeria oelandica*; two mosses new to Norway. *Lindbergia*, **9**, 81–85.

Coker, P.D. (1988) Some aspects of the biogeography of the Høyfjellet with special reference to Høyrokampen, Bøverdal, Southern Norway. MPhil thesis, University College, London.

Dahl, E. (1956) Rondane; Mountain vegetation in South Norway and its relation to the environment. Skrifter utgitt av det Norske Videnskaps-academi i Oslo. *Matematisk-naturvidenskapelig klasse*, **3**, 1–374.

Dahl, E. and Krog, H. (1973) *Macrolichens*. Universitetsforlaget, Oslo.

Dirkse, G.M. (1995) The fourth Dutch forest survey, a stratified random approach to vegetation sampling. *Colloques Phytosociologiques* XXIII, Bailleul, 1994, pp. 410–416.

Flora Europaea (1964–1980) Parts 1–5. Cambridge University Press, Cambridge.

Gauch, H.G. (1982) *Multivariate Analysis in Community Ecology*. Cambridge University Press, Cambridge.

Hill, M.O. (1979) *TWINSPAN – A FORTRAN Program for Arranging Multivariate Data in an Ordered Two-Way Table by Classification of the Individuals and Attributes.* Cornell University, Department of Ecology and Systematics, Ithaca, New York.

Kent, M. and Coker, P.D. (1992) *Vegetation Description and Analysis – a Practical Approach.* Belhaven Press, London.

Landmark, K. (1950) *Geologiske undersøkelser i Luster-Bøverdalen*. Bergen Univ. Årbok, 1948.

Lid, J. (1963) *Norsk og Svensk Flora*. Det Norske Samlaget, Oslo.

McVean, D.N. and Ratcliffe, D.A. (1962) *Plant Communities of the Scottish Highlands*. HMSO, London.

Mucina, L., Ĉík, V. and Slavkovský, P. (1991) Trend surface analysis and splines for pattern determination in plant communities. In: E. Feoli and L. Orloci (eds) *Computer Assisted Vegetation Analysis*. Kluwer Academic Publishers, Dordrecht, pp. 355–371.

NGU (1968) *Tektonisk Kart over Omradet Skjåk/Sygnefjell. 1:100,000*. Norges Geologiske Undersøkelse, Oslo.

Nyholm, E. (1954–1969) *Illustrated Moss Flora of Fennoscandia. II. Musci. Parts 1–6*. Gleerups, Lund.

ter Braak, C.J.F. (1987) *CANOCO. A FORTRAN Program for Canonical Community Ordination by Partial Detrended Canonical Correspondence Analysis, Principal Components Analysis and Redundancy Analysis*. TNO Institute of Applied Computer Science, Wageningen.

Whittaker, R.H. (1978) *Classification of Plant Communities*. Junk, The Hague.

Whitten, E.H.T. (1970) Orthogonal polynomial trend surfaces for irregularly-spaced data. *Journal of the International Association for Mathematics in Geology*, **2**, 141–152.

APPENDIX 1

Copies of TWINSPAN, TREND and other analytical software for use on IBM-compatible PCs can be obtained from Biological Software, 23 Darwin Close, Farnborough, ORPINGTON, Kent BR6 7EP, UK, while CANOCO and CANOPLOT are available from Microcomputer Power, 111 Clover Lane, Ithaca, NY, USA. Both firms issue catalogues.

APPENDIX 2: ANALYSIS PARAMETERS

TWINSPAN

All defaults except as listed below.

500 Site Data Set

366 species in final tabulation
8 levels of division (128 possible end groups)*
9 pseudospecies cut levels set at 0, 5, 10, 20, 30, 40, 50, 60, 70%

100 Site Data Set

171 species in final tabulation
5 levels of division (32 possible end groups)*
9 pseudospecies cut levels set at 0, 5, 10, 20, 30, 40, 50, 60, 70%

CANOCO

500 Site Data Set – Indirect Gradient Analysis – DCA Option

All defaults applied except downweighting of rare species permitted.

100 Site Data Set – Direct Gradient Analysis Using CCA Option

All default values

CANOPLOT was only used for the display of the 100 species CCA output (Figures 8.5 and 8.6). The resolution of the program output is insufficient to cope with 500 sites.

The constancy table (Table 8.1) is derived from the 100 site data, and the vegetation map (Figure 8.4) is based on the same sites, with additional information from aerial photographs where particular areas transferred on the base map could be assigned to appropriate communities.

The *constancy class* ciphers (I to V) represent the percentage occurrence of a particular species in a set of relevés:

Constancy class	I	II	III	IV	V
Percentage	1–20	21–40	41–60	61–80	81–100

A species occurring in 14 sites out of 20 (70%) would be in constancy class IV, whereas a species occurring in just 3 sites (15%) would be in class I.

A *character-species* is one which is relatively restricted to the sites included in a particular phytosociological syntaxon, and therefore characterises it and provides a good indication of its environment. The moss *Mniobryum albicans* is a character-species of cold springs in the montane zone of Western Europe and Scandinavia, just as the lichen *Solorina crocea* is a character-species of ground subject to cryoturbation. These species rarely, if ever, occur in any other type of montane habitat.

* This criterion was established as the limit of interpretability for the data set.

9 Determining the Composition of the Blanket Bogs of Scotland Using Landsat Thematic Mapper

E. REID[1] **and N. QUARMBY**[2]

[1] *National Strategy, Scottish Natural Heritage, Caspian House, Clydebank Business Park, Clydebank, Glasgow, UK*
[2] *I.S. Ltd, Simonsway, Manchester, UK*

This chapter describes the current techniques used by the Scottish Blanket Bog Inventory (SBBI) for evaluating the extent, distribution and condition of the blanket bogs of Scotland. Analysis of Landsat Thematic Mapper data together with ancillary data has enabled six peatland classes to be mapped to a moderately high degree of accuracy. These can be described in terms of National Vegetation Classification categories. Statistical analysis of the classes has demonstrated their ecological validity.

INTRODUCTION

At both patch and landscape scales, resource managers and policy-makers require habitat information in order to increase their understanding of particular ecological processes. Land managers require information, for example, on hydrology, nutrient cycles, food webs, disturbance, damage, condition, vegetation and habitat context, in the form of maps. Contextual habitat mapping (considering a site relative to surrounding areas) is considered today to be a preferred element in the assessment and management of habitats at a local, regional and national scale. Consequently a need has arisen to develop inventory programmes which can fulfil some of these contemporary requirements.

In Britain there is a range of inventory methods currently available for determining the extent, composition and condition of land cover, habitat and plant community types. Three examples varying in scale are:

- Land-based survey, for example mapping the boundaries of vegetation types such as those classified under the National Vegetation Classification (NVC; Rodwell, 1991), usually to a scale of 1:10 000.
- Mapping NVC types, habitat features or land cover classes with the aid of Air Photo Interpretation (API) at a scale of 1:25 000 – 1:10 000.
- Small-scale inventory methods using satellite imagery to characterise land cover and habitat types.

Vegetation Mapping: From Patch to Planet. Edited by Roy Alexander and Andrew C. Millington.
© 2000 John Wiley & Sons Ltd.

This chapter describes the current techniques used by the Scottish Blanket Bog Inventory (SBBI) for evaluating the extent, distribution and condition of the blanket bogs of Scotland. In particular it illustrates the use of satelllite image processing allied with Geographic Information Systems (GIS) and field survey techniques in the Western Isles of Scotland (the island of Lewis and Harris).

Blanket mire is defined as a wetland which supports a surface vegetation which is normally peat-forming (Lindsay *et al.*, 1988). Blanket bog is listed as a 'priority' habitat (active bog), under the EC Habitats Directive (Table 9.1 – EC 1994) and under the UK Biodiversity Action Plan (Cm 2428, 1994). It is the most extensive semi-natural habitat found on land in the UK, and presents formidable challenges for an inventory programme. In Scotland this habitat covers just over 1.0 million hectares (1.4 m ha in Great Britain) and contains an archive of human, climatic and vegetation history ranging back in time through some 5000 – 10 000 years (Figure 9.1). Survey and inventory of this internationally important habitat in Britain, covering 13% of the global peatland resource (Lindsay *et al.*, 1988; Ratcliffe and Thompson, 1988), has been recognised by government conservation agencies as difficult because of the remoteness, the very great extent, and its apparent featureless nature.

SMALL AND INTERMEDIATE-SCALE SURVEY

Before outlining the use of satellite imagery for habitat characterisation, we mention other techniques available for surveying and assessing blanket bogs.

Small-scale Survey – NVC Mapping

The National Vegetation Classification system (NVC – Rodwell, 1991) is being used extensively for mapping particular blanket mire vegetation types throughout specified sites in Scotland. This provides a standard and directly comparable method for classifying vegetation and is used extensively by ecologists throughout Britain. The NVC, however, was not intended as a means for mapping large areas of semi-natural habitats, particularly where mosaics of vegetation types occur. It was, rather, designed to classify vegetation on the basis of a limited sample of plant relevés (quadrats – 2 × 2 m square). The scale of application of the NVC classification must therefore be appreciated when mapping at a scale which is beyond the limits of a 2 × 2 m quadrat, as problems can arise when identifying a homogenous stand of vegetation and drawing map boundaries around this stand (i.e. polygons) (Figure 9.2). Indeed in most contemporary surveys of blanket bogs, using this classification system, the resultant map, when displayed at a scale of 1: 10 000, often depicts mosaics which represent a mixture of small homogenous stands of one NVC type and homogenous stands of another NVC type. This information, although it is representative of the habitat, often leads to interpretation difficulty when presented on a map, as the complexity of the mosaics is difficult to quantify and display, particularly if two or more NVC types are found within one polygon.

The NVC in addition is limited as it cannot always classify or evaluate previously

Table 9.1 'Active' Peatland – National Vegetation Classification Types (after European Commission, 1994)

NVC Type	Description
M1	*Sphagnum auriculatum* bog pool community
M15	*Scirpus cespitosus–Erica tetralix* wet heath
M17	*Scirpus cespitosus–Eriophorum vaginatum* blanket mire
M18	*Erica tetralix–Sphagnum papillosum* raised and blanket mire
M19	*Calluna vulgaris–Eriophorum vaginatum* blanket mire
M20	*Eriophorum vaginatum* blanket and raised mire (active ONLY – M20a for raised bogs)

Figure 9.1 Blanket bog distribution in Scotland (boundaries derived from the British Geological Survey 1:50 000 drift series – where peat is greater than 1 m in depth)

unsampled vegetation types, habitat condition, site hydrology or micro-topography, all of which are important for assessing the scientific and conservation interest of peatlands. For these reasons peatland vegetation is inadequately represented in published NVC maps (see Rodwell, 1991; Horsfield *et al.*, 1996). Furthermore, NVC mapping requires a substantial capital investment in the form of personnel and time needed to cover large areas of peatland (one surveyor can map 100–150 ha day^{-1} of blanket bog).

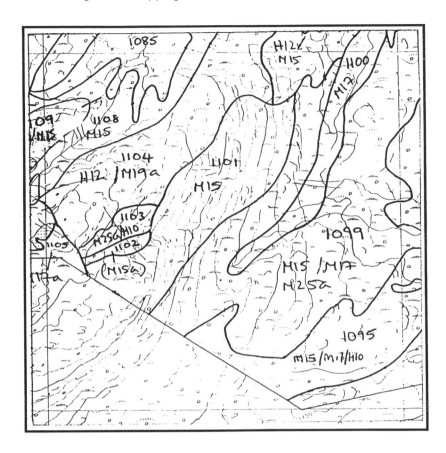

H10 – *Calluna vulgaris–Erica cinerea heath*
H12 – *Calluna vulgaris–Vaccinium myrtillus* heath
M15 – *Scirpus cespitosus–Erica tetralix* wet heath
M17 – *Scirpus cespitosus–Eriophorum vaginatum* blanket mire
M19a – *Calluna vulgaris–Eriophorum vaginatum* blanket mire, *Erica tetralix* sub-community
M25a – *Molinia caerulea–Potentilla erecta* mire, *Erica tetralix* sub-community

Figure 9.2 An extract from a typical SBBI NVC field map showing a 1 km square (1:10 000) displaying NVC pure communities/types and mosaics of types within a polygon

Intermediate Scale – Air Photo Interpretation

There are several examples of this type of survey, for example:

• The Landcover of Scotland (LCS – MLURI 1993)
• The National Countryside Monitoring Scheme (NCMS) (Tudor *et al.*, 1994)

API combined with GIS can provide a means of assessing the distribution and composition of land cover types deemed to be of conservation importance (e.g. heather moorland, Thompson *et al.*, 1995). API is also an extremely useful tool for estimating change over time in the appearance of the countryside. The categories of land cover definition or 'feature types' (Tudor *et al.*, 1994), however, are quite coarse for peatlands (e.g. heather moorland, unimproved grassland, blanket mire etc.). They do not give much information on the ecological variability within habitats and provide inadequate information on habitat 'condition'. This is a result of the limited spectral differentiation possible with air photography, particularly if the photography available is monochrome. In the case of blanket bog, the variability across plant communities dominated by different sedges, grasses, mosses and shrub species is critical to concluding on condition and management requirements.

SATELLITE IMAGERY AND THE SCOTTISH BLANKET BOG INVENTORY (SBBI): BACKGROUND

In 1995 Scottish Natural Heritage (SNH) identified the scale of remote sensing platforms which could be used to map different types of semi-natural habitat. Blanket mire was identified as being well suited to large-scale mapping by Landsat and SPOT satellites. This was supported by earlier research by Reid *et al.* (1994).

The SBBI uses a methodology developed specifically for mapping the extent, variability, condition and hence conservation interest of the blanket mire habitat throughout Scotland. The method uses data recorded by the Thematic Mapper (TM) Sensor from the Landsat 5 satellite. The significant difference between the approach adopted by the SBBI and other remote sensing methods, used in Britain, is that it is designed to map blanket mire condition and extent only, and thus makes no attempt to map other land cover types. Techniques have been developed to interpret the spectral characteristics of TM imagery allied with different features of blanket bog.

The SBBI favoured the data and technique developed, because:

• The data and processing are relatively inexpensive compared with air photography and ground-based survey.
• The data cover a wide area throughout Scotland – each TM image covers 185 km × 185 km, and TM imagery is available and more applicable for vegetation mapping than SPOT.
• The data can be combined with soils information (British Geological Survey) within a GIS in order to highlight the location of blanket bogs.
• The resolution (30 m square pixels) is appropriate for mapping such an extensive habitat.

- The maps indicate gaps in detail (e.g. NVC types, particularly condition classes) which can be targeted for ground survey.
- Major savings in time can be made with our methodology. If alternative ground survey were employed, then to acquire an inventory of the blanket bog resource would take approximately 80 person years.

The specific aims of the SBBI analysis and interpretation of each TM image are threefold:

- First, to identify and depict the different types of blanket bog habitat.
- Second, to determine the range of variation of different types of peatland vegetation/characteristics in order to identify high-quality peatland areas of interest.
- Third, to relate the results specifically to the criteria as set by the SSSI Guidelines (JNCC, 1994), and the EC Habitats Directive with particular reference to the Annex I type 'active blanket bog'.

The SBBI plan for completing the above is focused on the characterisation of blanket mire (Figure 9.3).

METHODOLOGY STEPS (USING THE ISLAND OF LEWIS AND HARRIS AS AN EXAMPLE)

Ground Survey and Ancillary Information

The satellite data used were six Thematic Mapper image bands acquired by the Landsat 5 satellite on 6 June 1992. Survey sites were targeted on the Landsat TM image using local knowledge, the British Geological Survey 1: 50 000 and 1: 625 000 drift series, an 'unsupervised classification' approach (this is termed the 1st stage classification) and air photography. The classification used an ISODATA algorithm described in more detail below.

Within these targeted sites, boundaries (polygons) were produced around all of the classes represented on a classmap, resulting from the unsupervised classification. These polygons were transferred (produced as vector data) onto 'raw' satellite imagery (i.e using a combination of bands 5, 4 and 3) where homogenous areas (blocks) of colour and tone were identified for ground reference survey collection. Where a visual match between the homogenous 5,4,3 composite and the classified vector boundary were identified, then these areas were targeted for survey (Plate 2 (Figure 9.4)). These blocks were reproduced as hard copy at a scale of 1: 10 000 and manually transferred on to air photography at the same scale using acetates. The photography was used to aid in navigation.

For this Landsat scene a total of 139 ground reference areas (blocks) were visited (Figure 9.5). A survey data sheet was completed for each area containing information on, for example, NVC type, condition, micro-topography, erosion features and vegetation structure (Table 9.2). In total, 30 attributes were collected for each area surveyed. The attribute recording framework (field survey methodology) is specifically designed to account for the ecological variation across blanket bog. It is also designed to

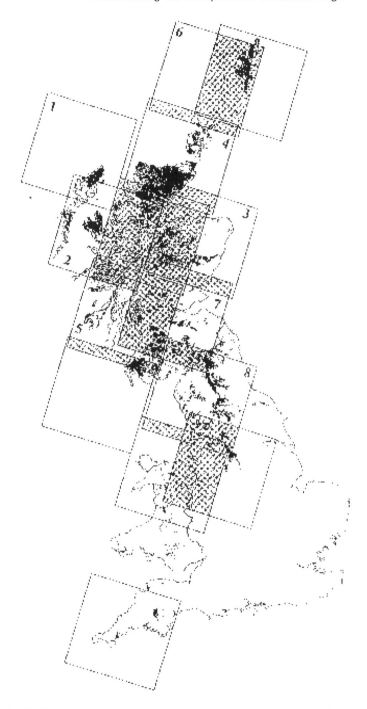

Figure 9.3 SBBI plan for completing an inventory of Scotland's blanket bogs (The order of priority of scenes may change depending on resource availability)

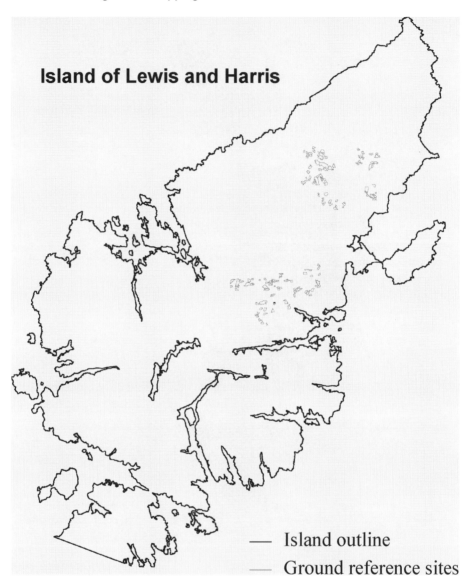

Island of Lewis and Harris

—— Island outline
—— Ground reference sites

Figure 9.5   Location of ground reference survey sites

Table 9.2 Vegetation data (scores in bold = those recorded in the field)

	% scores				
Shrub cover (ratio)	0	1	2	**3**	4
Herb cover (ratio)	0	**1**	2	3	4
Moss cover (ratio)	0	**1**	2	3	4
Dominant herb	*Eriophorum vaginatum*				
Dominant shrub	*Calluna vulgaris*				

consider the difference in scale between the variables on the ground and the resolution of the image. For example ground vegetation structure information is recorded as ratio scores of vegetation cover *throughout the block*.

In Table 9.2 the score for the vegetation cover of shrub, herb and moss species is 3:1:1, where these are in the following percentage range:

0 = zero cover
1 = 1–25% cover
2 = 26–50% cover
3 = 51–75% cover
4 = 76–100% cover

Following the completion of field survey, selected parameters from the survey sheets (such as those recorded above) were amalgamated into a spreadsheet to aid in the interpretation of the data, statististical analysis, image classification and the production of spectral class descriptions.

Ancillary Data Input

Existing SSSI vector data and the British Geological Survey (BGS) peatland areas (indicating peat over 1 m in depth) were imported as vector ancillary layers in the same file containing the satellite data. A digital elevation model (DEM) for Lewis was purchased from the Ordnance Survey. This information was supplied in 20 km × 20 km tiles at a 1:50 000 scale. Each pixel within the image represented an area of 50 m × 50 m on the ground. Each tile was placed into a seamless mosaic layer within the satellite image database. The tile was resampled to a 30 m × 30 m pixel size to match the pixel size of the satellite data, using a cubic convolution procedure.

The 139 ground reference blocks identified on photographs/satellite image enlargements, were digitised on screen using the original satellite image as a reference image.

Satellite Data Processing – Classification and Interpretation

To aid the classification process, a number of masks were produced to restrict the analysis to peatlands. For example, British Geological Soil deposit data, held in the GIS, were used to highlight the presence or absence of blanket bog (Figure 9.6).

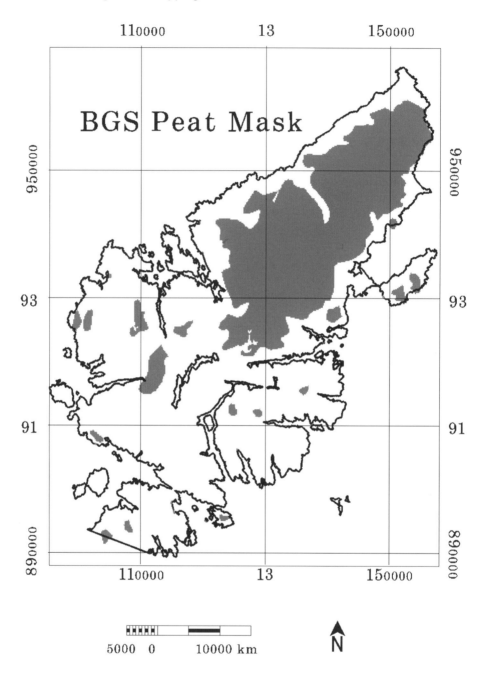

Figure 9.6   Example of an image mask – highlighting blanket bog

In addition, a number of statistical analyses (e.g. Principal Component Analysis) were performed on the imagery to reduce the size of the data set and in order to remove any correlated or confusing spectral features in the imagery.

The final classification of the image produced output, for Lewis, in three stages:

- An initial unsupervised ISODATA classification was used to identify a mask of the gross peat area. The ISODATA algorithm used is based upon a cluster analysis. The software was asked to identify up to 16 classes within the peat mask area below 200 m. This was the same number of classes as specified in the pre-survey image processing. The input data were the first three principal component images. The software selects 16 seed points within the data, and computes the optimum locations of the cluster centres by iteration. The areas were classified to produce 15 'peat' classes. These spectral classes were examined and aggregated to produce seven peat classes and one non-peat (rock) class.
- Areas outside the peat class mask (i.e. outside blanket bog) but below 200 m were classified into eight categories which were subsequently amalgamated into three classes (grassland, woodland and sandy dunes/soils). No accuracy analysis was performed on these non-peatland classes.
- Areas above 200 m were assigned to a single class. It was considered that at altitudes above 200 m (on Lewis) the vegetation would be upland in character on thinner peats, and as such would not be considered as active blanket bog. In addition the BGS peat soils data did not indicate that there was deep peat over 200 m. In total, the final classification contained 12 classes.

A 'sieve' filter was applied to the classification to remove some of the small isolated pixels or polygons. Polygons which were smaller than 8 pixels in size were removed and replaced by the class value belonging to the largest neighbouring polygon. A 'sieve' filter was preferred to a mode filter since the latter will remove narrow polygons and peninsula features in polygons, creating a blocky effect.

PRODUCTION OF CLASS DESCRIPTIONS

Each class required an ecologically meaningful description to convey the nature of the class and the range of variation, if any, within that class. To aid interpretation selected data from the survey sheets were amalgamated into a summary table arranged in order of spectral class. The table describes and lists various physical attributes such as shrub and herb cover and erosion intensity. This was extremely useful in comparing *general* block characteristics with the classes produced by the ISODATA classification. Other essential information used in preparation of the spectral class descriptions included the application of ecological knowledge, past survey data, block descriptions, slides of the vegetation and local knowledge of the peatland areas.

Quantitative Analysis – Accuracy of the Classification and Ground Information

Quantitative analysis of the peatland classification involved:

- A test of the robustness of the classification using a Transformed Divergence test on the separability of classes.
- An assessment of the accuracy of the classification in comparison with the ground reference data.
- Statistical analysis using multivariate and univariate Analysis of Variance (ANOVA) and Canonical Variates analysis (CVA) to ascertain the link between the classes and ecological field observations.

Class Separability

The separability measure used was a Transformed Divergence test. This test compares the means and covariance of each class pair and determines their statistical separability. Values range from 0 to 2 with values greater than 1.9 indicating that a class is very separable. The average separability of all classes was 1.84. This indicates that the classes were all fairly well separated and are robust.

Classification Accuracy

To test the accuracy of the classification, each ground reference block description was manually examined and labelled according to the spectral class or classes *which would have been predicted* from the ground reference survey. For example a block described as microbroken ground with smooth patches and a mixture of the M17 vegetation type would be manually labelled as class 1. Blocks containing a single class label were selected and assigned an attribute representing the spectral class number for that polygon/block (i.e. blocks containing a mixture of classes were excluded). In total 73 peatland blocks with a single class label were identified. These were converted from vectors to rasters and filled with the value of the class for each ground reference polygon/block. This provided a ground reference image.

A comparison was made between all the pixels in the ground reference image (i.e. developed by the prediction above) and the corresponding pixels in the classified image at the same spatial location. The comparison produced a contingency matrix indicating the proportion of correctly identified pixels for each spectral class, and the proportion of pixels allocated incorrectly for each class. An overall accuracy figure was calculated by weighting the correctly identified proportion by the number of pixels belonging to that class in the ground reference image.

The contingency table produced by comparing the ground reference data with the classification image is shown as Table 9.3. The diagonal from top left to bottom right shows the proportion of correctly identified pixels for each class. The horizontal axis shows the errors of omission. For class 1, 76.7% of pixels have been correctly classified, 17.4% of pixels have been incorrectly labelled as belonging to class 2. The overall accuracy was 76.98%.

Table 9.3 Contingency table comparing ground reference data with the classification image

Class	No. pixels	\multicolumn{9}{c}{Percent pixels classified by code}

Class	No. pixels	0	1	2	3	4	5	6	7	8
1	3816	0.4	**76.7**	17.4	2.2	0.3	1.7	0.8	0.1	0.6
2	1860	0	2.4	**82.5**	5.1	0.4	9.5	0	0.1	0
3	596	3.5	0.5	12.9	**63.3**	3.9	15.6	0.2	0	0
4	836	1.8	0	0.6	4.8	**89.8**	1.4	0	0	0
5	852	0.5	1.8	20.8	3.4	5	**67.5**	0	1.1	0
6		0								
7	460	2.6	0	0.7	0.4	1.5	21.3	0.4	**70.9**	2.2
8	136	0	0	4.4	0.7	0	0	24.3	0	**70.6**

Ecological Validity of the Classification

The unsupervised classification approach requires the image interpreter to supply a name to each class after the classification has been carried out. It is not certain that the classes produced will be meaningful in terms of ground observations. It is therefore helpful to use a statistical analysis to determine whether the classes are ecologically meaningful.

Field measurements (e.g. Table 9.2) of the proportion of shrub cover, herb cover, moss cover, bare peat cover and erosion had been made to characterise each ground reference block. As with the classification accuracy assessment, only ground reference blocks with a single class label were assessed. Classes 1, 2, 3, 4, 5 and 7 could be compared for statistical purposes. The other two classes contained too few samples to provide meaningful results.

In essence a general linear model can be set up of the form $Y = XB + E$, where, Y is the matrix of dependent variables, in this case these are five proportions of cover types, X is the categorical variable representing the spectral class, B is a matrix of regression coefficients, and E is a matrix of random errors.

The data matrix containing the six variables for each ground reference block was entered into the SYSTAT statistical analysis software and the following hypothesis was set:

- The null hypothesis for the statistical tests is that there is no significant difference in the proportions of cover type for a given spectral class.
- The alternative hypothesis, therefore, is that there is a significant difference in the proportions of cover types (e.g. erosion cover, shrub cover, moss cover etc.) for a given spectral class.

The tests produce several statistics which are of value in assessing the link between the spectral class and the proportions of cover type. These are Univariate Analysis of Variance (ANOVA), Multivariate Analysis of Variance (MANOVA) and canonical coefficients. The ANOVA statistic tests each cover type separately against the spectral class to see if the variation in spectral class can be explained by a corresponding variation in the proportion of a given cover type. The MANOVA statistic tests for all

Table 9.4 Univariate Analysis of Variance (ANOVA) results

Variable	E	Probability
SHCOV	6.893	0
HERBCOV	8.774	0
MOSSCOV	2.289	0.056
BARECOV	8.045	0
EROSION	5.753	0

of these variables together, to see if the variation in spectral class can be explained by a corresponding variation in the proportion of all the cover types as a whole. The canonical coefficients can be used to derive canonical variates for each spectral class. This is determined by multiplying the matrix of coefficients for each canonical covariate (one coefficient per cover type) by the least squares mean of each cover type (Chatfield and Collins, 1980). This was carried out for the first two canonical variates. The plot of the first two canonical variates, together with a significance limit can then be used to determine whether the spectral classes are separable in terms of the proportions of cover type.

Analysis involving ANOVA established that variations in the field measurement scores of shrub and herb cover, erosion and bare peat were all clearly linked to each of the spectral classes within the 95% confidence limits. Table 9.4 shows the probability that the null hypothesis is true for each variable. A value less than 0.05 indicates that the alternative hypothesis should be accepted at the 95% confidence level. At a 90% confidence level the alternative hypothesis should also be accepted for the moss cover variable. The MANOVA test indicated that when all the field measurement variables were assessed together the alternative hypothesis could be accepted at the 95% confidence level. The final classification is therefore a reflection of real spatial variability in vegetation/physical cover type.

The resulting canonical variates analysis plot for the six classes tested is shown in Figure 9.7. It can be seen that classes 3, 4 and 7 are easily separable in terms of their cover type characteristics. Classes 1 and 2 overlap significantly. Class 5 overlaps with class 2, and to a lesser extent with class 1.

The discrimination of classes 1, 2 and 5 in terms of image interpretation is based primarily on the proportion of moss cover and erosion. Class 1 degrades to class 5 as the proportion of erosion increases. Class 2 is a transitional state between class 1 and class 5. While the ANOVA tests demonstrate that the proportion of erosion can be used to differentiate between classes, the discriminatory power of the proportion of moss cover is less effective. It is therefore not surprising that there should be overlap between classes 1, 2 and 5.

In order to link the field observations of National Vegetation Classification types within a block to the spectral classification a simple proportion of types per spectral class was calculated. It was not possible to perform any further statistical analysis of the NVC data since there was more than one NVC observation per ground reference block, i.e the NVC is inherently hierarchical and thus NVC classes and sub-communities are correlated. For example NVC type M17a (*Scirpus cespitosus–*

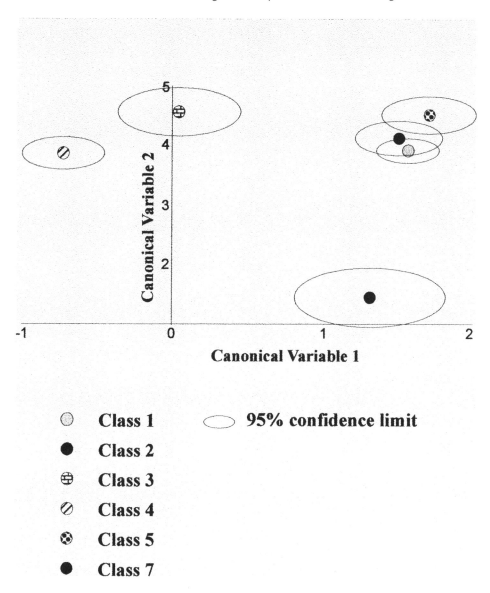

Figure 9.7   Canonical Variates Plot

Eriophorum vaginatum blanket mire – Sub-community: *Drosera rotundifolia–Sphagnum* spp.) is a subcommunity of type M17 (*Scirpus cespitosus–Eriophorum vaginatum* blanket mire), and many of the species found in M17a are also found in M17. In addition, for this area of Scotland, the vegetation was difficult to assign to many different NVC types as the vegetation was mainly variable within the one type, M17, which represented a large majority of the vegetation keyed across the island.

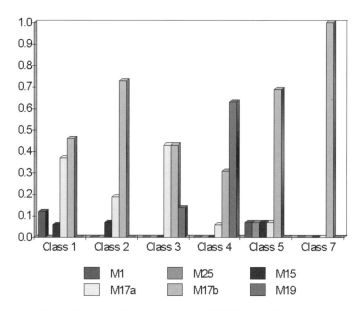

Figure 9.8   Chart showing the proportion of NVC type for each spectral class

For each class, the proportion of each NVC category has been plotted on Figure 9.8. The y-axis shows the proportion of each NVC type per class. To facilitate a comparison of the spectral classes, the M17 observations were equally divided between the M17a and the M17b categories with the aid of the field descriptions.

For classes 1, 2 and 5, which were not well discriminated by the proportion of moss cover, it is clear that the proportion of M17a decreases from class 1 through to class 5. Also classes 2 and 5 both contain higher proportions of M17b vegetation categories, and class 1 contains a significantly higher proportion of the M1 vegetation community. All of these classes are therefore represented in the figure as mixtures of NVC communities.

Final Classification Map and Descriptions

Using the results from the above analysis, final habitat descriptions were developed which are related to SSSI guidelines and the EC Habitats Directive (HD). A short summary report is produced for each class (colour) across the peatlands in the satellite scene. This provides:

- Class name and *dominant* NVC type in class.
- Relationship to *SSSI guidelines* – habitat structural elements, dominant microtopography, most common NVC type/types.
- Relationship to *HSD Annex 1 habitat* (i.e. Table 9.1 – active blanket bog).
- *Vegetation characteristics* – description in detail of the vegetation within this class.

Table 9.5 Areal distribution of each land cover class for the island of Lewis

Spectral class	Hectares	% of area
1 M17 mixture	14 716	7
2 M17 microbroken	29 610	15
3 M17a smooth, wet	13 653	7
4 M19	21 339	11
5 M17b eroded	32 022	16
6 Marginal peatland	7 092	4
7 Bare peat	22 165	11
8 Rocky ground	11 134	5
9 Grassland	11 272	6
10 Woodland	1 544	1
11 Sandy soil/dunes	2 670	1
12 Land above 200 m	23 151	11
Other	10 113	5

M17 = *Scirpus cespitosus Eriophorum vaginatum* blanket mire (sub-communities: M17a = *Drosera rotundifolia–Sphagnum* spp. and M17b = *Cladonia* spp.) M19 = *Calluna vulgaris–Eriophorum vaginatum* blanket mire.

- *Physical features* – description of the physical land features described by this class (e.g. degree of erosion, bare peat etc.).

As a result of the image processing of the island of Lewis and Harris, seven peatland classes have been determined. Table 9.5 shows the area for each spectral class for the island of Lewis. Plate 3 (Figure 9.9) shows the peatland classification and Plate 4 (Figure 9.10) shows the area of active blanket bog. The majority of the peatland of Lewis and Harris is composed of active bog. The active bog classes (classes 1–5 and 7) account for 67% of the total land area.

FUTURE APPLICATION

The SBBI technique of peatland assessment based on satellite imagery has become a valuable and cost-effective tool for mapping and quantifying, at a coarse scale, large expanses of blanket bog in Scotland. It could also offer considerable potential for mapping blanket bog habitats in other parts of the EU, and provides a valuable means of monitoring change between different vegetation types through time.

By applying this kind of mapping, SNH can identify and then target, for further detailed ground survey, areas of high natural heritage importance. Large areas of Scotland's blanket bogs have not been studied in any detail – some have only been walked over in more recent times. This technique therefore provides the first opportunity to determine the full extent and composition of blanket bogs in Scotland. The SBBI is currently concentrating on those parts of Scotland which are likely to have substantial areas of active blanket bog but which have been unmapped or poorly surveyed.

REFERENCES

Chatfield, C. and Collins, A.J. (1980) *Introduction to Multivariate Analysis*. Chapman and Hall, London.

Cm 2428. (1994) *Biodiversity: The UK Action Plan, Volume 2, Annex F, Lists of Key Species, Key Habitats and Broad Habitats*. Command 2428. HMSO, London.

European Commission (1994) *Manual for the Interpretation of Annex 1 Priority Habitat Types of the Directive 92/43/EEC*. European Commission, Directorate-General Environment, Nuclear Safety and Civil Protection. Nature Protection and Soil Conservation (Habitats 94/3 FINAL), Brussels.

Horsfield, D., Thompson, D.B.A. and Tidswell, R. (1996) *Revised Atlas of Upland NVC Plant Communities in Scotland*. Information and Advisory Note, Scottish Natural Heritage, Perth.

JNCC (1994) *Supplement to The Guidelines for the Selection of Biological Sites of Special Scientific Interest: Bogs, Nature Conservancy Council (1989)*. Joint Nature Conservation Committee, Peterborough.

Lindsay, R.A., Charman, D.J., Everingham, F., O'Reilly, R.M., Palmer, M.A., Rowell, T.A. and Stroud, D.A. (1988) *The Flow Country – The Peatlands of Caithness and Sutherland*. Nature Conservancy Council, Peterborough.

Macaulay Land Use Research Institute (1993) *The Landcover of Scotland* 1988 *Final Report*. The Macaulay Land Use Research Institute, Aberdeen.

Ratcliffe, D.A. and Thompson, D.B.A. (1988) The British uplands: their ecological character and international significance. In: M.B. Usher and D.B.A. Thompson (eds) *Ecological Change in the Uplands*. Special Publication Series of the British Ecological Society, No. 7, Blackwell Scientific Publications, Oxford, pp. 9–36.

Reid, E., Mortimer, G.N., Thompson, D.B.A. and Lindsay, R.A. (1994) Blanket bogs in Great Britain: an assessment of large-scale pattern and distribution using remote sensing and GIS. In: P.G. Edwards, R.M. May and N.R. Webb (eds) *Large Scale Ecology and Conservation Biology*. 35th Symposium of the British Ecological Society, Blackwell Scientific Publications, Oxford, pp. 229–246.

Rodwell, J.S. (ed.) (1991) *British Plant Communities. Vol. 2: Mires and Heaths*. Cambridge University Press, Cambridge.

Thompson, D.B.A., Hester, A.J. and Usher, M.B. (eds) (1995) *Heaths and Moorland Cultural Landscapes*. HMSO, Edinburgh.

Tudor, G.J., Mackey, E.C. and Underwood, F.M. (1994) *The National Countryside Monitoring Scheme: The Changing Face of Scotland, 1940s to 1970s. Main Report*. Scottish Natural Heritage, Perth.

10 An Assessment of the Land Cover Map of Great Britain within Headwater Stream Catchments for Four River Systems in England and Wales

A.M. BROOKES,[1] **M.T. FURSE**[2] **and R.M. FULLER**[3]

[1] *EOS, Farnham, UK*

[2] *Institute of Freshwater Ecology, Dorset, UK*

[3] *Institute of Terrestrial Ecology, Monks Wood, Huntingdon, UK*

Field surveyed and remotely sensed data (from the Land Cover map of Great Britain – LCMGB) were compared for small headwater stream catchments within four regions of England and Wales in the context of research investigating agricultural impacts on stream macroinvertebrates. Resulting values for direct correspondence varied (from 37.6% to 65.0%) between the four catchments and gave a combined figure of 45.8% for all four catchments. Systematic non-correspondence between the data sets was investigated and rules formulated to facilitate the production of compromise land cover maps and hence revised correspondence figures. These revisions resulted in a general increase in correspondence though the degree of change varied between both cover classes and catchments. Results suggest that although the LCMGB does not represent a suitable alternative source for land cover data at the scale of the individual headwater catchment site, it provides an invaluable source of hitherto unavailable data at the regional scale of the whole river system.

INTRODUCTION

Mapping vegetation from a land use or land cover perspective is an important requirement of a range of environmental applications, including land use planning, landscape monitoring, natural resource management and habitat assessment for species conservation. The increasing use of computerised Geographical Information Systems (GIS) for these types of applications over the past decade has widened the scope for the acquisition, handling and analysis of land cover and land use data.

The choice between data collection in the field or by remotely sensed means (air photo or satellite) has been available for many years but has always been limited by the question of scale. Field survey data and air photos are costly and time-consuming to collect and so are limited in the geographical extent to which they can be applied. There have been few attempts to map vegetation on a national scale in the field: the national Land Use Surveys of the 1940s (Stamp, 1947) and 1960s (Coleman, 1961) took

Vegetation Mapping: From Patch to Planet. Edited by Roy Alexander and Andrew C. Millington.
© 2000 John Wiley & Sons Ltd.

tremendous effort (much of the work being done by schoolchildren) and years to complete and have never been digitised or repeated (Fuller *et al.*, 1994c). More recent national surveys have adopted a sampling strategy instead, such as the Countryside Surveys, conducted by the Institute of Terrestrial Ecology (ITE) at a six-yearly interval (Barr *et al.*, 1993), which use a stratified sample of 1 km squares for field surveying to provide representative coverage of Great Britain.

Satellite imagery provides an alternative that is more readily accessible, is geographically extensive and comparatively cheap but until recently was not available at a fine enough resolution to be applicable at less than regional scales. Improvements in remote sensing technology over the past decade have resulted in classified image data becoming available at resolutions of less than 30 m, which are suitable for vegetation mapping at a local scale. The first national digital map of land cover at the field parcel scale, known as the Land Cover Map of Great Britain (LCMGB), was produced by ITE in 1993, using classified LANDSAT Thematic Mapper data output at a resolution of 25 m (Fuller *et al.*, 1994a).

The choice between field mapped and remotely sensed data involves more than the question of availability. Fundamental differences in the way the data are represented and in the levels of error and uncertainty affect their suitability for subsequent modelling and analysis (Goodchild, 1994). Field mapping is prone to a range of errors (particularly differences in interpretation between surveyors) and ought to incorporate quality control measures to quantify errors of this type (Cherrill and McClean, 1995). Image data are similarly prone to errors during correction, interpretation and classification. An awareness of data quality is important when considering the level of accuracy required to meet the purposes for which the data were acquired.

In this chapter, field surveyed and remotely sensed data (from the Land Cover Map of Great Britain) are compared for small headwater stream catchments within four regions of England and Wales. The requirement for the land cover data is within the context of research investigating agricultural impacts on stream macroinvertebrates. The comparison of the data sets complements earlier assessments of the LCMGB done on a national basis (Fuller *et al.*, 1994a) and within a single region (Cherrill, 1994; Cherrill *et al.*, 1995). The aims are:

- to look for direct correspondence between the field surveyed and remotely sensed data on a local and regional scale;
- to investigate and try to accommodate systematic non-correspondence between the data sets and produce compromise land cover maps for each stream catchment; and
- to assess the LCMGB as a data source for future work in catchments where field surveyed data are not available.

DATA

Field Survey Data

The main source of the field surveyed land cover data was from Stage 3 of a project conducted by the Institute of Freshwater Ecology (IFE) for the National Rivers

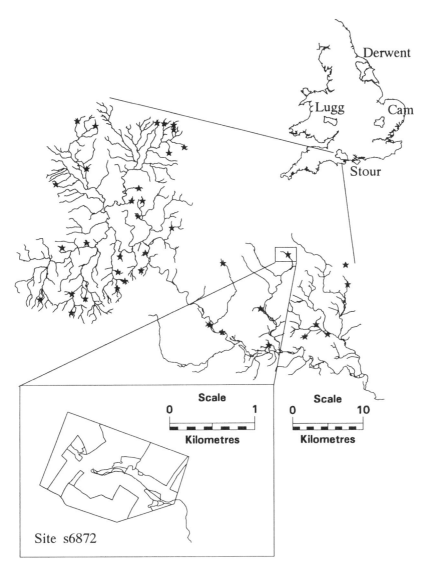

Figure 10.1   Location maps of study areas and headwater sites on the river Stour. Site map based upon an Ordnance Survey Map with permission of the Controller of Her Majesty's Stationery Office

Authority (NRA) entitled The Faunal Richness of Headwater Streams (Furse *et al.*, 1993, 1995). Four river systems were chosen for biological sampling and field mapping on the basis of their geographical and agricultural disparateness and size (Furse *et al.*, 1995): these were the Cam (Cambridgeshire), the Derwent (Yorkshire), the Lugg (Hereford and Worcester) and the Stour (Dorset) – Figure 10.1. The number and location of headwater sites were selected according to the size of the river system and to represent a range of major vegetated land cover types, resulting in 15 sites on the

Cam, 39 each on the Derwent and Lugg and 38 on the Stour (Furse *et al.*, 1995). Each study site consisted of a biological sampling point in the stream and the estimated catchment area on the ground which drained to that point. All the stream sites lay within 2.5 km of the source (the working definition of 'headwater'), most within 1 km, and so the catchment areas were small, ranging from 1 to 3 km^2.

Field survey booklets were prepared by marking the estimated catchment boundaries of each site onto photocopies of OS 1:10 000 maps (Furse *et al.*, 1995). Field surveying and biological sampling of the sites took place simultaneously in the summer of 1993. The field surveyors mapped the land cover/use of all the field parcels within the catchment using a detailed hierarchical classification system based on that used by the Countryside Surveys of Great Britain (Barr *et al.*, 1993). For the purposes of the Faunal Richness study, the field survey maps were reclassified into the simpler 25 class system of the Land Cover map (see below). Permission was obtained by the NRA from the Ordnance Survey to digitise from the field booklets and this was done using ARC/INFO to create polygon coverages of the field parcels within each catchment and line coverages of the parcel boundaries. Since the field surveyors were not required to map or label the types of field boundaries between parcels (i.e. hedge, wall, fence etc.) these features were not represented in the polygon coverages. In the line coverages, the features were just labelled road, footpath, track or field boundary. Secondary codes were given to boundaries which the field surveyors had marked as either new to the map or no longer present. The polygon coverages were used in the Faunal Richness project to calculate the total area within each catchment of each cover type.

Remotely Sensed Data

The remotely sensed data were obtained for the same areas of the country by extraction from the Land Cover Map of Great Britain. The LCMGB was produced using an automated classification of composite Landsat TM data based on summer and winter imagery acquired between November 1987 and November 1990 (Fuller *et al.*, 1994b). The LCMGB uses a classification system of 25 cover types (Table 10.1) which include urban, suburban, bare ground and arable land (tilled land) plus 18 semi-natural vegetation types, comprising three woodland, four heathland, three wetland, seven grassland and one bracken class.

METHODS

Data Preparation

In ARC/INFO the 131 polygon coverages were converted to field surveyed (FS) grids with a cell size of 25 m using the POLYGRID command. In order to ensure that the FS cells were geometrically aligned to the National Grid and thus matched the LCMGB cells, it was necessary to read the tic points of each polygon coverage (which were on 1 km or 500 m intersections of the National Grid) and set the analysis window in GRID to the tic extent before using the POLYGRID command. (POLYGRID other-

Table 10.1 LCMGB classification system showing class numbers and names

0	Unclassified
1	Sea/estuary
2	Inland water
3	Beach/flats
4	Saltmarsh
5	Grass heath
6	Mown/grazed turf
7	Meadows/semi-natural grass
8	Rough/marsh grass
9	Moorland grass
10	Open shrub moor
11	Dense shrub moor
12	Bracken
13	Dense shrub heath
14	Scrub/orchard
15	Deciduous woodland
16	Coniferous woodland
17	Upland bog
18	Tilled land
19	Ruderal weed
20	Suburban/rural development
21	Continuous urban
22	Inland bare ground
23	Felled forest
24	Lowland bog
25	Open shrub heath

wise defaults to using the bottom left corner of the minimum bounding rectangle of the polygon coverage as the starting point of the raster grid which is not likely to coincide with the National Grid.)

In the ARC/INFO GRID module, remotely sensed (RS) grids of the same area as each FS grid were created from the LCMGB by setting the analysis window and mask to that of the FS grid.

Comparison Procedures

Direct Comparison

For each catchment, the FS and RS grids were compared cell by cell and an output grid created containing ones for all the RS cells whose class value exactly corresponded with the FS class and zeros for all non-corresponding cells. The percentages of ones and zeros in the correspondence grids were calculated and written to a results table to give the initial figure of direct correspondence between the data sets. A residuals grid was also created to show the class values of the RS grid which did not correspond.

Correspondence matrices of the spread of FS cells among the 25 classes against the same for the RS cells were generated for all catchments combined in each of the four river systems using a C program and ASCII versions of all the grids. From these it was possible to identify the most commonly confused classes and deduce likely reasons for the differences.

Revised Comparison

Where explanations for non-correspondence between classes were possible these were formulated into rules which could be applied during a repeat comparison of the data to produce a 'compromise' grid and a revised correspondence figure. A revised residuals grid was generated by comparing the new compromise grid against the original residuals grid.

Comparison Against Linear Features

From the remaining non-corresponding cells in the RS grids some patterns could be seen which appeared to represent linear features. These were not represented in the FS data because the field surveyors were not required to label or map the types of field boundaries between parcels. Although incomplete and somewhat patchy, this potential source of boundary data from the RS data, which could add diversity to the rather simplified FS mapped data, was considered useful to the research project and an attempt was made to include it in a new version of the compromise grid.

The line coverages of boundary information from the field survey booklets were converted to raster grids with a cell size of 10 m. A further comparison was made in GRID, this time between the RS grid, the compromise grid and the line grids, for each catchment, to find cells in the RS grid which coincided with linear features and appeared to represent a linear feature. This latter condition was difficult to implement but a crude test was made to see whether the cell in question differed from any of its eight immediate neighbours, using the FOCALVARIETY function. If it had a variety of greater than 1, then the cell was either at a boundary or within one cell of a boundary.

Some discrimination was made between the different line feature codes and the class values. For example, if the cell value of a line was a road but the RS cell value was not in the urban, suburban or bare ground classes, then the new compromise cell value was forced into the suburban class. Overriding conditions were applied to certain classes, namely that any inland water cell in the compromise grid could not be overwritten by a boundary cell, and that RS cells in the arable or unclassified categories could not become boundary cells. After applying these conditions, the RS cell value was put into the revised compromise grid wherever it coincided with tracks and footpaths and existing or new field boundaries. For field boundaries no longer on the map and all other cells which did not coincide with a linear feature or had a neighbourhood variety of 1, the output cell value was kept the same as the original compromise grid.

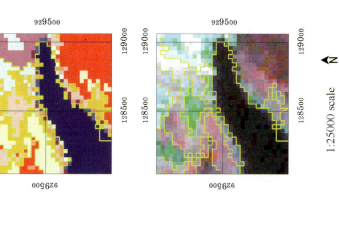

Site 12 (Achmore): Unsupervised classification and 5,4,3 false colour composite with vector boundaries from the classification superimposed.

1:25000 scale

Plate 2 – Figure 9.4 A typical field survey site

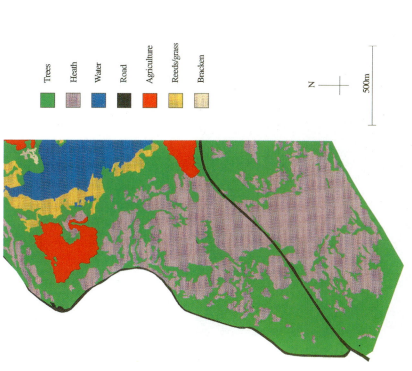

Trees

Heath

Water

Road

Agriculture

Reeds/grass

Bracken

N

500m

Plate 1 – Figure 7.6 Vegetation map of the Dinnet National Nature Reserve study area, 1986

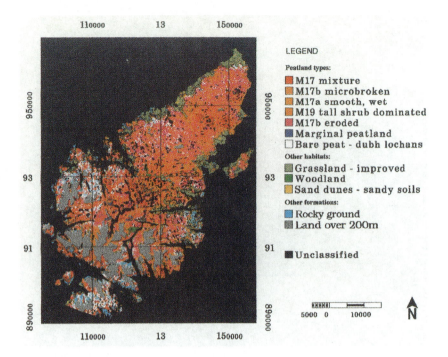

Plate 3 – Figure 9.9 Classified peatland map of Lewis and Harris

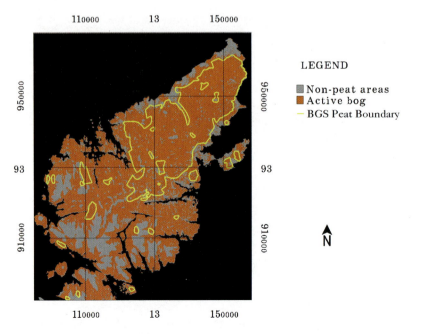

Plate 4 – Figure 9.10 The distribution of Active Blanket Bog on Lewis and Harris

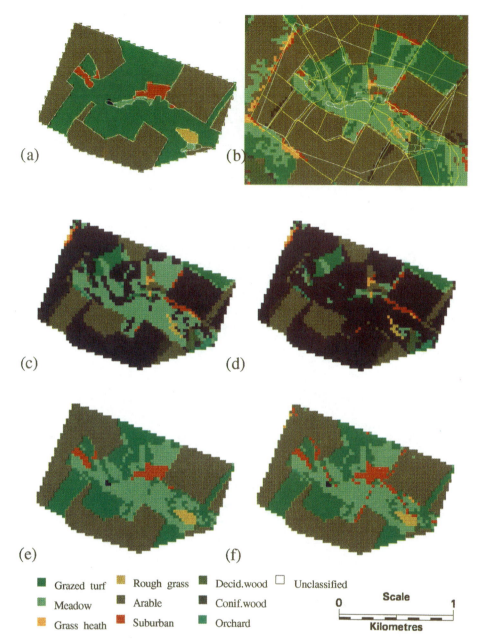

Grazed turf Rough grass Decid.wood Unclassified
Meadow Arable Conif.wood
Grass heath Suburban Orchard

Scale
0   1
Kilometres

Plate 5 – Figure 10.3 Comparison of datasets for a site on the Stour: (a) FS grid with polygon boundaries overlaid, (b) RS grid with polygon and line boundaries overlaid, (c) residuals grid showing cells that directly correspond coloured dark blue and non-corresponding cells from the RS grid in their own colours, (d) revised residuals grid showing improved correspondence after applying rule, (e) compromise grid (see text), (f) revised compromise grid after comparison with the line grids (see text). Reproduced by permission of the Institute of Terrestrial Ecology, based upon an Ordnance Survey Map with permission of the Controller of Her Majesty's Stationery Office

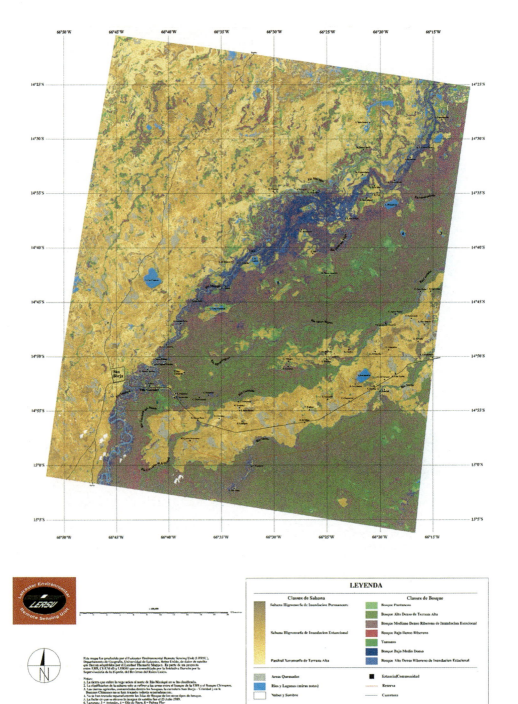

Plate 6 – Figure 11.4 Map of the vegetation classification of the reserve

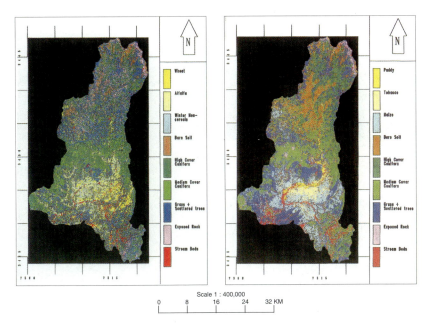

Plate 7 – Figure 12.1 The Landsat image before and after topographic adjustments. Note that the relief effects are reduced after reducing the pixel value from a slope to its horizontal equivalent through solar radiation modelling

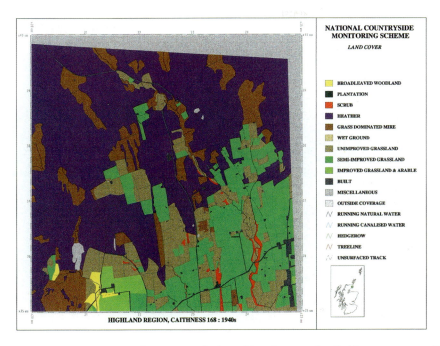

Plate 8 – Figure 13.2 A GIS-generated plot of land cover for Caithness square 168 (1940s)

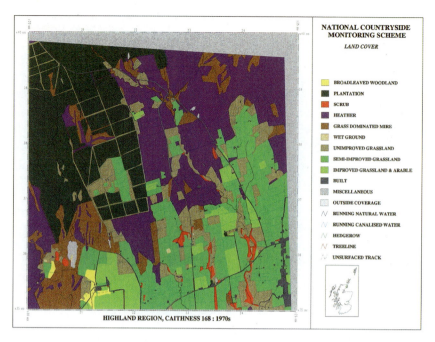

Plate 9 – Figure 13.3   A GIS-generated plot of land cover for Caithness square 168 (1970s)

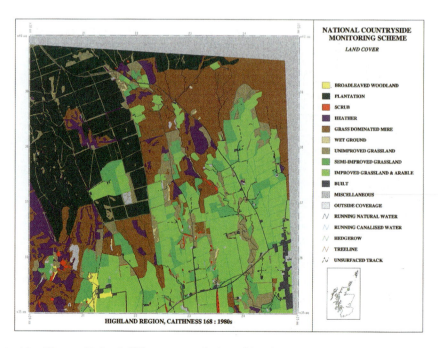

Plate 10 – Figure 13.4   A GIS-generated plot of land cover for Caithness square 168 (1980s)

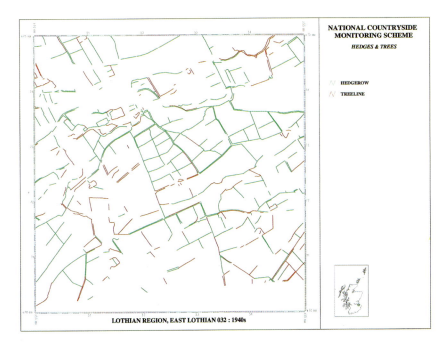

Plate 11 – Figure 13.7 A GIS-generated plot of treelines and hedgerows for Lothian square 32 (1940s)

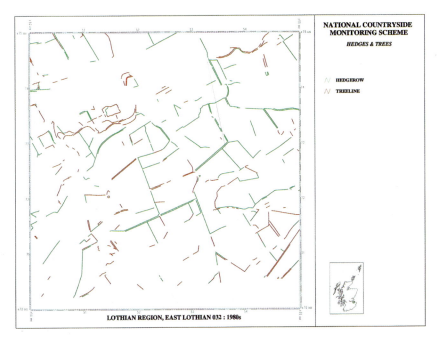

Plate 12 – Figure 13.8 A GIS-generated plot of treelines and hedgerows for Lothian square 32 (1980s)

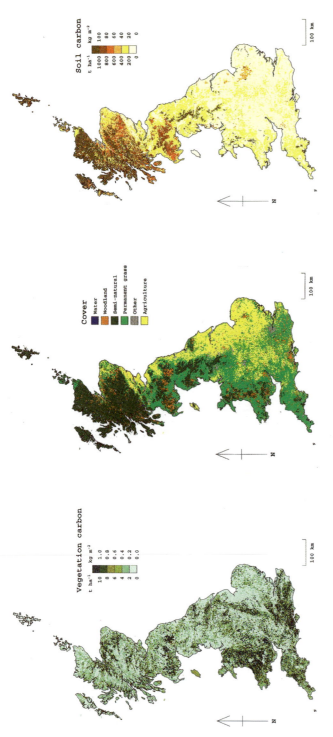

Plate 13 – Figure 15.1 Distribution of carbon in vegetation based on allocating the total vegetation carbon in all the squares of an ITE land class, estimated from survey data, proportionally to each square using the relative occurrence and location of different vegetation cover types from the ITE Land Cover Map

Plate 14 – Figure 15.2 Distribution of main cover groups from ITE Land Cover Map used in estimating soil carbon for Great Britain

Plate 15 – Figure 15.3 Distribution of soil carbon in Great Britain based on soil data from MLURI for Scotland, SSLRC soil data for England and Wales, Forestry Authority bulk density data for Scottish peats and ITE Land Cover Map

RESULTS

Class Distribution

The distribution of class values in the RS data is more diverse with a wider spread of class values than the FS data (Figure 10.2). Differences in the land cover composition among the four river systems are also noticeable. The catchments on the Cam are predominantly arable with over 60% of the cells from both data sources in class 18 and few other classes represented. The Stour catchments are dominated by agricultural grass (classes 6 and 7) with woodland and arable classes accounting for the majority of the remaining cells. The Derwent and Lugg catchments have a greater proportion of their land cover in the heathland, semi-natural and woodland classes (Figure 10.2).

For the four rivers combined, 93% of the total number of cells in the FS data fall into only 6 out of the 26 classes whereas the same classes account for only 63% of all the RS cells (Table 10.2a). Out of the top six classes in both data sets there are four classes in common (classes 6, 18, 15 and 16) but their rank order differs (Tables 10.2a and 10.2b). The most marked differences between the FS and RS data in the respective proportions of the total number of cells are for classes 6 and 7. Overall, more than twice as many cells are classed as 6 in the FS grids than the RS grids and over 11 times more RS cells are classed as 7 than FS cells.

Direct Comparison

From the correspondence matrices for all sites combined within each river system, the calculated percentages of cells which are classified as the same in both data sets are shown in Table 10.3. There are considerable differences between the four rivers, with correspondence ranging from 37.6% for the Stour catchments to 65.0% in the Cam. The combined percentage correspondence is 45.8% for all four rivers (Table 10.3).

In the individual classes, the best direct correspondence figures come from classes 18, 16, 11 and 6 (Table 10.4), although there is a wide discrepancy between the figures in classes 11 and 6 for the two data sources. The comparatively low figures overall illustrate the amount of classification difference which existed.

At the scale of the individual site catchments, the figures for direct correspondence also varied widely (Table 10.5). A maximum of 95.5% direct correspondence occurred at one site on the Cam, where mean correspondence was highest overall, but, at the other end of the scale, minimum correspondence figures were very low (6.8% at one site on the Lugg) – Table 10.5. Even after the revised comparison, the minimum direct correspondence figures remained low – Table 10.5, bracketed numbers.

Plate 5 (Figure 10.3) illustrates the comparison of the data sets for site s6872 on the Stour. Figure 10.3a is the FS grid with the polygon field boundaries overlaid and Figure 10.3b is the RS grid of the same area, taken from the LCMGB, overlaid with the polygon and line boundaries. Figure 10.3c is the residuals grid showing all the cells that directly correspond shaded in dark blue and those that do not shaded in their own colour from the RS grid.

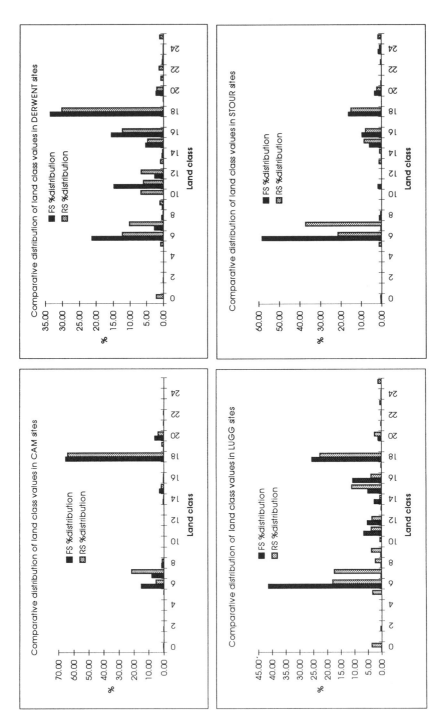

Figure 10.2 Bar charts showing the distribution of class values in both the FS and RS data, for all four rivers

Table 10.2(a)   Actual and cumulative percentages of the top six land cover classes in the FS data. Figures in parentheses are the percentages for the same classes in the RS data

Rank	Class name	Class number	Actual %		Cumulative %	
			FS	(RS)	FS	(RS)
1	Mown/grazed turf	6	38.2	(15.9)	38.2	(15.9)
2	Tilled land	18	30.1	(27.8)	68.2	(43.7)
3	Coniferous wood	16	10.2	(6.9)	78.5	(50.6)
4	Dense shrub moor	11	6.3	(2.6)	84.7	(53.2)
5	Deciduous wood	15	5.1	(7.3)	89.8	(60.5)
6	Suburban	20	2.8	(2.5)	92.6	(63.0)

Table 10.2(b)   Actual and cumulative percentages of the top six land cover classes in the RS data. Figures in parentheses are the percentages for the same classes in the FS data

Rank	Class name	Class number	Actual %		Cumulative %	
			RS	(FS)	RS	(FS)
1	Tilled land	18	27.8	(30.1)	27.8	(30.1)
2	Meadow/semi-natural grass	7	22.7	(2.0)	50.4	(32.1)
3	Mown/grazed turf	6	15.9	(38.2)	66.4	(70.3)
4	Deciduous wood	15	7.3	(5.1)	73.7	(75.4)
5	Coniferous wood	16	6.9	(10.2)	80.6	(85.6)
6	Bracken	12	2.7	(2.1)	83.3	(87.7)

Table 10.3   Direct correspondence between FS and RS class values. Figures in parentheses are the results obtained after the revised comparison

River system	No. of study sites	Total no. cells	No. with direct correspondence	% direct correspondence
Cam	15	19 984	12 986 (14 588)	65.0 (73.0)
Derwent	39	38 188	19 457 (24 068)	51.0 (63.0)
Lugg	39	39 740	16 395 (23 458)	41.3 (59.0)
Stour	38	48 141	18 080 (33 918)	37.6 (70.5)
All combined	131	146 053	66 918 (96 032)	45.8 (65.8)

Table 10.4 Direct correspondence between the RS and FS data in each class for
all catchments combined, expressed as a
percentage of the total number of cells in
that class

Class number	% direct correspondence in **FS** data	% direct correspondence in **RS** data
0	1.2	0.1
1	–	–
2	0.0	0.0
3	–	–
4	–	–
5	–	0.0
6	**33.1**	**79.3**
7	41.3	3.7
8	2.1	1.7
9	24.8	4.0
10		– 0.0
11	**32.1**	**76.9**
12	27.5	21.5
13		– 0.0
14	6.4	9.9
15	39.9	27.9
16	**57.4**	**85.0**
17		– 0.0
18	**70.4**	**76.2**
19	0.0	0.0
20	15.6	17.9
21	0.9	0.0
22	0.0	0.3
23	0.0	0.0
24	23.4	40.5
25	–	0.0

Table 10.5 Minimum, maximum and mean direct correspondence percentages for
the individual headwater site catchments within each river system. Figures in parentheses are the results obtained after the revised comparison

| River system | % direct correspondence | | | |
	Minimum	Maximum	Mean	S.D.
Cam	21.6 (35.2)	95.5 (95.5)	67.0 (73.3)	21.8 (16.4)
Derwent	9.5 (17.5)	91.3 (96.4)	46.7 (60.2)	23.8 (21.9)
Lugg	6.8 (20.8)	83.0 (88.0)	40.5 (57.5)	18.8 (17.0)
Stour	10.3 (11.1)	79.7 (96.7)	34.3 (69.8)	16.8 (23.3)

Classification Differences

Most of the class differences occurred within the eight classes represented in the top ranks of both data sets.

Mown/grazed turf (6) – Meadow (7): There was considerable overlap between classes 6 and 7 with 41.3% of the FS data in class 6 represented as 7 in the RS data. The distinction between the classes is essentially based on the length of the grass which may vary several times within one season depending on cutting/grazing regimes and can vary within fields at any one time depending on grazing patterns. For the revised comparison, some within-field variation was introduced into the compromise grid by making a rule whereby the RS class 7 cells took precedence over FS class 6 cells where they co-occurred.

Open shrub moor (10) – Dense shrub moor (11): The confusion between these classes is explained by the classification systems of the two data sources. The shrub moorland class of the field survey system is subdivided by the LCMGB system into classes 10 and 11. In the translation of the FS data into LCMGB classes all shrub moorland parcels were put into class 11, resulting in only 32.1% of these directly corresponding with the RS data, but an additional 26.5% were represented as class 10 in the RS data. To include this distinction in the compromise grid a rule was made in the revised comparison which accepted the RS class 10 cells in place of FS class 11 where they co-occurred.

Deciduous woodland (15) – Coniferous woodland (16): Overlap in the woodland classes occurred particularly where FS class 16 was represented as 15 in the RS data. This confusion was interpreted as a result of the satellite detecting patches of deciduous or mixed species in a coniferous stand which were not detected or mapped by the field surveyors. This variation was included in the compromise grid by a rule which took the RS class 15 value in preference to the FS class 16.

Tilled land (18) – Managed grass (6 and 7): The time delay of 3–4 years between the two surveys clearly gave rise to the confusion between arable and managed grass cover types as a result of agricultural rotation taking place in the intervening time. For all the catchments combined, 17.6% of the FS cells classed as 18 occurred as 6 or 7 in the RS data, which, when added to the proportion that directly correspond (70.4% – Table 4), accounts for a total of 87.9% of the distribution of the FS cells in class 6. Similarly, for class 18 in the RS data 13.0% were mapped as 6 in the FS data accounting for 92.3% of the cells when added to those directly corresponding. The more recent field surveyed classification was kept in the revised comparison.

Felled forest (23) – Coniferous woodland (16): Another instance of the time delay between the surveys causing confusion was apparent where cells in class 16 in the RS data were mapped as 23 in the FS data. Not surprisingly, none of the FS cells mapped as 23 were classed the same in the RS data but 41.8% were classed as 16. Again the more recent classification was kept in the compromise grid.

Unclassified data (0): A small proportion of the FS data had to be assigned to class 0 where the field surveyors had omitted to label a particular parcel. In the revised comparison procedure, the RS value for any unclassified cells was assigned to the compromise grid to supplement the data.

Remaining classes: Most of the remaining non-correspondence was spread among several classes making it more difficult to account for on a systematic basis and so no further rules were formulated for the revised comparison which favoured the RS data over the FS data. One obvious mis-classification in the RS data was apparent where arable cells in the FS data were classed as suburban, urban or bare ground, but this accounted for only 4.9% of the FS cells in that class. Other more general causes of difference between the two surveys are discussed below.

Revised Comparison

The revised comparison produced a new map of the land cover/use for each catchment as a result of applying the rules described above. The compromise grid (Figure 10.3e) contained mostly the same class values as the original FS grid for that catchment except where a rule changed it to that of the RS grid. The revised residuals grid showed the improved correspondence between the FS and RS data when these classes had been taken into account. The increase in correspondence for each catchment depended on the proportion of the grid in the affected classes; figures ranged from 0% to a maximum increase in direct correspondence of 78.4% of the cells at a single catchment.

The bracketed figures in Table 10.3 show the increase in correspondence achieved on a per-river basis and for all cells combined. Catchments on the Stour showed the largest increase in correspondence (from 37.6% to 70.5%) because of the high proportion of land cover in the agricultural grass categories. The lower improved correspondence figures for the Derwent and Lugg reflect the higher proportion of cells in a more varied range of classes in their study catchments.

Comparison with Linear Features

In most catchments there were a few cells in the RS data which coincided with the digitised linear features and matched the conditions defining a boundary cell and so were included in the revised compromise grid (Figure 10.3f). However, these boundary feature cells rarely formed a contiguous line but more often occurred as single cells or occasionally in short runs. Although they did introduce some variability to the FS data at the field margins and provided explanation for another small proportion of the non-corresponding RS cells, their usefulness is limited because of the patchy nature of the results.

DISCUSSION

Land cover/use mapping by whatever method always results in a simplification of what is on the ground and so no single survey can be treated as 'ground truth'. The results of any survey are heavily dependent on a combination of the underlying purpose for which the data are collected, the choice of classification system and the final style of representation. The two surveys compared here differ on all three points, which can help to explain some of the causes for unidentified non-correspondence.

Added to this is the question of errors in the data which undoubtedly contributed to the problem.

The quality of the LCMGB has previously been assessed against field survey data on a national basis using a sample of the 1 km square field data collected for the 1990 Countryside Survey (Fuller *et al.*, 1994a; Wyatt *et al.*, 1994). Similar methods were used to measure direct correspondence between cells from 143 1 km squares. The direct correspondence obtained for that survey (making no allowance for differences in class definitions, spatial resolution and time differences between the surveys) was 46%, a figure identical to the 46% obtained overall in the current assessment (Table 10.3). This demonstrates that although the field survey for the regional assessment is at least three years later than the first, there is no noticeable reduction in the correspondence between the data sets.

In the national assessment, further comparisons were performed which made allowances for differences in classification definition, boundary cells and time differences to improve the correspondence figures. This led to an overall correspondence of 76% including boundary cells and 82% excluding them (Fuller *et al.*, 1994a). This compares with the revised correspondence measured in the current regional assessment of 66% after making allowances for classification definitions and time differences.

Another regional assessment of the LCMGB in comparison with field survey data has recently been completed for the Tyne catchment in the north of England (Cherrill *et al.*, 1995). Similar conclusions were drawn with regard to differences due to methodological approach and the classification systems. In the Tyne study, the comparison was made with particular emphasis on the mapping of dwarf shrub vegetation and coniferous forests.

The distinction between land cover and land use mapping is apparent when comparing the underlying aims of the LCMGB survey and the field surveys for the Headwater project. The Headwater field survey had both land use and land cover in its aims, with the bias towards land use because of the interest in the management practices and human interference likely to affect the watercourses. From this stance, field parcels would have been mapped by associating the relevant code to the whole parcel on the assumption that the whole field was being used in the same way, although on the ground the cover may have been patchy within the parcel. The LCMGB, on the other hand, is purely a land cover map which means that if the cover types have a different spectral signature within a single field then the classification will reflect this.

The translation of the FS classification into the simpler 25 class system forced the data through an extra stage of interpretation – which was not performed by the same people that carried out the original field survey – and inevitably resulted in loss of information and the introduction of some errors through mis-classification and mis-interpretation. There is no formalised translation from the Countryside Survey classification system to the LCMGB 25 class system and, although many classes are easily translated, some do not satisfactorily fit into any single category: for example, the category scattered trees (for which there was no density information) was translated into pasture/amenity turf (class 6), where field surveyors indicated the presence of livestock although the same area in the LCMGB would often be classed as woodland. This type of interpretation problem was apparent between several classes for which there was no obvious or consistent choice for translation.

The raster-based representation of the LCMGB imposes a minimum mappable unit onto the data and forces the shape of features to conform to this structure. By contrast, the vector-based representation of conventional mapping, assumed by the FS data, is not subject to these constrictions but still suffers from simplification and other limitations imposed by cartographic conventions. Conversion of the digitised FS polygon coverages into raster grids forced the data through both sets of limitations. Some small features were inevitably lost in the process, especially ponds, resulting in the 0% direct correspondence in class 2 (inland water) between the data sets.

Other potential errors when comparing the two data sets are those of spatial registration. In a previous comparison of the LCMGB with field maps of 1 km squares, the average displacement was found to be 0.8 cells (20 m) (Fuller *et al.*, 1994a). In the case of the Headwater field booklets, spatial errors may have crept into the data during any of the following stages of preparation, collection and transfer:

- photocopying and joining of the original map sheets to create the catchment maps;
- photoreducing the maps to create the field booklets;
- estimating the position of land cover boundaries in the field and sketching them;
- creating a single composite map of the field booklet by tracing onto OHP acetate;
- digitising the composite map and transforming to National Grid coordinates;
- subsequent editing of digitised coverages to enable polygon topology to be built.

No spatial correction was applied to either data set during the comparisons, although for some individual catchments there was an obvious geometric misregistration of one or two cells in a particular direction.

CONCLUSIONS

In this chapter the comparison of the LCMGB with a set of field survey data has served to demonstrate the ways in which the two sets of data differ and attempted to produce a compromise land cover map by taking obvious causes of differences into account. Classification differences accounted for much of the non-correspondence observed, in which case reducing the number of classes compared would undoubtedly increase the correspondence measured. That option was not employed here, in order to retain as much detail as possible, but may be considered in future work using these data. Another option for future work is to apply an error model to predict the certainty of a cell belonging to any one class, such as that described by Goodchild (1994).

Neither data set is seen as an end in itself within the context of the overall research project since both sources of data will be used in the subsequent analysis stages, and so this comparison is not meant as a judgement on either.

On the scale of the headwater site catchment areas, the correspondence results presented here demonstrate that the LCMGB is not directly comparable to field surveyed data, particularly at sites dominated by semi-natural vegetation types, and so is not a suitable alternative data source of land cover data for new unmapped headwater catchments. At the regional scale of the whole river system, however, the overall correspondence figures were comparable to the national assessment of the

LCMGB. At this scale, the LCMGB therefore represents an extremely accessible and invaluable source of hitherto almost unobtainable land cover data.

ACKNOWLEDGEMENTS

This study was carried out as part of a PhD research fellowship funded by the Freshwater Biological Association. We would like to thank Dave Morris at the Institute of Hydrology for allowing use and reproduction of the digital river network and the National Rivers Authority for obtaining copyright permissions for the field survey data.

REFERENCES

Barr, C.J., Bunce, R.G.H., Clarke, R.T., Fuller, R.M., Furse, M.T., Gillespie, M.K., Groom, G.B., Hallam, C.J., Hornung, M., Howard, D.C. and Ness, M.J. (1993) *Countryside Survey 1990: Main Report*. Department of the Environment, London.

Cherrill, A.J. (1994) A comparison of three landscape classifications and investigation of the potential for using remotely sensed land cover data for landscape classification. *Journal of Rural Studies*, **10**(3), 275–289.

Cherrill, A.J. and McClean, C. (1995) An investigation of uncertainty in field habitat mapping and the implications for detecting land cover change. *Landscape Ecology*, **10**(1), 5–21.

Cherrill, A.J., McClean, C., Lane, A. and Fuller, R.M. (1995) A comparison of land cover types in an ecological field survey in northern England and a remotely sensed land cover map of Great Britain. *Biological Conservation*, **7**, 313–323.

Coleman, A. (1961) The second land use survey: progress and prospect. *The Geographical Journal*, **127**, 168–186.

Fuller, R.M., Groom, G.B. and Jones, A.R. (1994a) The Land Cover Map of Great Britain: An automated classification of Landsat Thematic Mapper data. *Photogrammetric Engineering and Remote Sensing*, **60**(5), 553–562.

Fuller, R.M., Groom, G.B. and Wallis, S.M. (1994b) The availability of Landsat images of Great Britain. *International Journal of Remote Sensing*, **15**(6), 1357–1362.

Fuller, R.M., Sheail, J. and Barr, C.J. (1994c) The Land of Britain, 1930–1990: a comparative study of field mapping and remote sensing techniques. *The Geographical Journal*, **160**(2), 173–184.

Furse, M.T., Winder, J.M., Symes, K.L., Clarke, R.T., Gunn, R.J.M., Blackburn, J.M. and Fuller, R.M. (1993) *The Faunal Richness of Headwater Streams: Stage 2 – Catchment Studies, Volume 1 Main Report*. National Rivers Authority R&D Note 221, Bristol.

Furse, M.T., Symes, K.L., Winder, J.M., Clarke, R.T., Blackburn, J.M., Gunn, R.J.M., Grieve, N.J. and Hurley, M. (1995) *The Faunal Richness of Headwater Streams: Stage 3 – Impact of Agricultural Activities, Volume 1 Main Report*. National Rivers Authority R&D Note 392, Bristol.

Goodchild, M.F. (1994) Integrating GIS and remote sensing for vegetation analysis and modeling: methodological issues. *Journal of Vegetation Science*, **5**, 615–626.

Stamp, L.D. (1947) *The Land of Britain: the Final Report of the Land Utilisation of Britain*. Geographical Publications, London.

Wyatt, B.K., Greatorex-Davies, N.G., Bunce, R.G.H., Fuller, R.M. and Hill, M.O. (1994) *Comparison of Land Cover Definitions. Countryside 1990 Series: Volume 3*. Department of the Environment, London.

11 Mapping Humid Tropical Vegetation in Eastern Bolivia

J. WELLENS,[1] A.C. MILLINGTON,[1] W. HICKIN,[1] R. ARQUEPINO[2] AND S. JONES[1]

[1] *Leicester Environmental Remote Sensing Unit (LERSU), Department of Geography, University of Leicester, UK*
[2] *Centro Universitario Estudios Medio Ambiente y Desarollo (CUEMAD), Universidad Mayor de San Simon, Casilla, Bolivia*

This chapter describes how Principal Component Analysis (PCA) and unsupervised classification techniques were applied to Landsat Thematic Mapper satellite data to produce a vegetation map for the Reserva de la Biósfera de Beni 'Estación Biológica del Beni' in eastern Bolivia. Due to the strong spectral differences between savannas and forests, separate image processing pathways were developed for each of these broad vegetation types. However, in both areas the level of inundation is a critical factor in determining the types of vegetation present. In the savanna, PCA alone was used to analyse and map the hydromorphically controlled ecotone between hydromorphic and xeromorphic extremes. For the forest areas, PCA was used in combination with unsupervised classification to discriminate seven different forest types. Good levels of discrimination were achieved between forests that are subject to different levels of inundation, but it was not possible to subdivide the non-inundated *terra firme* forests. The separate maps that were produced for forest and savanna areas were merged to create a provisional map of vegetation units in the reserve. This was verified using field observations, video data acquired during an over-flight of the reserve, and also by discussion with a group of vegetation scientists familiar with the reserve. The verification procedures led to minor modifications to the vegetation categories mapped.

INTRODUCTION

The last 15 years have seen a growth in recognition of the importance of tropical forests in the global climate system, in biogeochemical cycling and in maintaining biological diversity. This has led to more protected areas (e.g. national parks, non-exploitative reserves, biosphere reserves, protected headwaters) being established. However, baseline surveys for protected areas are often non-existent, thereby raising questions about the suitability of their geographical locations and the ability to recognise the impact of human-induced and natural environmental change in protected areas after they have been designated. Such data are lacking for most protected areas in Bolivia at the present time despite government investments in the conservation sector during the 1990s (Baudoin, 1995).

This chapter describes a project in which a combination of Landsat Thematic Mapper image data, field observations and hand-held aerial video footage was used to

Vegetation Mapping: From Patch to Planet. Edited by Roy Alexander and Andrew C. Millington.
© 2000 John Wiley & Sons Ltd.

provide a baseline vegetation map for the Reserva de la Biósfera 'Estación Biológica del Beni' (EBB) in Bolivia (Salinas, 1995). The overall aim of the research is to combine this map with information about animal distributions to create habitat maps and models. The resulting vegetation and habitat maps will be used to monitor changes in plant communities and animal habitats in the future.

THE RESERVA DE LA BIÓSFERA 'ESTACIÓN BIOLÓGICA DEL BENI'

The 135 000 ha Reserva de la Biósfera 'Estación Biológica del Beni' is located in the Department of Beni in the eastern lowlands of Bolivia (Figure 11.1). The area comprises low-lying, alluvial plains at an altitude of about 220 m a.s.l., the relative relief is about 5 m. The area is crossed by rivers that debouch from the Andes. The tectonically controlled migration of these rivers, which has mainly occurred since the late Tertiary and continues at present (Hanagarth, 1993), is marked by a network of abandoned channels and other fluvial features within the reserve which, in turn, have affected the spatial distribution of plant communities. The establishment of the reserve in 1982 resulted from a 'debt-for-nature swap' (Walsh, 1987). It has since been accorded Category 1: Scientific Reserve Status by IUCN and has also been designated a Man and the Biosphere Reserve by UNESCO. As a consequence of the relatively long history of international links, the reserve has a better infrastructure and management plan than any other protected area in Bolivia.

Approximately 90 000 ha of the reserve comprises lowland tropical *terra firme* and inundated forests. Edaphic, humid grassland savannas (known locally as *pampa*) surround the forest block. A significant proportion of the forest and *pampa* are seasonally flooded, and in the *pampa* grassland types range from short, dry grasslands to wetlands with tall, swamp grasslands. In a landscape ecology context, the forest block is connected to a large forest area to the south, the Bosque Chimanes, by four forest corridors (all of which are stream corridors) and numerous forest islands (which act as patches) within the *pampa*. The Bosque Chimanes is more-or-less contiguous with lowland tropical forests further south and the Andean lowland montane and foothills forests to the west (Figures 11.1 and 11.2). Although the reserve is in the humid tropics and receives > 2200 mm of rainfall annually, there is a distinct dry season between May and October. Within the reserve over 100 species of mammals and 400 species of birds have been identified, 19 of which feature in the IUCN Red Data lists. These include black-spider monkey (*Ateles paniscus*), marsh deer (*Odocoileus dichotomus*), jaguar (*Felis onca*) and black alligator (*Melanosuchus niger*). The reserve and its surrounding area is the homeland of an indigenous Amerindian tribe – the Chimane – who comprise about 800 people scattered throughout 21 settlements. Most of these settlements are situated along the Río Maniqui, to the north of the reserve, and thus are mainly within the reserve's sustained multiple use zone. Recently, 30 000 ha of the reserve have been designated indigenous Chimane territory with associated rights over the natural resources in the area.

The international status of the reserve has resulted in a great number of ecological and zoological studies since its inception (e.g. Cabot *et al.*, 1986; Garcia and Tarifa,

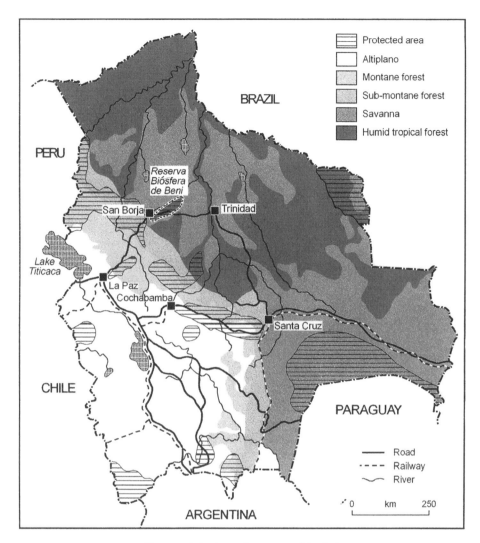

Figure 11.1   Location map of Bolivia

1987; Rocha, 1990; Dallmeier *et al.*, 1992). This research has focused on the more accessible periphery of the reserve. The central parts of the reserve are uninhabited and comprise forests and wetlands that are difficult to access. This central area forms the reserve's core zone and has been set aside for strict protection. The surrounding sustained multiple use and buffer zones mainly comprise a matrix of *pampa* with islands and corridors of forest.

The *pampa* has traditionally been used for cattle ranching, and the use of fire as a grazing management tool is commonplace. These fires threaten the spatial distribution of the forest islands and corridors. The forest corridors have also come under threat from deforestation processes since the construction of a highway between San Borja

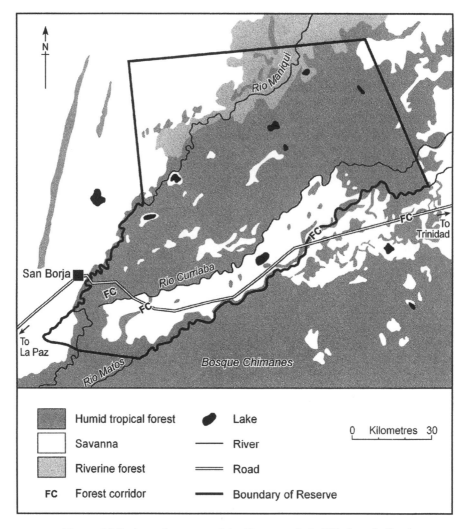

Figure 11.2   Location map of the Reserva de la Biósfera de Beni

and Trinidad (Figure 11.2). Deforestation has resulted from both land clearance for cultivation (mainly by in-migrants from the Altiplano), and as a result of illegal logging by small-scale operators removing trees such as mahogany (*Swietenia macrophyla*). There are concerns about the impact that deforestation may have on further fragmentation of this landscape of forest corridors and islands within the *pampa* and how this will affect the movement of animals, especially those with arboreal habitats, between the reserve and the Bosque Chimanes.

Prior to this project, the only vegetation map of the reserve had been produced at a scale of 1:266 000. This was based on a manual interpretation of a photographic copy of a false-colour composite Landsat MSS image (Miranda, 1991). This image had not been geometrically corrected and the map lacked critical cartographic information. The only copies available to the reserve's staff appeared to be dyeline prints that were cartographically distorted. The aim of the research reported here was to produce a 1:100 000 scale, geometrically correct map of the vegetation units in the reserve with a brief ecological commentary on each unit.

SATELLITE DATA ACQUISITION AND PROCESSING

Satellite data have shown great potential for mapping the current extent of the tropical forest biome in the neotropics (Hill and Foody, 1994; Moran *et al.*, 1994; Paradella *et al.*, 1994; Brondizio *et al.*, 1996), for monitoring changes in forest extent (e.g. Nelson and Holben, 1986; Malingreau *et al.*, 1989), for assessing forest productivity, and for monitoring topical forest disturbance and successional regimes (Mausel *et al.*, 1993). The use of satellite data offers a major advantage over field-based mapping techniques in protected tropical forest areas, such as EBB, which often have poor accessibility and are relatively isolated.

Map production was based on the most recent Landsat TM scene that was available when the project started. This image was acquired in July 1989 during the dry season. Unfortunately, approximately 10% of the north-east of the reserve fell outside the image and was not mapped. However, the image did cover the majority of the lowland forests and *pampa*, three of the four forest corridors and the northern part of the Bosque Chimanes. The lack of a cloud-free, wet season image meant that a multitemporal mapping approach using several TM scenes was not possible. This is unfortunate as multitemporal approaches to land cover map production have generally proved more accurate than single date approaches (Townshend and Justice, 1986; Millington *et al.*, 1992; Stone *et al.*, 1994).

Although the image was six years old when the project began, it was felt that this would not be a problem for mapping the forests in the core zone. Their protected status, combined with access difficulties, means that very little anthropogenic disturbance would be likely to have occurred in this area. Nonetheless, there are two areas where changes have occurred in the forests. These are:

1. The largest of the forest corridors, to the east of San Borja, which has experienced significant clearance for cultivation due to its location outside the initial boundary of the reserve and close to the San Borja–Trinidad highway (it is now within the reserve).
2. Along the Río Maniqui, which forms the northern boundary of the reserve, where there was a substantial shift in the course of the river in 1995.

In contrast to the relatively stable forest areas, the *pampa* is very dynamic. Changes in the spatial extent of vegetation units can occur over a few hours due to burning or grazing in addition to the seasonal hydrological changes in the region. Thus the Landsat TM image only provided information about the status of the *pampa* at the time of the satellite overpass on 23 July 1989.

In addition to the satellite data, we recorded a video and accompanying commentary on the vegetation during an over-flight of the reserve in a light aircraft. A standard hand-held camcorder was used to video the flight through one of the aircraft's windows. The flight was at an altitude of approximately 500 m above the ground and lasted for two hours. It followed a predetermined flight line of approximately 300 km that was selected with the aid of the existing map and the park guards' *a priori* information. This was done in order to include as many different plant communities as possible. As there was no on-board GPS, comprehensive notes were made about the route using the aircraft navigation instrumentation. These notes, in combination with prominent landscape features and the spatial patterns of the savanna–forest boundary, enabled the over-flight route to be drawn onto a hard-copy of the TM image. This was digitised and used as a vector overlay during image processing. Based on this overlay, and the field of view of the camcorder at the flying altitude, we estimate that the video coverage is approximately 30% of the total area of the reserve.

GEOMETRIC CORRECTION

It was essential both for map production and for future monitoring of the reserve that the image was geometrically corrected. A set of ground control points (GCPs) was selected from readily identified features on the imagery (e.g. road junctions, airstrips, ranch buildings) and marked up on hard-copy imagery. Due to the nature of the landscape in this region, the GCPs were restricted to the west and south of the image where roads and ranches are concentrated. These locations were visited in September 1995 and latitude and longitude coordinates established using a Garmin GPS in non-differential mode. Twelve of these GCP locations were used for geo-rectification. These gave a RMS error of 0.76 pixels, equating to an error of about 23 m in the south and west of the image. Due to the sparsity of GCPs in the east and north of the image, geo-rectification errors in these areas will be > 23 m.

IMAGE PROCESSING STRATEGY

As the main objective was to produce a map of forest and *pampa* vegetation units, an image processing strategy was developed that was based on separate image processing pathways to produce maps of forest units and maps of savanna units (Figure 11.3). These maps were later merged, along with maps of water, cloud and cloud shadow, and burn scars to produce the final vegetation map of the reserve and its surrounding region. This strategy was adopted because initial inspections of the imagery high-lighted a very large spectral contrast between the forest and savanna communities, and also between water (both lakes and rivers), cloud and cloud shadows.

The initial step was to separate the areas of forest and *pampa*. This was done by examining various false-colour composite images until the band combination with the best visual contrast between forest and *pampa* was found (TM bands 2, 4 and 5). A density slice of this band combination, created using a red–green–blue clustering algorithm, was used to produce masks of the *pampa* and the forest regions. In addition, masks of water, clouds and cloud shadows were produced. It was found that these features could be readily identified using Principal Component (PC) 3 of a Principal

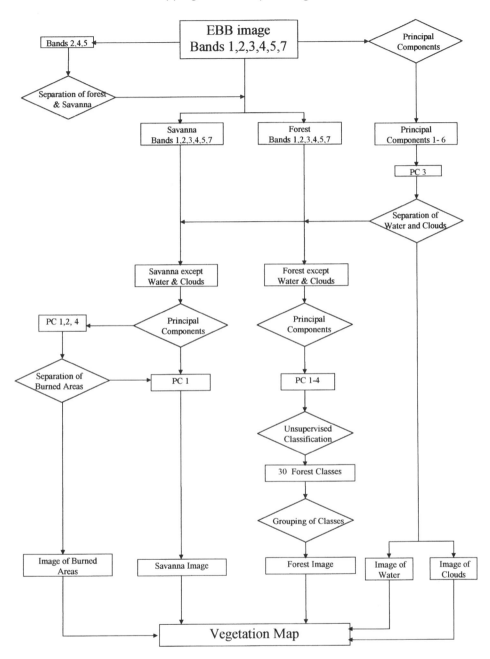

Figure 11.3 The image processing pathway

Component Analysis (PCA) carried out on the six reflective bands of TM. Again masks were created using density slicing. This ensured that the pixels representing these features would not influence the image analysis procedures used to map vegetation units. All of the cloud features fall outside the reserve and therefore we decided not to interpret the vegetation obscured by the cloud shadows. The result of these initial processing steps was to create the following four image masks.

- A *pampa* mask. This masks out *pampa* vegetation units and shows only forested areas plus any water, clouds and cloud shadows in the forest area.
- A forest mask. This masks out forest vegetation units and shows only areas of *pampa* plus any water, clouds and cloud shadows in the *pampa*.
- A water mask. This masks out lakes and rivers over the entire image.
- A cloud and cloud shadow mask. This masks out clouds and cloud shadows over the entire image.

The remainder of the image processing focused on two images. First, the *pampa* image – a six-band image which contained only *pampa* landscapes and had forest, water and cloud and cloud shadows masked out. Secondly, the forest image – a six-band image that contained only forest landscapes and had *pampa*, water and clouds and cloud shadows masked out.

MAPPING *PAMPA* VEGETATION UNITS

The *pampa* formations in EBB range from dry, short-grass savanna, often containing an open canopy of fire-tolerant woody species and termitaria (a landscape known locally as *sartenejal*), through to hydromorphic tall grasslands in topographic lows. The latter communities are often inundated for more than eight months and, in some cases, remain flooded throughout the year. The extreme inundation experienced in these hydromorphic grasslands gives them the appearance of a transition between swamp and savanna, and they are often referred to locally as *pampa–yomomo*. Sartenejal and *pampa–yomomo* represent the two extremes of a hydromorphic gradient in which topography influences the depth and length of flooding.

   PCA was applied to the six-band *pampa* image. The explanation of variance in the *pampa* image follows the typical pattern of PCA (Table 11.1) with most variance explained by PC1 (75%). The loadings on the eigenvectors indicate that most of the information in PCA 1 comes from TM bands 4, 5 and 7 (Table 11.2) indicating that it is an integrated albedo image. An analysis of the spatial variations in albedo, in the context of fieldwork undertaken during the 1995 dry season, indicated that the main control on albedo is the moisture content of the soil and vegetation. Areas with the highest albedo are found in the *sartenejal* landscape, whilst the *pampa–yomomo* has a much lower albedo. Thus the PC1 image appears to provide a good differentiation between areas subject to different moisture levels and, therefore, different lengths of inundation. However, it was not possible to calibrate the data values in the PC1 image to actual moisture levels since the highly dynamic nature of the *pampa* means it is impossible formally to link the image data acquired in 1989 with field data acquired at a later date.

Table 11.1 Eigenvalues produced from PCA of the six-band savanna image

Principal component	Eigenvalue
1	240.6
2	66.9
3	8.9
4	4.0
5	1.3
6	0.6

Table 11.2 Loadings on the eigenvectors for the six-band savanna image

Band	Principal component					
	1	2	3	4	5	6
1	0.13	−0.10	−0.51	−0.26	−0.80	0.09
2	0.10	0.02	−0.31	−0.22	0.18	−0.90
3	0.16	−0.12	−0.47	−0.52	0.55	0.41
4	0.34	0.91	−0.17	0.13	0.01	0.09
5	0.84	−0.19	0.46	−0.21	−0.07	−0.03
7	0.36	−0.33	−0.43	0.74	0.14	0.03

It has already been acknowledged that the subtle topographic changes in the *pampa*, which are a function of river channel migration and the resulting sedimentation patterns (Hanagarth, 1993), lead to gradual change in *pampa* communities along a hydromorphically controlled ecotone. Rather than attempt to divide the *pampa* vegetation using arbitrary cut-off points in the PC1 image, it was decided to acknowledge the continuum in the *pampa* vegetation by showing the gradation from the wettest areas to the driest areas in 220 levels defined by PC1 values. However, there was a complication in basing the hydromorphic gradient solely on PC1. The areas with low PC1 values indicate fire scars as well as very wet areas. This is immediately obvious from the spatial distribution of pixels with low PC1 values, as those related to fire scars are restricted to the dry *pampa* communities. Burning is carried out every dry season to encourage the growth of new shoots and, consequently, the burn scars from 1989 were found on the imagery but not during fieldwork in 1995. It was decided to process burnt areas separately and identify such areas as fire scars on the final map. A FCC of PC1, PC2 and PC4 provided the best differentiation between wet *pampa* communities and fire scars, and a fire scar mask was created from this FCC. This mask was applied to the PC1 image in order to produce the maximum contrast in the non-burnt *pampa*.

MAPPING FOREST VEGETATION UNITS

The use of standard per-pixel image classification techniques has not proven an accurate method for mapping tropical forest communities because of the strong spatial heterogeneity in the reflectance values from the canopy (Tuomisto *et al.*, 1994, 1995; Hill, 1996). The irregular nature of the forest structure means that the surface sensed by the satellite sensor may vary from tall emergent trees to an open canopy gap within the space of a few metres. This implies that the signal received by the satellite sensor is a function of a series of complex interactions between solar radiation and these surfaces. Thus, the resulting image has a number of elements that can be summarised as illuminated and shaded canopy, and illuminated and shaded ground (usually soil or water). As a consequence neighbouring pixels within the image frequently have very different reflectance values due to the wide variability in the proportions of the four main components of each pixel. This provides a speckling effect that hinders effective classification. To help overcome these problems, PCA was applied to the forest image prior to unsupervised classification. The explanation of variance in the PCAs from the forest image is provided in Table 11.3 and again follows the typical pattern of PCA: PC1 only explains 58% of the variance. The loadings on the eigenvectors are given in Table 11.4. They indicate that most of the information in PC1 and PC2 (which together explain 94% of the total variation in the original imagery) are from TM bands 4 and 5. These bands lie in the red and near-infrared regions of the electromagnetic spectrum, which are well known for their ability to detect differences in the spectral characteristics of vegetation. Examination of the PC1 and PC2 images showed that they were able to distinguish between forest canopies in the reserve on the basis of their reflectance characteristics. The PC3 and PC4 images, although only accounting for 3.9% of the overall variance, contained important information about the spatial distribution of the different forest units. In summary, two major spatial controls on forest distribution are apparent on these four PC images:

- Large differences in the reflectance of forested areas along the Río Maniqui in the north of the reserve and all other forest types to the south of the Río Maniqui.
- Broad sinuous, channel-like tracts of forest within the main forest block.

The spectral characteristics of the floodplain forests found along the current course of the Río Maniqui are very different to those of the other forest types in the reserve. This can be explained by the fact that the gallery forests along the Río Maniqui have a distinct structure and different species than the *terra firme* forests to the south. Hanagarth (1993) argues that the riparian vegetation in this region is typical of other floodplain forests adjacent to the Andes (Salo *et al.*, 1991; Seidenschwarz, 1986). Four main types of vegetation are found: *Cecropia*-dominated woodlands, *Gynerium sagittatum* mattoral, short-stature woodlands dominated by *Salix humboldtiana* and *Tessaria integrifolia*, and communities of grasses, herbs and perennials. A strong control on this vegetation is the flooding regime of the river (Hanagarth, 1993). However, part of their spectral distinctiveness is due to the fact that the Río Maniqui is a whitewater river containing significant amounts of sediment that has its origins in the Andes. When these forests are flooded during the wet season, light-coloured sediments are deposited

Table 11.3 Eigenvalues produced from PCA of the six-band forest image

Principal component	Eigenvalue
1	51.7
2	32.8
3	2.7
4	1.2
5	0.8
6	0.4

Table 11.4 Loadings on the eigenvectors for the six-band forest image

Band	Principal component					
	1	2	3	4	5	6
1	0.09	−0.17	−0.69	0.44	−0.54	−0.01
2	0.10	−0.13	−0.35	−0.02	0.47	0.80
3	0.08	−0.22	−0.45	−0.08	0.61	−0.60
4	0.84	0.54	−0.06	−0.06	0.00	−0.04
5	0.50	−0.70	0.41	0.30	0.02	0.00
7	0.15	−0.35	−0.16	−0.84	−0.35	0.03

on the vegetation as the annual flood retreats. Subsequently, in the dry season, a high proportion of the reflectance from many pixels will be from recently deposited alluvium on the forest floor and on the lower branches and trunks of trees. This comprises light-coloured sands with high reflectance in visible and near-infrared (VISNIR) channels of TM. As the rivers migrate and former floodplain forests become non-flooded (or flooded to a lesser extent), less sediment is deposited each year and organic matter will accumulate because of the low-lying, waterlogged conditions. This organic matter accumulation will suppress the soil reflectance component of these forests (Galvão and Vitorello, 1998), thereby providing strong contrasts between the overall reflectance of forests which are in the main area of sediment accumulation, the contemporary channel, and those in areas with a lower sedimentation regime.

The sinuous forest communities correspond spatially to river courses in the Bosque Chimanes and in the southern part of the reserve. In most cases they are palaeochannels of the nine phases of the Río Maniqui identified by Hanagarth (1993). The forest areas associated with these rivers are generally much wider than the floodplains that would be expected from the size of the rivers. This is in line with Miranda's (1991) estimate of 50% of the reserve's forests being inundated for up to four months each year. It is argued that they represent forests on the seasonally inundated floodplains of palaeo-rivers that had greater discharges than those at the present time. Their distinctive spectral response is again due to the interactions between canopy structure and electromagnetic radiation. However, in this case it is not freshly deposited sediment that causes the strong contrast in reflectance, as these are sediment-poor blackwater rivers unlike the sediment-rich Río Maniqui. The main difference between these forests and the *terra firme* forests adjacent to them is that the trees along these seasonally inundated areas are not under moisture stress in the dry season, unlike many trees in the *terra firme* forests. Consequently middle infrared (MIR) absorption will be high. Evidence for this is provided by the high eigenvector loadings of PC1 to PC4 in TM bands 5 and 7 (Table 11.4). The canopy in these forests is also more fragmented than in the *terra firme* forests and a greater proportion of the reflectance from individual pixels emanates from standing water or moist soil than in the *terra firme* forest. In addition, because of the fragmented nature of the canopy, shaded areas will account for a greater proportion of pixels than in *terra firme* forests and the overall reflectance in TM bands 1–3 will be lower.

The differentiation between seasonally inundated forests and those experiencing very little or no inundation (*terra firme* forests) is relatively clear using PCA. However, within these two broad forest types, ecotonal variations due to hydromorphic gradients are evident within inundated forests, within *terra firme* forests, and between inundation and *terra firme* forests. It was impossible to apply the same image processing procedure to map the spatial distribution of vegetation units across ecotonal gradients in the forested area as was used for the *pampa*. In the *pampa* there was only one gradient – the hydromorphic gradient – to consider, whereas in the forests there are a number of environmental gradients. In the inundated forests these gradients are influenced by factors such as soil moisture, the presence or absence of standing water and substrate properties (especially sediment colour), while in the *terra firme* forests the vertical structure and, in particular the number of canopy strata, appear to be the major factors. Consequently, the first four PCs were submitted to an unsupervised classification using the Iterative Self-Organising Data Analysis Technique (ISODATA) algorithm (Tou and Gonzalez, 1974) which separated the forest area into 30 classes. This is considerably more categories than exists in the reserve, but the aim at this stage in the image processing chain was to produce an image which summarised the PCA and which could be subsequently interpreted. Each of these classes was interpreted separately in terms of:

- its spatial distribution compared to the other 29 classes;
- the minimum, maximum, mean and standard deviation of the six TM bands within each of the classes; and
- the video footage from two transects; (i) along the forest–savanna boundary to the south of the reserve, and (ii) between Laguna Piraquinas and Estancia Venezia to capture the variation across the Río Maniqui floodplain.

On the basis of these interpretations the 30 classes were merged into eight forest classes (later revised to seven classes). Brief details of these classes are provided in Table 11.5.

MAP PRODUCTION AND VERIFICATION PROCEDURES

The *pampa* and forest image maps were merged. Subsequently, the areas of fire scars, water and cloud features, obtained from the relevant masks, were combined with the map of forest and *pampa* vegetation to produce a provisional map of vegetation units

Table 11.5 Vegetation units on final map

Vegetation classification	Description
Savanna Continuum	
Sabana higromorfa de inundación permanente	Permanently flooded hydromorphic savannas
Sabana higromorfa de inundación estacional	Seasonally flooded hydromorphic savannas
Pastizal xeromorfo de terraza alta (*Sartenejal*)	Dry savanna on raised ground with a rough micro-relief created by termite mounds. Contains open woody fire-resistant trees
Forest Classes	
Bosque pantanoso	Poorly drained, marshy forests that experience inundation for more than 8 months each year. The forest is dense and low (10–15 m) and has a high abundance of lianas and *Heliconia* sp.
Bosque alto denso de terraza alta	*Terra firme* forests comprising 3–4 strata with an average upper canopy height of 30–40 m
Bosque mediano denso ribereño de inundación estacional	Medium density, seasonally flooded riverine forest – with an average canopy height of 30–40 m but with low species diversity (dominated by Palmaceae) as a result of the impeded drainage
Bosque bajo denso ribereño	Riverine forest generally below 15 m with a high density of Palmaceae and lianas
Yomomo	Marshy depressions dominated by a floating mat of Cyperacae that reaches a height of 1.5 m above the water
Bosque bajo medio denso	Medium density low forest generally below 15 m found on recently deposited terraces of the Río Maniqui
Bosque alto denso ribereño de inundación estacional	High density, seasonally flooded, riverine forest – with an average canopy height of 30–40 m and low species diversity. Palms are abundant and the forest floor is rich in forbs. The forest contains a greater cover of emergents than the medium dense version

in the reserve and its surrounding area. Given the issues of accessibility over much of the reserve, three strategies were adopted to verify the map. First, it was analysed and discussed by a group of scientists with extensive experience of the Bolivian vegetation (and particularly of EBB) during a workshop held in March 1996. Secondly, field checking which focused on the interpretations of the forest and *pampa* units was carried out in the buffer and sustainable multiple use zones during May 1996. Finally, qualitative comparisons were made with the video data acquired during September 1995. The main changes to the provisional map that arose from these checking procedures were:

- The recognition of a new class – *bosque pantanoso*. This occurs around wetlands in the forested area, and had been previously mapped as wet savanna.

- A minor transitional class between *terra firme* forest and inundated forest was merged into the *terra firme* category.

The final map incorporating these changes is shown in Plate 6 (Figure 11.4) and brief details of the vegetation classes are provided in Table 11.5.

CONCLUSIONS

This work has demonstrated that standard image processing techniques can be successfully applied to single-date VISNIR image data to produce a reasonable level of discrimination between both forest and savanna communities in the humid tropics. The image processing techniques employed, particularly the PCA, appear to discriminate successfully between the major vegetation communities found within EBB and are especially useful for discriminating between the different savanna formations. PCA used in combination with unsupervised classification techniques was also able to differentiate between the different forest types found in the reserve. However, while these techniques appear to differentiate successfully the *terra firme* forest from the inundated forest classes, it was not possible to subdivide the *terra firme* forest category. *Terra firme* forests account for approximately 45% of the forests in EBB and occupy the oldest river terraces on the better-drained, consolidated alluvium to the south of the reserve. Their floristic composition is highly complex with over 100 species of tree having been identified. In support of our inability to distinguish subdivisions of *terra firme* forest on the basis of canopy reflectance (which equates to forest structure), Dallmeier *et al.* (1992) consider that it is impossible to differentiate them floristically in terms of a characteristic group of species. Attempts have been made to identify 'guilds' of typical large trees, and initial studies suggest that there are at least 13 different 'guilds' of *terra firme* forest, all of which are characterised by large trees and emergents (Dallmeier *et al.*, 1992). The fact that the PCA and unsupervised classification techniques were unable to differentiate between the different 'guilds' of *terra firme* forest is therefore understandable, since it is floristic rather than the structural characteristics that define them. The vegetation classification presented in Table 11.5 was considered to be a good representation of the distribution of vegetation in EBB by the group of experts who reviewed the vegetation map in 1996. However, this impression is entirely qualitative. Furthermore, field verification within the buffer and sustainable multiple use zones of the reserve also indicated good qualitative correspondence between the map and field observations. Given the problems of access to the core zone of the reserve, it is difficult to establish any qualitative assessment of the overall accuracy of the vegetation map. To a certain extent our findings are in contrast to those of Tuomisito *et al.* (1995) and Hill (1996) in lowland Peru, who suggest standard image processing methods cannot be applied successfully to humid tropical forests. It is possible that our findings mean that it is possible to map humid tropical forests successfully in some environments but not others. The area we have investigated in Beni has a pronounced dry season and a well-developed floodplain topography, whereas the area that Tuomisito *et al.* (1995) and Hill (1996) studied in Peru is significantly more humid than this environment and vegetation–soil–topography relationships may be less clear in such environments.

A further point in terms of practical application is that the map produced from this exercise represents a significant improvement on the previous vegetation map produced for EBB in two ways:

- it is geometrically rectified and can therefore be imported into a GIS, and
- it can be updated by the digital analysis of further satellite data in the future and change metrics can easily be calculated.

Finally, in addition to the image processing techniques employed, we have demonstrated the utility of simple hand-held video footage acquired from light aircraft as a source of data for image interpretation and image verification. This is clearly a cost-effective method of data acquisition for vegetation mappers working in difficult or poorly accessible terrain.

ACKNOWLEDGEMENTS

We gratefully acknowledge the funding for this research from the Darwin Initiative for the Survival of the Species (grant 162/4/154). The imagery was purchased under NERC grant GST/02/706. We would like to thank France Gerard (Institute of Terrestrial Ecology, Monks Wood) for collecting some of the ground control data. Our thanks also go to Carmen Miranda and her colleagues in La Paz and El Porvenir who have made our research both possible and stimulating. The feedback from colleagues during a workshop held under the auspices of Darwin Project 162/4/154 in Cochamamba in 1996 has led to the improvement of the original version of the map and we thank them for sharing their insights with us.

REFERENCES

Baudoin, M. (1995) Conservation of biological diversity in Bolivia. In: S. Amend, and T. Amend (eds) *National Parks without People? The South American Experience.* IUCN, Gland and Quito, pp. 95–102.

Brondizio, E., Moran, E., Mausel, P. and Wu, Y. (1996) Land cover in the Amazon Estuary: linking of the Thematic Mapper with botanical and historical data. *Photogrammetric Engineering and Remote Sensing,* **62**, 921–929.

Cabot, J., Serano, P., Ibanez, C. and Braza, F. (1986) Lista preliminar de aves y mamiferos de la reserva Estacion Biologica del Beni. *Ecologia en Bolivia,* **8**, 37–44.

Dallmeier, F., Foster, R., Romano, C., Rice, R. and Kabel, M. (1992) *Gui para el usuario de las parcelas experimentales de biodiversidad.* Programa de Diversidad Biological del Instituto Smithsonian/Programa del Hombre y la Biosfera, Washington, DC.

Galvão, L.S. and Vitorello, I. (1998) Role of organic matter in obliterating the effects of iron on spectral reflectance and colour of Brazilian tropical soils. *International Journal of Remote Sensing,* **19**, 1969–1980.

Garcia, J.E. and Tarifa, T. (1987) Primate survey of the Estacion Biologica del Beni, Bolivia. *Primate Conservation,* **9**, 97–100.

Hanagarth, W. (1993) *Acerca de la Geoecológia de las Sabanas del Beni en el Noreste de Bolivia.* Instituto del Ecología, La Paz, Bolivia.

Hill, R.A. (1996) Identifying and mapping tropical vegetation and its phenology using satellite imagery: case studies from Peruvian and Brazilian Amazonia. Unpublished PhD thesis, University of Wales, Swansea.

Hill, R.A. and Foody, G.M. (1994) Separability of tropical rain-forest types in the Tambopata–Candamo Reserved Zone, Peru. *International Journal of Remote Sensing*, **15**, 2687–2693.

Malingreau, J.P., Tucker, C.J. and Laporte, N. (1989) AVHRR for monitoring global tropical deforestation. *International Journal of Remote Sensing*, **11**, 855–867.

Mausel, P., Wu, Y., Li, Y., Moran, E.F. and Brondizio, E.S. (1993) Spectral identification of successional stages following deforestation in the Amazon. *Geocarto International*, **4**, 61–71.

Millington, A.C., Styles, P.J. and Critchley, R.W. (1992) Mapping forests and savannas in Sub-Saharan Africa from Advanced Very High Resolution Radiometer imagery. In: P.A. Furley, J. Proctor and J.A. Ratter (eds) *Nature and Dynamics of Forest–Savanna Boundaries.* Chapman & Hall, London.

Miranda, C. (ed.) (1991) *Plan de Manejo de Reserva de la Biósfera Estación Biológica del Beni.* Artes Gráficas, La Paz, Bolivia.

Moran, E.F., Brondizio, E., Mausel, P. and Wu, Y. (1994) Integrating Amazonian vegetation, land-use and satellite data. *BioScience*, **44**, 329–338.

Nelson, R. and Holben, B. (1986) Identifying deforestation in Brazil using multi-resolution satellite data. *International Journal of Remote Sensing*, **7**, 429–448.

Paradella, W.R., Da Silva, M.F., De, A., Rosa, N. and Kushigbor, C.A. (1994) A geobotanical approach to the tropical rain forest environment of the Carajás Mineral Province (Amazon Region, Brazil), based on digital TM-Landsat and DEM data. *International Journal of Remote Sensing*, **15**, 1633–1648.

Rocha, O. (1990) Lista preliminar de aves de la Reserva de la Biósfera 'Estación Biológica del Beni'. *Ecológica en Bolivia*, **12**, 57–68.

Salinas, E. (1995) Beni Biosphere Reserve and Biological Station: education and development. In: S. Amend and T. Amend (eds) *National Parks without People? The South American Experience.* IUCN, Gland and Quito, pp. 103–116.

Salo, J.S. and Kalliola, R.J. (1991) River dynamics and natural forest regeneration in Peruvian Amazon. In: A. Gómez-Pompa, T.H. Whitmore and M. Hadley (eds) *Rain Forest Regeneration and Management.* UNESCO, Paris, pp. 261–270.

Seidenschwartz, F. (1986) Vergleich von Flußufergesellschaften mit Wildkrautvegetation im tropischen Tiefland von Peru. *Amazonia*, **6**, 79–111.

Stone, T.A., Schlesinger, P., Houghton, R.A. and Woodwell, G.M. (1994) A map of the vegetation of South America based on satellite imagery. *Photogrammetric Engineering and Remote Sensing*, **52**, 397–399.

Tou, J.T. and Gonzalez, R.C. (1974) *Pattern Recognition Principles.* Addison-Wesley, Reading, Massachusetts.

Townshend, J.R.G. and Justice, C.O. (1986) Analysis of the dynamics of African vegetation using the Normalised Difference Vegetation Index. *International Journal of Remote Sensing*, **7**, 1435–1445.

Tuomisito, H., Linna, A. and Kalliola, R. (1994) The use of digitally processed satellite images in studies of tropical rain forest vegetation. *International Journal of Remote Sensing*, **15**, 1595–1610.

Tuomisito, H., Roukolainen, K., Kalliola, R., Linna, A., Danjoy, W. and Rodriguez, Z. (1995) Dissecting Amazonian biodiversity. *Science*, **269**, 63–66.

Walsh, J. (1987) Bolivia swops debts for conservation. *Science*, **237**, 597–598.

12   Mapping Vegetation in Complex, Mountainous Terrain

A.C. MILLINGTON[1] **and S. JEHANGIR**[2]

[1]*Department of Geography, University of Leicester, UK*
[2] *Punjab Forest Service, Islamabad, Pakistan*

We consider the difficulties that may be encountered in mapping vegetation and land cover change in dissected, mountain environments. The research we report on is based in the Himalayas of northern Pakistan. The specific problems addressed in the research are (i) overcoming the effects of highly dissected terrain on the spectral properties used for mapping vegetation and land cover; and (ii) the production of vegetation and land cover maps using archive image and ancillary data and their verification. In addressing the first problem we considered the issue of shadowing and have attempted to solve this with a method based on modelling spatial and temporal variations in solar radiation across the study area. A spectral transformation, which included shadow matching, an adjustment factor grid, masking snow, ice and cloud, and multiplication of each image band by an adjustment factor, was used.

Land cover maps were produced for 1979 and 1989. These were 'verified' using a combination of field surveys (at 126 sites) and interviews with landowners. The map accuracies were 82.9% (1979) and 76.3% (1989). The major land cover change was not, as expected, a dramatic reduction in forest cover, but a change in the types of forest cover. In fact there was only a 7.7 km^2 reduction in forest area over the entire 1040 km^2 area that was analysed. However there was a significant change in the forest cover categories; the area of forest with $> 50\%$ canopy cover declined by 81.2 km^2 from 1979 to 1989, whilst the area of forest with canopy cover ranging from 25–50% increased by 90.9 km^2

INTRODUCTION

The uses to which vegetation and land cover maps are put are many and varied (cf. Chapter 18). In this chapter we describe how land cover was mapped in northern Pakistan and used to provide inputs to a Geographical Information System (GIS)-based distributed hydrological model. Rather than simply describing the production of the vegetation maps, we will concentrate on three issues that have wider relevance outside the study area. These are:
1. Overcoming the effects of highly dissected terrain on the spectral properties of remotely sensed data used for mapping vegetation and land cover.
2. The production of vegetation and land cover maps using archive image and ancillary data and their verification.
3. The use of land cover maps to produce inputs to hydrological models.
    The study area is the Siran Basin in the western Himalayan foothills of Pakistan. It

Vegetation Mapping: From Patch to Planet. Edited by Roy Alexander and Andrew C. Millington.
© 2000 John Wiley & Sons Ltd.

lies between 34°14'17"N and 34°47'56"N and covers an area of 1040 km². Elevation ranges from 870 to 4285 m a.s.l. In the upper catchment, the slope angles reach 56°, but in the lower basin more open, gently sloping terrain can be found. Rock types are extremely varied – there are sedimentary, igneous and metamorphic rocks in the catchment as well as Quaternary and Recent sediments. The complex terrain and variety of rock types gives rise to 21 soil associations. The area is under the influence of both the south-west Indian Monsoon and Mediterranean weather systems (Ives and Messerli, 1989). Climate types vary from subtropical at low elevations to alpine at high elevations. Mean annual precipitation increases along a gradient from approximately 750 mm in the south-west to 1500 mm in the north-east (Jan, 1992). Approximately two-thirds of the precipitation occurs during the summer monsoon, the wettest months being July and August. The remaining rain falls in winter. Humidity is lowest from April to June.

The complex terrain, variety of soil associations and strong climatic gradients, combined with human activities, lead to a variety of vegetation and land cover types. The sequence of altitudinal vegetation zones outlined in Table 12.1 is complicated by the land cover types that result from disturbance of natural vegetation by the mixed agro-pastoral farming systems that dominate rural life in northern Pakistan. In the low-lying plains within the Siran Basin, irrigated double-cropping cultivation dominates land use. On the lower and middle slopes, irrigated and rain-fed wheat, maize and rice cultivation are common, although recently there has been a growth in tobacco cultivation. There is little cultivation above 2450 m a.s.l.

Forested land falls under one of three legal categories. Reserved forests are government owned and were declared free of public rights and concessions at the time of settlement (1871). However, during the second period of settlement (1904–05) limited public rights to grazing, fodder collection, grass cutting and collection of dead wood were acknowledged. *Guzara* forests are under private ownership, with their use being restricted to meeting subsistence requirements for timber, fuelwood, grazing and grass cutting. Protected forests are formerly privately owned forests which were taken back into government protection in 1959.

Hydrological interest in the Siran Basin emanates from the fact that the Siran is a major input to the Tarbela Dam. This is a key structure in the Indus irrigation system and is the world's largest earth dam. The rivers in the basin also supply irrigation water for arable agriculture within the catchment. The suspended sediment load of these rivers is potentially extremely high for two reasons, first, because of the naturally high erosion rates within the catchment, and, secondly, because of the potential effects of land cover change and deforestation within the catchment on erosion. This chapter focuses on the derivation of land cover maps from two different years and their uses in providing inputs to a GIS-based hydrological model developed for the Siran Valley (Jehangir, 1995).

DATA SOURCES

Any research into vegetation and land-cover mapping in mountainous areas using remotely sensed data or a GIS requires digital elevation data. In the Siran Basin study

Table 12.1 Altitudinal vegetation zones in the western Himalaya (after Jan, 1992)

Zone	Elevation range (m a.s.l.)	Vegetation description
Alpine pasture	> 3200	Alpine grass–herb meadows, with stunted and contorted *Juniperus* spp. at treeline
Moist temperate mixed conifer forests	2290–3200	Pure stands of *Abies pindrow*, *Cedrus deodar* and mixed stands of *A. pindrow–Picea smithiana–Picea wallichiana*. In moist areas patches of broadleaved trees, e.g. *Acer pictum, Aesculus indica, Juglans regia* and *Quercus incana* are found. Above 2590 m a.s.l. mixed fir and spruce forest dominates, while below this altitude *P. wallichiana* co-dominates with *Cedrus deodar*
Chir pine–Blue pine forests	1675–2290	Mixed stands dominated by chir pine (*P. longifolia*) and blue pine (*P. wallichiana*). At lower elevations within this zone *P. longifolia* dominates, while above 1980 m a.s.l. *P. wallichiana* dominates. Between 1675 and 1980 m a.s.l. species dominance is controlled by aspect
Chir pine	900–1675	Subtropical zone which is dominated by chir pine (*P. longifolia*)
Scrub forest	300–900	

digital elevation data were obtained by digitising contour lines from 1:63 650 topographic maps. Severe restrictions are placed on the use of post-1950 small-scale topographic maps of Pakistan because of security concerns. None were made available to this study as it was undertaken outside Pakistan. Therefore, topographic maps (Sheets F-43/1 to /3, and F-43/5 to 8) published in 1941 by the Survey of India were obtained from the British Library (London). The contour interval on these maps is 50 feet (15.24 m) on gentle slopes and 100 feet (30.48 m) on steep slopes. It was assumed that the contour information from these maps would still be valid 50 years after publication, except for areas where there had been landslides and in the vicinity of river channels that may have migrated. However, information about land cover, the built-up areas of towns and villages, and the communications network could only be used as an historical record. Digitising was accomplished in ARC/INFO ADS version 3.4 and a Triangular Irregular Network (TIN) was derived using ARCTIN.

Two Landsat images were used for land cover mapping. These were a TM image acquired on 10 July 1989 and a MSS image acquired on 5 March 1979. The MSS image was purchased from the EROS Data Centre and the 1989 image supplied by SUPARCO (the Pakistan Space and Upper Atmosphere Research Council). We tried to obtain earlier digital imagery to compare to the 1989 image. However, all of the digital MSS imagery acquired for this scene prior to 1979 had been destroyed and only prints were available. Both images had been pre-processed for system errors (Bernstein and Ferneyhough, 1975; Bernstein, 1983). Further 'pre-processing' of the TM image had to be undertaken due to the variable number of scan lines in bands 2–5 and 7. This resulted in some lines being cropped at the top and bottom of the image. In addition, as band 1 was unreadable, it was not used in the image analysis. The full range of data used in land cover mapping and in the GIS-based hydrological modelling are listed in Table 12.2.

VEGETATION AND LAND COVER MAPPING IN HIGHLY DISSECTED TERRAIN

The major problems associated with land cover mapping and monitoring in mountainous terrain are:

- shadows caused by sun–topography interactions;
- anisotropic reflectance;
- snow and ice cover; and
- enhanced cloud cover due to orographic precipitation processes.

To a certain extent, the latter two problems can be overcome by choosing imagery with little snow or ice, and with no, or minimal, cloud cover. However, as will be shown later, choosing snow-, ice- and cloud-free images is not always possible because the data archives are limited and, even if snow-, ice- and cloud-free images are available they may not be from the most optimal seasons for land cover mapping. Moreover, the ability to obtain imagery throughout the entire growing season is limited because of the occurrence of snow, ice and cloud. Even if imagery free of these problems is acquired, the problem of shadowing remains. Howard (1991) argued that shadowing is a cause of major spectral inaccuracies when mapping vegetation and land cover in mountainous areas. Kimothi and Jadhav (1998) argue that, in months with high sun elevation, IRS LISS-1 data is suitable for land cover mapping because of the local acquisition time in the Himalayas. However, this does not overcome the shadow problems if a multitemporal approach to land cover mapping is used, because months with low sun elevation must be used; or, as is the case in this study, archive data sets from other sensors are used to compare land cover trends over time. Walsh (1987) has provided a thorough review of the influence of topography on the spectral response of satellite image data. The effects of anisotropic reflectance are known (Justice *et al*, 1980; Smith et al., 1980; Hugli and Frei, 1983; Colby, 1991; Colby and Keating, 1998) but rarely considered. We did not consider anisotropic reflectance in this research.

Landsat MSS and TM image data of the Siran Basin have potentially serious radiometric inaccuracies caused by the large relative relief which is manifest as shadowing. The two land cover classifications derived from the Landsat image data yielded misleading results for all north-facing slopes. This was particularly acute in the maps derived from the 5 March 1979 image. When the data were submitted to clustering algorithms, all of the land cover types on the north-facing slopes formed a single cluster, with a low standard deviation, due to the lack of spectral separability between the pixel Digital Number (DN) values regardless of the land cover types they represented. Supervised classification of the land cover types was consequently hindered by the difficulties of locating training areas on slopes in shadow. Therefore, overcoming shadowing was a major task for land cover mapping in the Siran Basin.

An objective method of reducing the topographic effects in an image involves adjusting pixel DN values for aspect and gradient prior to image classification (Howard, 1991). Such a spectral transformation can be performed to eliminate, or at least significantly reduce, the inaccuracies caused by either sun angle or topographic relief. The topographic information required in such a procedure requires a geo-rectified

Table 12.2   Data used in research

Primary data source	Parameters derived from data source	Source
Topographic maps (1:63 650)	DEM, slope angle and aspect, hydrological network	Survey of India (1941)
Geological maps (1:250 000)	Map of rock types (9 types)	Soil Survey of Pakistan (1988)
Soil association maps (1:250 000)	Map of soil associations (21 associations), soil texture map (derived from soil association and geological maps), soil depth map (derived from the soil association map)	Soil Survey of Pakistan (1988)
Forest tenure map	Map of legal status of 'forest' land	NWFP Forest Dept, Siran Forest Development Project
Landsat imagery (MSS image 5 March 1979, TM image 10 July 1989)	Land cover maps	EROS Data Center (1979 image), SUPARCO (1989 image)

Digital Elevation Model (DEM). Clearly accurate geo-registration of the image data for superimposition over a DEM is an essential prerequisite for this type of spectral transformation (Nguyen and Ho, 1988). Civco (1989), Yang and Vidal (1990) and Walsh *et al.* (1994) have adopted this approach to correct Landsat data for the topographic redistribution of solar radiation.

The approach we used was also based on solar radiation modelling for the entire catchment. It was based on the assumption that the total insolation across a catchment can be mapped at a particular point in time for a particular day (Iqbal, 1983). In this study the particular point in time represents the time of the satellite overpass. Image rectification was confined to the Siran Basin and its immediate surrounding area as topographic maps were only available for that area. Twenty-nine points (mainly sharp bends and confluences in rivers, and road junctions) were chosen for rectification from the imagery. They were located on the topographic maps and their latitude and longitude coordinates obtained. During the warping process some of these points were eliminated in reducing the initial RMSE of the rectification from seven to between one and two (Table 12.3).

Many remote sensing studies have stressed the importance of integrating digital elevation data with image data (e.g. Parachini and Folving, 1994). One way in which digital elevation data can be used is to reduce the relief effect on spectral responses (Howard, 1991). This can be done in two ways: either by post-classification enhancement (Cibula and Nyquist, 1987) or by spectral transformation during pre-processing. Hoffer (1979) used elevation values and two derivatives of a DEM, slope angle and aspect, with a layer of spectrally distinct classes to enhance the results of a land cover classification. Walsh *et al.* (1990) superimposed several topographic layers with Landsat TM data to characterise the hydrological structure of a catchment. A method based

Table 12.3   Accuracy of image registration

Image type and acquisition date	Mean RMSE	
	pixels	metres
MSS (5 March 1979)	1.18	66.1
TM (10 July 1989)	1.67	50.1

on photointerpretation (the analysis of tone, texture and terrain) was adopted by Strahler (1981) after discriminating between well-illuminated regions and those in shadow. Nguyen and Ho (1988) draped a false-colour composite over a three-dimensional surface generated from a Digital Terrain Model (DTM) prior to manual enhancement.

The approach used in this research was to model the spatial and temporal insolation patterns over the catchment (Jehangir, 1995). For the spectral transformation of Landsat data it was assumed that for a window in the visible and near-infrared (NIR) parts of the spectrum, the reflectance from any surface is proportional to the insolation. This implies that F_t, the factor which projects the total radiation received on the horizontal surface to any sloping surface (Duffie and Beckman, 1974), can be used to reduce the reflectance from that sloping surface to that on the equivalent horizontal surface. Although a particular land cover will have widely different spectral reflectance properties on different gradients and aspects (i.e. spectral anisotropy), all locations must have the same reflectance after being reduced to their horizontal equivalents. Therefore, after removing the shadowing effects caused by topographic relief, the pixel DN values will represent land cover properties (vegetation and soil) not a combination of vegetation, soil and topographic influences. The spectral transformation procedure used in this research involved four steps:

- shadow matching
- generation of an adjustment factor grid
- snow, ice and cloud masking
- multiplication of the image bands by the adjustment factor

Shadow Matching

Before integrating the DEM with the remotely sensed data using radiation modelling, areas of shadows were identified on FCC images. These areas were spatially matched with potential shadow areas on a synthetic reflectance image – a shaded relief image – derived for the image acquisition dates and overpass times as suggested by Howard (1991). This was done visually. It enabled areas of low reflectance that were not due to shadowing to be eliminated from the spectral transformation procedures.

According to the Lambertian laws of reflectance and a vertical observer's angle, the reflectance R at a point on an image is given by:

$$R = \max(0, k \cos \alpha) \tag{12.1}$$

where k is a coding coefficient (255 on a 16-bit display), and α is the angle between the surface and the sun direction (Nguyen and Ho, 1988).

The sun azimuth and elevation information were obtained in the image header files. These data were used to generate the shaded relief image used for matching shadow areas with those potential shadow areas identified on Landsat FCCs (Plate 7 (Figure 12.1a)).

The relief adjustment factor used to adjust the pixel values for gradient and aspect is the reciprocal of the total solar radiation factor (F_t) at any particular point in the catchment at the time of overpass. The total radiation on an arbitrarily oriented surface has three components: beam, diffuse and reflected (Duffie and Beckman, 1974; Iqbal, 1983; Whiteman, 1990; Moore et al., 1991) and is given by:

$$R_t = R_b F + R_d (1 + \cos\beta) / 2 + (R_b + R_d)(1 - \cos\beta) \rho / 2 \qquad (12.2)$$

where R_t = total radiation received on the sloping surface; R_b = direct (beam) radiation on a horizontal surface; F = the ratio between the potential radiation on the sloping surface and a horizontal surface; R_d = diffuse radiation on the horizontal; β = slope angle (%); and ρ = albedo.

The total radiation ratio (F_t) between the sloping surface and a horizontal surface is

$$F_t = (R_b F/R_h) + (R_d (1 + \cos\beta) / 2R_h) + (1 - \cos\beta) \rho / 2 \qquad (12.3)$$

where R_h = total radiation received on the horizontal surface.

The effects of shading from direct sunlight by surrounding sites in enclaves were not accounted for by this method (Dozier and Bruno, 1981; Moore et al., 1991). However, a visual inspection of the DEM revealed very few enclaves and those that did exist were very small in area.

Snow, Ice and Cloud Masking

The parts of the Siran Basin that were freshly covered in snow were found to have the maximum reflectance in all four MSS bands (5 March 1979) irrespective of gradient or aspect. Ice-covered areas had the maximum reflectance in the visible bands, but were very low in the NIR for all aspects and gradients. These areas could not be corrected by the same method that was used for the areas of vegetation, bare soil and rock exposures. These areas were dealt with by masking the areas of snow and ice in all bands before multiplying the remaining pixels with the relief factor. Classification of the original MSS data into 25 classes yielded six classes representing areas of snow and ice. These classes were combined and masked before adjusting for relief. In addition, areas of clouds and cloud shadows were also masked before adjustment for relief. This meant that these areas were not adjusted for shadow effects. Snow and ice were restricted to the highest elevations in the catchment in both images. Therefore it was felt that not correcting for ice and snow would have little overall influence on the spectral transformation to correct for shadowing, as shadowing was restricted to the lower slopes and floors of deep valleys.

Image Multiplication by the Relief Factor

The final stage in the image adjustment procedure was the multiplication of the DN values in the different bands by the relief factor. This was done for all areas except those that had been masked out. The relevant image statistics for the MSS image (5 March 1979) are given in Table 12.4.

LAND COVER MAPPING

The production of land cover maps from the Landsat MSS and TM imagery used standard procedures and combined both image processing and GIS operations (Figure 12.2). In this chapter we will only focus on the land cover classification scheme, the integration of image and elevation data to refine the land cover maps, and accuracy assessment procedures.

In the classification scheme used for land cover mapping in this study all of the main land cover types associated with the major land uses and the main 'undisturbed' vegetation types were included as they all contribute to the hydrological activities within the catchment. The land cover and vegetation types that potentially occur in the Siran Basin were identified from topographic maps, land use maps and reports, and reports of forest maps. An appreciable proportion of the basin (approximately 8%) lacks any soil or vegetation cover; these areas mainly comprise rock exposures and the gravel trains of braided river channels. At elevations > 3500 m a.s.l. much of the land is covered by snow throughout the year. The land cover and vegetation types comprise various agricultural land cover types, forests and pastures. The range of potential cover types that could be found in the basin were used to construct a matrix of the major land cover strata. Two major land cover strata, cultivation and forest, were subdivided into Level I classes (Table 12.5). For the cultivated land cover strata the subdivisions were based on the main crops grown in the area and a bare soil class to account for fallow fields. Subdivision of the forest strata was slightly more complex. Initially three classes were used. These were based on arbitrary canopy cover thresholds of > 50%, 25–50% and < 25%. However, it was very difficult to distinguish known areas of the lowest cover category from areas of grassland (pasture). This is because areas of very open, degraded and regenerating forest often have a similar spectral response to pastures with occasional trees (Lo and Fung, 1986). Therefore it was decided to merge the grassland and low canopy cover forests into one class – grassland with scattered trees.

The training areas for supervised classification of Level I classes were chosen from topographic maps (1:63 650 scale) and land use maps (1:50 000 scale). As the images had been rectified to the same coordinate system as the maps, the training areas were easily identified on images composed of various combinations of principal components, and image maps of NDVI and PVI (Parachini and Folving, 1994). The training area data were saved as separate vector polygon coverages so that they could be copied onto the TM and MSS images. The minimum number of samples in each class was set at 50 to provide meaningful statistics (Su and Schultz, 1993). Data from all four MSS bands were used in the supervised classification, the bands were input with equal

Table 12.4 Comparison of the DN values of the Landsat MSS image (5 March 1979) for the area of the Siran Basin before and after spectral transformation for relief

Channel (nm)	DN values before spectral transformation		DN values after spectral transformation	
	Mean	S.D.	Mean	S.D.
400–500	26.78	26.11	28.99	29.89
500–600	31.84	27.33	33.65	29.84
600–700	42.99	24.60	45.58	27.66
700–1100	38.66	21.55	41.45	26.20

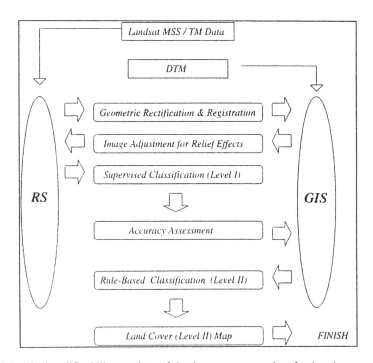

Figure 12.2 A simplified illustration of the image processing for land cover mapping

probability and a zero rejection threshold used so that no unclassified pixels were obtained. Five bands (TM bands 2–5 and 7) were used for the classification of the TM image. In the TM image, areas of cloud shadows were left unclassified.

In the upper parts of the basin winter snow cover obscured many underlying ground features in the MSS image which was acquired during March. At lower elevations, snow on the forest floor was found to alter the spectral response of the forest at this time of the year. There was a requirement to produce maps of the summer land cover types from the MSS image (i.e. after snowmelt) to compare to the maps of 'summer'

Table 12.5   A matrix of the main land cover strata and Level I land cover classes for the Siran Basin. The categories in the right-hand column represent the Level I land cover classes that were used in the later aspects of land cover mapping

Major land cover	Code	Level I land cover class	Code
Cultivated	1	Wheat	11
		Alfalfa	12
		Winter non-cereal crops	13
		Paddy rice	14
		Tobacco	15
		Maize	16
		Bare soil	17
Forest	2	Coniferous forest (> 50% canopy cover)	21
		Coniferous forest (25–50% canopy cover)	22
Grassland	3	Pasture with scattered trees	31
Exposed rock	4	Bare rock exposures	41
River channels	5	Gravel and sand of dry river beds	51
Snow and ice	6	Snow and ice	61

land cover types from the TM image. To do this it was assumed that the land cover assignments of the areas that had been fully or partly covered by snow in 1979 were similar to those in 1989. No maps or reports are available which provide specific information on the land cover at high elevations in the late 1970s to check this assumption. However, we justify this assumption on the grounds that the areas covered in snow are generally at high elevation and, as these are more-or-less inaccessible and rarely used by farmers, they will have undergone little change in land cover type over the 10 years between 1979 and 1989.

A further problem when comparing the land cover maps is that there are areas covered by cloud and cloud shadow in the 1989 TM image that were not obscured in the 1979 MSS image. In these areas we compared the same geographical areas in the TM image and the topographic maps. The cloud-covered areas all corresponded to an 'agriculture–forest–grass' class on the topographic maps. Therefore they were assumed to correspond to class 31 (grass with scattered trees) and class 31 was used to replace cloud and cloud shadow in the 1989 land cover map.

The land cover map produced from the TM image was warped to 80 m resolution to interface with the GIS (which had been set at an 80 m cell size to correspond to the pixel size of MSS imagery) for the purposes of change detection. The land cover areas for the two maps (Figures 12.3 and 12.4) are given in Table 12.6.

Accuracy Assessment

Ground-based land cover verification is particularly important in areas where other sources of information are unreliable such as the Himalayas (Thompson and Warburton, 1985). Bearing in mind the climatic and logistical constraints on ground data collection in this region, and the nature of the crop calendars for different crops, it was

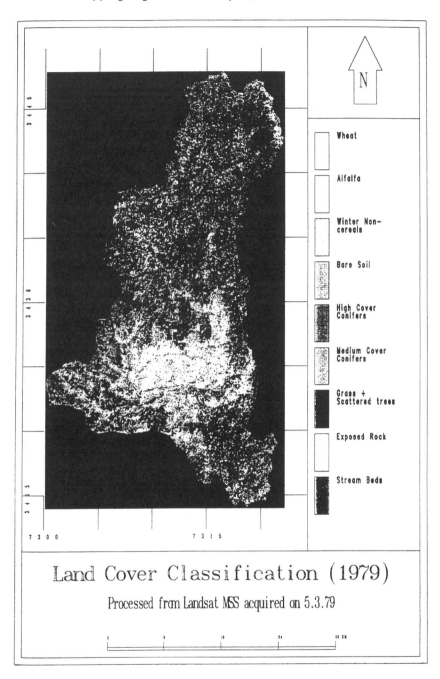

Figure 12.3 The land cover map for 1979. This map was derived by the supervised classification of Landsat MSS. Note the forest classes have been defined on the basis of canopy cover density, because different forest types are not spectrally distinguishable

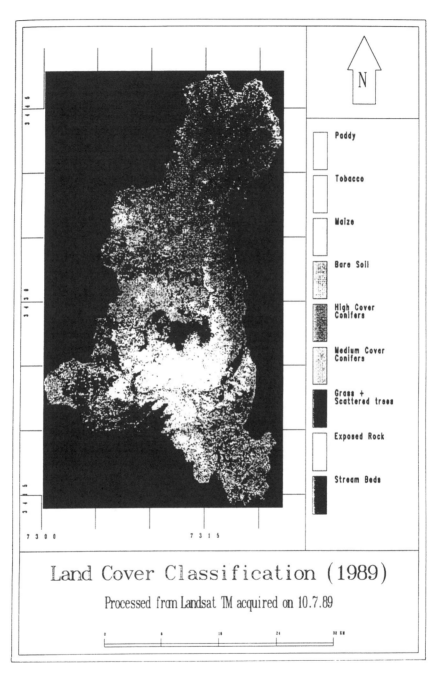

Figure 12.4   The land cover map for 1989 derived from Landsat TM. Note the changes which have taken place over 10 years, e.g. the areas where high and medium cover conifers have been replaced by very low cover conifers and/or grass. Also note the expansion of stream beds

Table 12.6   Areas of Level I land cover classes in 1979 and 1989 (in km^2)

Code	Land cover class	Land cover areas		Difference in areas
		1979	1989	
11	Wheat	64.8	0.0	64.8
12	Alfalfa	65.4	0.0	65.4
13	Winter non-cereal crops	78.4	0.0	78.4
14	Paddy	0.0	17.5	−17.5
15	Tobacco	0.0	56.5	−56.5
16	Maize	0.0	152.5	−152.5
17	Bare soil	155.9	135.4	20.5
1	*Total cultivated*	364.4	361.9	2.6
21	Coniferous forest (> 50% cover)	146.6	65.5	81.1
22	Coniferous forest (25–50% cover)	184.5	275.4	−90.9
2	*Total forest*	331.1	340.9	9.8
31	Pasture and scattered trees	265.2	256.9	8.3
41	Bare rock exposures	46.6	16.0	30.6
51	Gravel and sand of dry river beds	33.5	65.2	−31.7

decided to ground verify the maps derived from classified imagery in winter. This was done in December 1994 and January 1995, some five years after the last image was acquired and at a different time of the year to the 'summer' land cover class maps. The five-year gap between acquisition and verification was a logistical constraint imposed on the project by virtue of the fact that no later TM image could be acquired within the project's budget. Image verification at a different time of the year from image acquisition should not be a problem in non-agricultural areas (e.g. forests and grasslands) where inter-seasonal changes in vegetation do not affect the land cover classification assignments as much. In the cultivated areas there is a strong seasonal bias due to the double cropping season, with distinctive cropping patterns for the summer and winter seasons. This has already been noted in the different land cover classifications for the MSS and TM images that were derived from imagery acquired in different seasons (Table 12.6). However, this is less of a problem than might be anticipated because of the crop rotation within any one farmer's fields. The classification of a field to a particular crop in 1979 or 1989, regardless of season, does not mean that that crop will be grown in the field the next year. Therefore, verification of the agricultural land cover parcels concentrated on verification of the fields as 'cultivated' land rather than to a particular crop. The end result was any change in the overall cultivated area between 1979 and 1989.

The minimum number of sites visited for each major land cover stratum was set at 30. These were selected randomly within each stratum. Five main strata were identified:

- cultivated,
- forests,
- grasslands,
- rock exposures, and
- river channels.

The total number of sample sites scheduled to be visited was 150 (30 × 5). However, 15 of these sites were inaccessible due to their extremely high elevations, therefore only 135 sites were visited. The sizes of the sample sites to be checked were determined by the equation devised by Justice and Townshend (1981):

$$A = P(1 + 2L) \tag{12.4}$$

where P = the pixel dimension (in m), and L = the accuracy of location in terms of pixels.

Assuming an accuracy of 1.5 pixels, each sampling site measured 360 m × 360 m or 0.13 km^2. Given the total number of sample sites surveyed, 135, the total area checked in the field was 17.5 km^2 or 1.68% of the study area. Sample sites within each stratum were selected using a stratified random sampling scheme, the most efficient and preferred method in field verification (Rudd, 1971; Van Genderen et al., 1978; Card, 1982; Rosenfield, 1982; Congaltan, 1988). This was achieved by using the five strata as the stratification and selecting sample sites using latitude and longitude coordinates using random numbers. The sites were then marked on the topographic maps using a transparent grid. The sites were located in the field using a Magellan Navigator 3000 GPS. The dimensions for each site were measured using a compass and a 30 m tape. Nine of the 135 sites were not accessible, due to topographic difficulties or access being restricted by the owners. At these sites as many observations as possible were made using binoculars. The following data were collected for the 126 sites that were visited in the field (as against those that were observed with binoculars) – slope angle and aspect, altitude and ground cover (% cover and dominant species). This range of observations has been used for deforestation studies in ecologically similar environments in Nepal (Jordan, 1994). Other data collected were latitude and longitude of the mid-point, map sheet number, soil texture, soil depth and the rock outcrop (as a %). In addition a sketch map of the land cover was made and the actual land cover class was identified and, if people were available to interview, the best estimate of the actual land cover class of the site in 1989 and 1979. Confusion matrices were produced from the ground data and image data (Table 12.7).

The overall classification accuracies of the maps are 82.9% (1979) and 76.3% (1989). In both cases the highest accuracy for the main land cover strata are for the forests, 97.2% (1979) and 85.7% (1989), (Table 12.8). The least accurate strata in both maps were the river channels, 50 and 11.1%, respectively. In addition to the mathematical classification accuracy, the probability of a pixel being classified correctly within a 95% confidence limit using a binomial expansion technique was calculated. Using this technique the classification accuracies for the entire maps were 77.7% (1979) and 71.2% (1989). With the exception of the grass and scattered trees class, all land cover classes in the 1979 map had higher levels of accuracy than in the 1989 map.

Table 12.7 Confusion matrices for 1979 and 1989 land cover maps of the Siran Basin

(a) 1979

Land cover classes from image maps	Land cover classes from field survey										Total	Accuracy
	PAD*	TOB*	MZE*	BS	C50	C2550	PST	ROK	STR	FLW†		
Wheat (WHT)	12	0	0	0	0	0	0	0	0	0	12	100
Alfalfa (ALF)	0	11	0	0	1	2	1	0	0	0	15	91.66
Winter non-cereals (WNC)	0	0	17	0	1	0	2	0	0	0	20	85
Bare soil (BS)	0	0	0	14	0	1	0	0	0	0	15	93.33
Coniferous forest > 50% cover (C50)	0	0	0	0	19	1	0	0	0	0	20	95
Coniferous forest 25–50% cover (C2550)	0	0	0	1	1	14	0	0	0	0	16	87.5
Pasture and scattered trees (PST)	2	0	0	3	0	0	17	0	0	0	22	77.27
Bare rock exposures (ROK)	0	0	0	0	0	0	2	3	0	0	5	60
Dry river beds (STR)	0	0	0	0	0	0	1	0	5	4	10	50
Total	14	11	17	18	22	18	23	3	5	4	135	82.98

* The codes for the first three columns of the 1979 confusion matrix can be found in Table 12.7b.
† FLW = Fallow fields

(b) 1989

Land cover classes from image maps	Land cover classes from field survey										Total	Accuracy
	PAD	TOB	MZE	BS	C50	C2550	PST	ROK	STR	FLW†		
Paddy rice (PAD)	3	1	0	0	0	0	0	0	0	0	4	75
Tobacco (TOB)	0	15	0	0	1	0	2	0	1	0	19	78.54
Maize (MZE)	0	1	16	0	0	2	1	0	1	0	21	76.19
Bare soil (BS)	1	0	0	12	0	2	0	0	0	0	15	80
Coniferous forest > 50% cover (C50)	1	0	0	0	11	0	0	0	0	0	12	91.66
Coniferous forest 25–50% cover (C2550)	1	0	1	0	1	25	1	1	0	0	30	80
Pasture and scattered trees (PST)	3	0	1	0	0	0	20	0	0	0	24	83.33
Bare rock exposures (ROK)	0	0	0	0	0	0	0	1	0	0	1	100
Dry river beds (STR)	0	0	0	0	2	0	1	0	1	5	9	11.11
Total	9	17	18	12	15	29	25	2	3	5	135	76.29

† FLW = Fallow fields

Table 12.8 Mapping accuracy of the main land cover strata for the 1979 and 1989 land cover maps

	Land cover map of 1979			Land cover map of 1989		
	Total sample sites	Number of correct sites	Accuracy (%)	Total sample sites	Number of correct sites	Accuracy (%)
Cultivated	62	54	87.1	59	49	69.5
Forest	36	35	97.2	42	37	85.7
Grassland and trees	22	17	77.3	24	20	83.3

The high accuracy of the forest classes is mainly a consequence of their very distinct spectral response compared to all other classes. The mixed class – grass and scattered trees – was confused with many other classes and, in the field, comprised low-density coniferous forest stands, grasslands, abandoned agricultural areas, and fallow agricultural land. The low accuracy assessments of the exposed rock and river channel classes were probably due to the inclusion of village areas and roads. Population expansion in this area has meant that part of the apparent increase in the area of river channels is due to increased house building and road construction, as well as to the newly exposed stream beds due to channel migration. Snow cover also affected mapping accuracy. Snow under the forest canopies changed the spectral responses of the forest classes leading to mis-classification. There is also some residual hill shading that could not be corrected for in the image adjustments for shadowing.

Multi-layer, Rule-based Land Cover Classification (Level II)

Certain hydrological processes, particularly interception, evapotranspiration, infiltration and runoff, are strongly determined by the nature of the tree canopies. Therefore the canopy cover proportion used in this study was potentially useful in parameterising the hydrological model that was developed subsequently. However, the ground conditions, e.g. the presence or absence of ground vegetation and the amount of ground vegetation and litter, are also important influences on hydrological processes such as infiltration and surface runoff. For example, the two forest classes (> 50% and 25–50% canopy cover) can comprise stands with radically different ground conditions and therefore they could have very different hydrological responses to the same rainfall event.

The Level I forest classes were determined on the basis of the spectral properties of their canopies and contain no information about their species composition. Therefore, they are poorly defined in terms of parameterisation of hydrological models. The situation with respect to forest is more complex. Five types of forest are found within the basin (Jan, 1992) and, as these can be defined on an altitudinal basis, it is possible to use the digital elevation data as an additional data set to refine the accuracy of the land cover maps for hydrological purposes. The combination of spectral and digital elevation data has been used for vegetation and land cover mapping in Mediterranean ecosystems and in temperate forests (e.g. Cibula and Nyquist, 1987). Justice and Townshend (1981) and Curran (1985) both advocate the use of additional *a priori* data as an aid to increasing the accuracy of vegetation mapping. Therefore digital elevation data were used to subdivide the two Level I classes (21 and 22) in Table 12.5. Using the vegetation categories outlined by Jan (1992), the potential theoretical forest classes are provided in Table 12.9.

The multi-rule classification was carried out in ARC/INFO GRID (Version 6.1) using the two level I land cover maps (Figs. 12.3 and 12.4), the digital elevation model and an aspect map that had been produced from the DEM. The areas of the different forest types in the two land cover maps are provided in Table 12.10.

Changes in Land Cover and Vegetation between 1979 and 1989

The areal estimates of the main land cover types in the two maps revealed that there had been no major changes in the main land cover types (cultivated, coniferous forest and grassland) during the 1980s (Table 12.10), the differences in areas being -2.7, 9.7 and $-8.2\,km^2$, respectively. However, mapping using the Level II classification scheme indicated that there had been significant changes in the spatial distribution of the different types of forest between 1979 and 1989. All of these statistics are based on post-classification change detection as recommended by Estes *et al.* (1982).

The main changes are in the forest areas (Table 12.10). Some areas have remained more or less the same with respect to forest cover, but many areas have shown dramatic reductions in forest cover (e.g. moving from the $>50\%$ cover class to the 25–50% cover class) and some forest areas have disappeared altogether. Some areas with increased forest cover were also noted. Complete loss of forest cover has mainly occurred on the plains and on relatively gentle slopes during the 1980s. These results support Haigh's (1991) assertion that deforestation in the Himalayas is a matter of recent historical record. Deforestation has resulted from the fact that more land has been brought into cultivation, the ease of accessibility, and the settlement of refugees during the Russian occupation and subsequent civil war in Afghanistan (Allan, 1987). New plantations and areas of regeneration are also found within the basin and these, to a minor extent, have countered the loss of forest area and the thinning of forest cover. These result from tree planting programmes during the 1970s, watershed restoration projects and village-level projects to assist in the supply of food, fuel and fibre (Dixon and Perry, 1986; Khattak, 1987).

Forest changes must also be analysed in the context of forest ownership, which is perhaps the most important factor in forest management in Pakistan (Azhar, 1989). We have analysed the changes in forest cover by forest tenure type (Fig. 12.5). These data indicate that most deforestation between 1979 and 1989 occurred on privately owned land and, since the focus of afforestation and planting campaigns in the past has been on privately owned waste lands (Dixon and Perry, 1986; Khattak, 1987), most afforestation has also occurred on privately owned land.

Table 12.9   Level II classification of forest classes in the Siran Basin

Elevation range (ma.s.l.)		300–910	910–1675	1675–1980		1980–2290	2290–2590	2590–3200
Aspect		All aspects	All aspects	South-facing	North-facing	All aspects	All aspects	All aspects
Forest type (from Jan, 1972)		Scrub forest	Chir pine forest		Blue pine forest		Mixed blue pine and deodar forest	Mixed fir and spruce forest
Level II Classes	Canopy cover > 50%	211	212		213		214	215
	Canopy cover 25–50%	221	222		223		224	225

Table 12.10   Areal extent of Level II forest types in the Siran Basin in 1979 and 1989

Class code	Class descriptor	Area (km^2)		Difference (km^2)
		1979	1989	
211	Scrub forest	1.5	0.1	−1.4
212	Chir pine forest	87.4	10.2	−77.2
213	Blue pine forest	29.0	21.5	−7.5
214	Mixed blue pine and deodar forest	16.6	20.7	+4.1
215	Mixed fir and spruce forest	12.1	12.9	+0.8
Total forest cover with canopy cover > 50%		*146.6*	*65.4*	*−81.2*
221	Scrub forest	20.7	6.2	−14.5
222	Chir pine forest	119.2	194.2	+75.0
223	Blue pine forest	22.0	47.9	+25.9
224	Mixed blue pine and deodar forest	9.3	13.1	+3.8
225	Mixed fir and spruce forest	13.3	14.0	+0.7
Total forest cover with 25–50% canopy cover		*184.5*	*275.4*	*+90.9*
Total forest cover		*331.1*	*340.8*	*+7.7*

PARAMETERISATION OF A HYDROLOGICAL MODEL USING LAND COVER DATA

The main aim of the research project in which these land cover maps were produced and analysed was the development of a distributed hydrological model for the Siran Basin (Jehangir, 1995). Consequently, an important aspect of the vegetation and land cover classes was the extraction of hydrological parameters. This is particularly important for distributed hydrological modelling in large basins, such as the Siran, where data sources are sparse and accessibility to take field measurements is limited. In

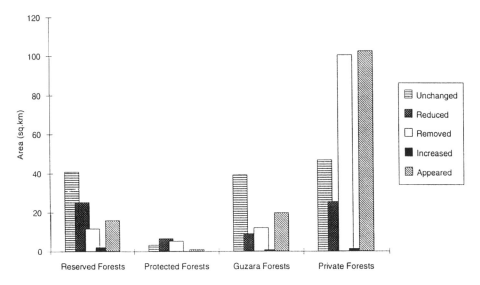

Figure 12.5 Changes in forest cover with relevance to forest ownership

the absence of experimental results, a further problem – common in modelling – is that the conversion of vegetation and land cover data to hydrological parameters is dependent on observation from other localities.

Two parameters that are required for the Penman–Monteith evapotranspiration model – crop or vegetation height and leaf area index (LAI) – could be estimated from ancillary data for the Level II land cover classes. They were estimated using conversion data provided by the Meteorological Office (1981) and Kirkby (1987). Seasonal and monthly variations in these parameters were adjusted using crop calendars for the region, and daily variations in the LAIs of seasonal crops were estimated using equations developed by the Meteorological Office (1981). In addition to crop/vegetation height and LAI, albedo was derived directly from TM and MSS data using the procedures outlined by Harris (1987). The estimated values for LAI and albedo for each Level II land cover class are given in Table 12.11.

CONCLUSIONS

We derive three main sets of conclusions from this research. First are those observations that relate to land cover mapping in complex, mountainous terrain. These are:

- the requirement to reduce the effects of shadowing before applying land cover mapping and monitoring techniques such as those that are outlined in this chapter;
- to account for snow and ice which might be lying on the ground;
- to account for the enhanced cloud cover common in these regions due to the orographic effects; and

Table 12.11   Estimation of albedo and LAI for Level II land cover classes for the Siran Basin

Land cover code and class	Albedo	Maximum LAI
11 Wheat	0.260	4.0
12 Alfalfa	0.199	
13 Winter non-cereal crops	0.232	
14 Paddy rice	0.189	
15 Tobacco	0.200	
16 Maize	0.201	
17 Fallow	0.213	
211 Scrub forest (> 50% cover)	0.127	6.0
212 Chir pine forest (> 50%)		5.0
213 Chir pine and blue pine forest (> 50%)		6.0
214 Blue pine forest (> 50%)		7.0
215 Mixed fir and spruce forest (> 50%)		8.0
221 Scrub forest (25–50% canopy cover)	0.175	4.0
222 Chir pine (25–50%)		3.0
223 Chir pine and blue pine forest (25–50%)		4.0
224 Blue pine forest (25–50%)		4.5
225 Mixed fir and spruce forest (25–50%)		5.0
31 Pasture and scattered trees	0.198	2.5
41 Bare rock exposures	0.198	0
51 Dry river beds	0.216	0

- to acknowledge that field observations for verification purposes are likely to be biased towards more accessible terrain.

In addition, spectral anisotropy should be considered and methods such as those employed by Colby and Keating (1998) be employed to reduce its effects.

The second group of conclusions concern the use of archive imagery in land cover mapping and change detection.

- Digital archive imagery is restricted by availability. In some cases digital data may even have been destroyed or have been stored in such a way that it is not readable.
- Security considerations may also affect the availability of images and maps. The ability to use 'old' maps is limited to relief data, but even then could be a problem in areas with very active geomorphological processes. In terms of the other data on topographic maps, they could provide historical land cover information.
- Verification of land cover maps produced from archive imagery is problematic and, unless relatively contemporary topographic or other maps are available, the only routes available are through contemporary reports and oral histories of local residents.

The final group of conclusions relate to the derivation of hydrological parameters from land cover maps. The main conclusion is that land cover maps can be used to provide spatially located parameters for hydrological models. However, to use this approach the land cover classes have to be converted to hydrological parameters. Often the conversion factors that are used are based on experimental work from other areas and may not be valid in different ecological and hydrological situations. An alternative approach would be to predict parameters directly from remotely sensed data using physical laws, as is the case for albedo.

ACKNOWLEDGEMENTS

We acknowledge the provision of the Landsat TM image data from SUPARCO (Karachi) and data obtained from the Surface Water Hydrology Project (WAPDA), Lahore; the Meteorological Office (Lahore); the Agro-met Office (Islamabad), the Pakistan Agricultural Research Council (Islamabad) and the Forest Development Corporation (Peshawar). SJ carried out this research under Quaid-I-Azam Scholarship No. EG2383(QA) from the Government of Pakistan. We would also like to thank the following people from the Geography Departments at Reading and Leicester who contributed towards this work – Rob Bryant, Sheila Dance, Bill Hickin, Allan Howard, Mitch Langford, Dave Orme, Richard Pope, Jane Wellens and Kevin White.

REFERENCES

Allan, N.J.R. (1987) Impact of Afghan refugees on the vegetation resources of Pakistan's Hindukush-Himalaya. *Mountain Research and Development*, **7**, 200–204.

Azhar, R.A. (1989) Communal property rights and depletion of forests in northern Pakistan. *Pakistan Development Review*, **28**, 643–651.

Bernstein, R. (1983) Image geometry and rectification. In: R.N. Coldwell (ed.) *Manual of Remote Sensing*. American Society for Photogrammetry, Falls Church, Virginia, pp. 873–920.

Bernstein, R. and Ferneyhough, D.G. (1975) Digital image processing. *Photogrammetric Engineering and Remote Sensing*, **41**, 1465–1476.

Card, D.H. (1982) Using known map category marginal frequencies to improve estimates of thematic map accuracy. *Photogrammetric Engineering and Remote Sensing*, **48**, 431–439.

Cibula, W.G. and Nyquist, M.O. (1987) Use of topographic and climatological models in a geographical database to improve Landsat MSS classification of Olympic National Park. *Photogrammetric Engineering and Remote Sensing*, **53**, 67–75.

Civco, D.L. (1989) Topographic normalisation of Landsat TM imagery. *Photogrammetric Engineering and Remote Sensing*, **55**, 1303–1310.

Colby, J.D. (1991) Topographic normalization in rugged terrain. *Photogrammetric Engineering and Remote Sensing*, **57**, 531–537.

Colby, J.D. and Keating, P.L. (1998) Land cover classification using TM imagery in the tropical highlands: the influence of anisotropic reflectance. *International Journal of Remote Sensing*, **19**, 1479–1500.

Congalton, R.G. (1988) A comparison of sampling schemes used in generating error matrices for assessing the accuracy of maps generated from remote sensing data. *Photogrammetric Engineering and Remote Sensing*, **64**, 593–600.

Curran, P.J. (1985) *Principles of Remote Sensing*. Longman, Harlow.

Dixon, R.K. and Perry, J.A. (1986) Natural resource management in northern Pakistan. *Ambio*, **15**, 301–305.

Dozier, J. and Bruno, J. (1981) A faster solution to the horizontal problem. *Computers and Geoscience*, **10**, 327–338.

Duffie, J.A. and Beckman, W.A. (1974) *Solar Energy Thermal Processes*. Wiley, London.

Estes, J.E., Stow, D. and Jensen, J.R. (1982) Monitoring land use and land cover changes. In: C.J. Johansen and J.L. Sanders (eds) *Remote Sensing for Resource Management*. Soil Science Society of America, Ankeny, Iowa, pp. 101–109.

Haigh, M.J. (1991) Reclaiming forest lands in Himalayas: notes from Pakistan. In: G.S. Rajwar (ed.) *Advances in Himalayan Ecology*. Today and Tomorrow Publications, New Delhi, pp. 199–210.

Hoffer, R.M. (1979) *Digital Processing of Landsat MSS and Topographic Data to Improve Capabilities for Computerised Mapping of Forest Cover*. LARS, Technical Report. Purdue University.

Howard, J.A. (1991) *Remote Sensing of Forest Resources – Theory and Application*. Chapman & Hall, London.

Hugli, H. and Frei, W. (1983) Understanding anisotropic reflectance in mountainous terrain. *Photogrammetric Engineering and Remote Sensing*, **49**, 671–683.

Iqbal, M. (1983) *An Introduction to Solar Radiation*. Academic Press, London.

Ives, J.C. and Messerli, B. (1989) *The Himalayan Dilemma – Reconciling Development and Conservation*. Routledge, London.

Jan, A. (1992) *Land Use Survey of Siran and Daur River Watersheds*. North West Frontier Records Inventory Series, 3. Pakistan Forest Institute, Peshawar.

Jehangir, S. (1995) Modelling the hydrological impacts of land cover change in the Siran Basin, Pakistan. Unpublished PhD thesis, University of Leicester.

Jordan, G. (1994) GIS modelling and model validation of deforestation risk in Sagarmatha National Park, Nepal. In: M. Price and D.I. Heywood (eds) *Mountain Environments and Geographical Information Systems*. Taylor & Francis, London, pp. 249–260.

Justice, C.O. and Townshend, J.R.G. (1981) Integrating ground data with remote sensing. In: J.R.G. Townshend (ed.) *Terrain Analysis and Remote Sensing*. Allen & Unwin, London, pp. 38–58.

Justice, C.O., Wharton, S.W. and Holben, B.N. (1980) *Application of Digital Terrain Data to Quantify and Reduce the Topographic Effect on Landsat Data*. National Aeronautics and Space Administration, Technical Monograph 81988, Greenbelt, Maryland.

Khattak, G.M. (1987) The watershed management in Manshera, Pakistan. In: L.S. Hamilton (ed.) *Forest and Watershed Development and Conservation*. Westview Press, Boulder, Colorado.

Kimothi, M.M. and Jadhav, R.N. (1998) Forest fire in the Central Himalaya: extent, direction and spread using IRS LISS-1 data. *International Journal of Remote Sensing*, **19**, 2261–2274.

Kirkby, M.S. (1987) *Computer Simulation in Hydrology*.

Lo, C.P. and Fung, T. (1986) Land-use and land-cover maps of central Guangdong Province in China from Landsat MSS imagery. *International Journal of Remote Sensing*, **7**, 1051–1074.

Meteorological Office (1981) *The Meteorological Office Rainfall and Evaporation Calculation System: MORECS*. Meteorological Office Hydrological Memo 45. Meteorological Office, Bracknell, UK.

Moore, I.D., Grayson, R.B. and Ladson, A.R. (1991) Digital terrain modelling: a review of hydrological, geomorphological and biological applications. In: K.J. Bevan and I.D. Moore (eds) *Terrain Analysis and Distributed Hydrological Modelling in Hydrology*. Wiley, Chichester.

Nguyen, P.T. and Ho, D. (1988) Multiple source data processing in remote sensing. In: J.P. Muller (ed.) *Image Processing in Remote Sensing*. Taylor & Francis, London, pp. 153–176.

Parachini, M.L. and Folving, S. (1994) Land use classification and regional planning in Val Melanco (Italian Alps): a study on integration of remotely-sensed data and DTM for thematic mapping. In: M. Price and D.I. Heywood (eds) *Mountain Environments and Geographical Information Systems*. Taylor & Francis, London, pp. 59–76.

Rosenfield, G.H. (1982) Sampling for thematic map accuracy. *Photogrammetric Engineering and Remote Sensing*, **48**, 131–137.

Rudd, R.D. (1971) Macro land planning with simulated space photographs. *Photogrammetric Engineering and Remote Sensing*, **37**, 365–372.

Smith, J., Lin, T. and Ranson, K. (1980) The Lambertian assumption and Landsat data.

Photogrammetric Engineering and Remote Sensing, **46**, 1183–1189.

Strahler, A.H. (1981) Stratification of natural vegetation for forest and rangeland inventory using Landsat digital imagery and collateral data. *International Journal of Remote Sensing*, **2**, 15–41.

Su, Z. and Schultz, G.A. (1993) A distributed runoff prediction model developed on the basis of remotely sensed information. *Proceedings EARSeL Specialist Meeting*, Dundee.

Thompson, M. and Warburton, M. (1985) Uncertainty on Himalayan scale. *Mountain Research and Development*, **5**, 115–135.

Van Genderen, J.L., Lock, B.F. and Vass, P.A. (1978) Remote sensing: statistical testing of Thematic Map accuracy. *Remote Sensing of Environment*, **7**, 3–14.

Walsh, S.J. (1987) Variability of Landsat MSS spectral response of forests in relation to stand and site characteristics. *International Journal of Remote Sensing*, **8**, 1289–1299.

Walsh, S.J., Cooper, J.W., Von Essen, I.E. and Gallagher, K.R. (1990) Image enhancement of Landsat Thematic Mapper data and GIS integration for evaluation of resource characteristics. *Photogrammetric Engineering and Remote Sensing*, **54**, 1135–1141.

Walsh, S.J., Butler, D.R., Brown, D.G. and Bian, L. (1994) Form and pattern in the alpine environment: an approach to spatial analysis and modelling in Glacier National Park, USA. In: M. Price and D.I. Heywood (eds) *Mountain Environments and Geographical Information Systems*. Taylor & Francis, London, pp. 189–216.

Whiteman, C.D. (1990) Observations of thermally developed wind systems in mountainous terrain. In: W. Blumen (ed.) *Atmospheric Processes over Complex Terrain*. Meteorological Monograph Volume 23, No. 45, American Meteorological Society, Boston, pp. 5–42.

Yang, C. and Vidal, A. (1990) Combination of digital elevation model with SPOT-1 HRV for reflectance factor mapping. *Remote Sensing of Environment*, **32**, 35–45.

Section 3

REGIONAL AND CONTINENTAL SCALE

13   Land Cover Changes in Scotland over the Past 50 Years

E.C. MACKEY and G.J. TUDOR
Environmental Audit Branch, Scottish Natural Heritage, Edinburgh, UK

The Wildlife and Countryside Act 1981 highlighted concern about the ways in which the structure and appearance of the countryside had been affected by changes in farming, forestry and other forms of land use. The National Countryside Monitoring Scheme (NCMS) was initiated in 1983 by the Nature Conservancy Council, and from 1986 onwards it has focused on Scotland. Now administered by Scottish Natural Heritage, this retrospective study of land cover is of key importance to understanding the extent and nature of changes which have taken place in Scotland since the late 1940s. The study is scheduled for completion in 1996 when results for the late 1980s will be assembled together with data for the early 1970s and the late 1940s.

The NCMS study of land cover change in Scotland has been made possible by the interpretation of aerial photographs and by innovations in the application of geographical information systems (GIS) for spatial analysis. This study describes losses and gains in vegetation cover, characteristic of a dynamically evolving landscape.

An unsupervised classification of Landsat MSS imagery for each local authority region, as mapped in 1984, generated up to five broad land classes or strata. Stratified random sampling of 5 km × 5 km or 2.5 km × 2.5 km squares within each local authority district was designed to detect a change of 10% or more in the extent of land cover features, with 95% confidence. Overall, the NCMS has mapped 467 sites in Scotland at a scale of 1:10 000 (with a minimum mappable area of 0.1 ha), which represents a 7.5% sample of the land area of Scotland.

Two case studies demonstrate the application of GIS for the analysis of land cover change, and allow analytical results to be compared visually with GIS-generated maps. In this context, each vegetation 'patch' is represented by a discrete GIS polygon. Linear features, such as hedges, are represented by lines. The NCMS method permits local-level vegetation patch mapping for the areas sampled and, by extrapolation, a more generalised representation of the relative proportions of land cover features at the regional and national scales. The definition and detail provided in the case studies provide useful insights into regional and national results.

Published results for the 1940s to 1970s comparisons have been employed throughout Scotland to inform strategic land resource planning, information which will be brought more up-to-date in 1996 when similar results for the late 1980s become available. This will mark the close of the NCMS programme, but it will also coincide with local authority boundary changes in Scotland. In order to make the greatest possible use of NCMS findings, and in an attempt to overcome its sample design restrictions, the potential for integrating NCMS with other GIS land cover data sets is being explored. However, the combination and comparison of geographical information from studies which vary in spatial extent, mapping resolution and feature classification, or which relate to different time periods, is not straightforward even in a GIS environment.

Technological advances are greatly extending capabilities for data acquisition and processing, paralleled by escalating demands for environmental information provision. In view of the considerable cost of land cover studies, ever greater consideration needs to be given to a broader range of interests in the collection and dissemination of geographical and ecological information, at the local, national and international levels. The consideration also of common standards for improving access to information and data exchange can only be of benefit in vegetation mapping from patch to planet.

BACKGROUND

The vegetation cover of Scotland is not static but exists in a dynamic state of change, as a consequence both of natural processes and of human activities (Thompson *et al.*, 1995). Natural change is, in the main, gradual and results from the combination of climatic fluctuations and long- and short-term geomorphological processes, as well as ecological dynamism and succession. The soils and vegetation of Scotland have evolved mainly in the warming climatic period since the retreat of the the glaciers of the last ice age, between 15 000 and 10 000 years ago. Human activities have also brought about change since that time, accelerated by the arrival of Celtic peoples some 2500 years ago with their new technologies of the Iron Age. Since then the vegetation of Scotland has been, and continues to be, very substantially altered. The so-called Highland 'clearances' of the nineteenth century made way for extensive sheep rearing, which continues to dominate Scottish hill farming. The broad tracts of wild, open land which are associated with remoteness today would have supported crofting communities in the past. Since 1850 or thereabouts, coinciding with the industrial and agricultural revolutions, the urban population of Scotland has expanded and the rural population has declined. By 1919, when the Forestry Commission was established, the once extensive forests of Scotland covered no more than 5% of the land area. Expansion of commercial forestry since then has been largely based on high-yielding, exotic conifers, stimulated by the need for reforestation to create a strategic timber reserve after the First World War (Scottish Natural Heritage, 1995).

Two major changes which occurred since 1945 were the collapse of the market for mutton and a steady decline in the price of wool. Hill sheep farming came to specialise in lamb production, for onward fattening on the low ground. This required an intensification of farming on the more fertile or more sheltered ground, with increased stocking densities to improve pasture utilisation. As the poorer land was, in general, unsuitable for lamb production it became available for commercial forestry which expanded onto former mire (peat forming vegetation), heather moorland and unimproved grassland. Where land was not sold for commercial forestry, as in much of the northern Highlands until the 1970s, the withdrawal of sheep and their shepherding created conditions for the expansion of the red deer population. Lowland arable farming has also changed. Prior to 1945 arable farming largely involved a rotation of crops and grass fallow, typically potatoes, cereals and grass. Technological advances and changing farm economics drove the intensification of lowland agriculture towards continuous cereal production and the ever-further reduction of labour inputs (Mutch, pers. comm.).

The National Countryside Monitoring Scheme (NCMS) is a retrospective study of land cover throughout Scotland from the late 1940s which provides new insights to understanding the extent and nature of changes which have taken place thereafter. The study, which was initiated in 1983 by the Nature Conservancy Council, was piloted in Cumbria (Nature Conservancy Council, 1987). It was fully implemented in 1986 as a national survey of land cover change, commencing in Grampian Region, and from then onwards the project has focused on Scotland. Since 1992 the NCMS has been administered by Scottish Natural Heritage (SNH) and was scheduled for completion in 1996.

METHODOLOGY

The NCMS is a sample survey which quantifies the extent of land cover change throughout Scotland between three post-war dates: the late 1940s (baseline date 1947), the early 1970s (baseline date 1973) and the late 1980s (baseline date 1988). This has been made possible by the existence of near-complete black-and-white aerial photographic coverage of Scotland, mainly at a scale of around 1:24 000, for the periods under comparison.

Scotland is characterised by considerable geographical variation in land cover (Ratcliffe and Thompson, 1988) and so each of the 12 Scottish administrative regions (as mapped on the Ordnance Survey Local Government Areas map of 1984) was stratified into broad land cover classes by means of an unsupervised classification of 1977/84 Landsat MSS imagery. The resulting strata vary in land cover composition from region to region, but within any region they approximate to upland, lowland, 'intermediate' and urban classes. A random sample was selected from each stratum, with a sampling rate designed to detect a change of 10% or more in the extent of land cover features with 95% confidence (Kershaw, 1988). The sample coverage of 467 selected squares (Figure 13.1) represents a 7.5% sample of Scotland by area.

The NCMS classification recognises 31 area and 5 linear features (Table 13.1). These were interpreted from aerial photography and mapped at 1:10 000 scale, with a minimum mappable area of 0.1 ha. The hand-drawn maps were then digitised and, because of the processing limitations of the early 1980s PC-based computer technology, the resulting vector data (line-based) was rasterised (grid-based). This generated data for analysis based on 10 m × 10 m pixels, each representing 0.01 ha on the ground. For each sample square it is straightforward in a GIS environment to re-group features for mapping and to calculate patch areas. By the electronic superimposition of sample squares for different time periods it is possible also to calculate land cover change, including the dynamics of gains and losses which occur between time periods, known as 'interchange'.

The outputs produced by Geographical Information System (GIS) software include matrices of land cover change for any two periods under comparison. The fully validated GIS-generated outputs for each district or region were made available to a suite of statistical programs, written in FORTRAN, which incorporated stratification details to calculate the estimates and standard errors. All regional results were then aggregated to provide Scotland-wide estimates, with accuracy in all cases indicated by standard errors and confidence intervals (Tudor et al., 1994b).

Field checking was carried out where there was uncertainty in the aerial photointerpretation. Results of a more comprehensive and independent accuracy assessment

Figure 13.1 The distribution of NCMS sample squares in Scotland showing the location of two case study areas

exercise, undertaken in 1995, will become available prior to the publication of the final results in 1996.

Difficulties which have been experienced in this study stem from the fact that many of the required analytical techniques were in their infancy when it commenced in the early 1980s. A study of agricultural change in the lowlands of Scotland (Langdale-Brown *et al.*, 1980) described a methodology for utilising aerial photography and stratified random sampling. In those pioneering days there was little experience to draw upon for determining appropriate mapping standards, or for optimising the

Table 13.1 The classification of NCMS features

	Land cover groups	Feature type	Feature code
1	**Grassland** (grassland types; semi-natural and agriculturally improved)	Unimproved Grassland Semi-improved Grassland Improved Grassland	UG SG G

2	**Mire** (areas where vegetation is growing in wet peat)	Blanket Mire:
		  heather dominated
		   grass dominated
		Lowland Raised Mire
			
			BM
			BMG
			LRM

3	**Heather Moorland** (includes all areas of heather other than heather-dominated blanket mire)	Heather Moorland	HM
4	**Arable** (areas under crops and cultivated areas)	Arable	A
5	**Woodland and Scrub** (areas of trees and scrub as well as areas of recently felled woodland)	Broadleaved Woodland Coniferous Woodland Tall Scrub Low Scrub Broadleaved Plantation Coniferous Plantation Mixed Woodland Young Plantation Felled Woodland Parkland	BW CW ST SL BP CP MW YP FW P
6	**Water** (areas of open and running water of greater than 10 m in width)	Standing Natural Water Running Natural Water Standing Man-made Water Running Canalised Water	SNW RNW SMW RCW
7	**Built** (built areas, urban and rural as well as roads and railways)	Built Land Transport Corridor	B TC
8	**Other** (semi-natural and artificially created feature types which are uncommon and do not fit into any of the above groups)	Bracken Wet Ground Marginal Inundation Quarry Rock/Cliff Bareground Recreation	BR WG MI Q RK BG R
9	**Linear** (recognisable features less than 10 m in width)	Hedgerow Without Trees Treeline (Including Hedgerow with Trees) Running Natural Water Running Canalised Water Unsurfaced Tracks	H T RNW RCW UT

classification of often similar features with regard to the reliability and replicability of interpreting the tonal and textural characteristics of variable-quality black-and-white aerial photography.

The first phase of the NCMS, which compared change between the late 1940s and the mid-1970s, was undertaken with custom-built GIS software (Marshall, 1985) running on stand-alone desk-top personal computers with concurrent CP/M operating systems. Since 1992 the project has upgraded to an Intergraph workstation environment, with its improved performance partly constrained by the requirement for downward compatibility with its predecessor system. It is envisaged that the full vector data set will ultimately reside on the SNH corporate network, within an ARC/INFO-based workstation environment. During the course of the study there will have been a migration across three GIS hardware–software platforms.

THE VISUALISATION OF LAND COVER CHANGE

Scotland-wide results from the NCMS for the late 1980s are not available at the time of writing but a preliminary comparison of the three time periods in the study has been possible for Central Region (Tudor and Mackey, 1995). While Central Region cannot be regarded as typical of Scotland as a whole, results showed that within the region a reduction in the extent of upland semi-natural features which had been recorded from the 1940s to the 1970s continued into the 1980s. A case study undertaken by the Macaulay Land Use Research Institute has also demonstrated this type of result in part of the Cairngorms for the years 1946, 1964 and 1988 (Scottish Office, 1992).

In the NCMS a vegetation 'patch' is represented by a discrete GIS polygon, classified according to the NCMS method. For each of the sampled areas, vegetation patches and linear features, such as hedges, can be mapped. More generalised views of vegetation cover at the regional level can also aid in illustrating land cover change across Scotland.

A CASE STUDY OF PATCHES: CAITHNESS

The continuing development and testing of geographical information processing in the SNH ARC/INFO environment during 1995 required the extraction of vector test data for the three NCMS time periods. The 5 km × 5 km 'tile' of data, selected by chance, serves to illustrate the application of GIS in studies of land cover change at a local scale. The Caithness case study area (Figure 13.1) can be viewed as GIS-generated maps for the three time periods (Plates 8, 9 and 10 (Figures 13.2, 13.3 and 13.4)). Vegetation changes are not only evident visually, but by automated measurement the GIS is also capable of quantifying patch areas and changes in area (Table 13.2). An interchange matrix (Table 13.3) provides further explanation of the types of change as interpreted from aerial photography, which can be viewed as an arrow diagram (Figure 13.5) to illustrate the direction and magnitude of change.

In the case study several distinct changes in vegetation cover from the late 1940s to the early 1970s, and thereafter to the late 1980s, are evident:

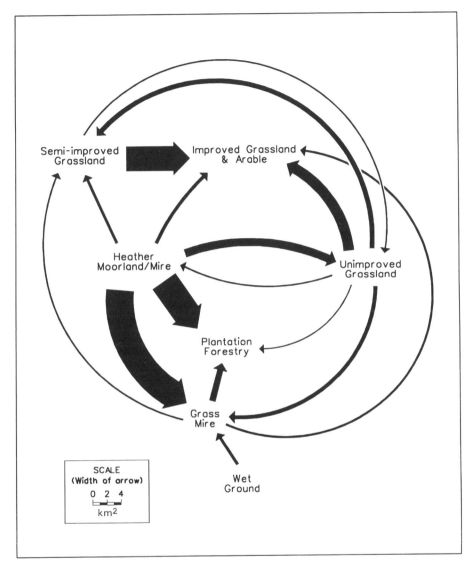

Figure 13.5 Interchanges of more than 0.1 km² between feature groups in the Caithness case study area (square 168) over the 1940s–1980s time period

- In the north-west of the case study area, a 5 km² block of heather-dominated mire was afforested after the late 1940s. This reduced the expanse of heather by nearly half, and covered one-fifth of the case study area in the 1970s. Despite a further expansion in the extent of young plantation along the northern boundary in the late 1980s, the overall plantation area remained essentially unchanged due to the opening-up of pockets of grassland within the afforested block.
- The extent of heather cover was further reduced between the 1970s and the 1980s. This was largely due to a change from heather-dominated to grass-dominated

Table 13.2 Caithness case study of areal features

Feature	Patches (no.)	Area		
		Total (ha)	%	Mean (ha)
1940s				
Broadleaved Woodland	5	14.6	1	2.9
Plantation Forestry	1	0.4	0	0.4
Scrub	17	17.8	1	1.0
Heather/Heather Mire*	20	1113.6	45	55.7
Grass Mire	25	193.8	8	7.8
Wet Ground	2	44.1	2	22.0
Unimproved Grassland	37	359.1	14	9.7
Semi-improved Grassland	39	435.3	17	11.2
Improved Grassland and Arable	7	44.1	2	6.3
Built	101	20.8	1	0.2
Miscellaneous	13	31.6	1	2.4
Outside Coverage	1	224.9	9	224.9
Total	268	2500.0	100	344.6
1970s				
Broadleaved Woodland	5	13.7	1	2.7
Plantation Forestry	50	507.3	20	10.1
Scrub	27	27.4	1	1.0
Heather/Heather Mire	51	550.0	22	10.8
Grass Mire	30	120.7	5	4.0
Wet Ground	9	53.4	2	5.9
Unimproved Grassland	63	352.0	14	5.6
Semi-improved Grassland	38	475.1	19	12.5
Improved Grassland and Arable	24	116.6	5	4.9
Built	129	21.9	1	0.2
Miscellaneous	15	36.7	1	2.4
Outside Coverage	1	224.9	9	224.9
Total	442	2500.0	100	285.2
1980s				
Broadleaved Woodland	11	10.6	0	1.0
Plantation Forestry	83	502.4	20	6.1
Scrub	42	9.4	0	0.2
Heather/Heather Mire	115	148.9	6	1.3
Grass Mire	54	515.8	21	9.6
Wet Ground	21	4.1	0	0.2
Unimproved Grassland	151	231.5	9	1.5
Semi-improved Grassland	67	163.6	7	2.4
Improved Grassland and Arable	101	620.2	25	6.1
Built	128	25.1	1	0.2
Miscellaneous	109	43.5	2	0.4
Outside Coverage	1	224.9	9	224.9
Total	883	2500.0	100	253.9

* 99% heather mire; 1% heather moorland.

Table 13.3 Caithness interchange matrix for summary groups: 1940s and 1980s (number of 10 m × 10 m pixels)

	1980s											
Feature Group	Broadleaved Woodland	Plantation Forestry	Scrub	Heather/ Heather Mire	Grass Mire	Wet Ground	Unimproved Grassland	Semi-improved Grassland	Imp. Grassland & Arable	Built	Miscellaneous	Total
Broadleaved Woodland	952		71	7	4		171	148	57	30	29	1469
Plantation Forestry							6	25		6	2	39
Scrub	1		231	78	36		956	176	231	6	58	1773
Heather/ Heather Mire	3	39 725	346	12 180	39 125	148	11 226	3347	4322	27	891	111 340
Grass Mire	1	8780	11	371	2673	127	2949	1736	2650	32	75	19 405
Wet Ground				474	3760		2	14	147		2	4399
Unimproved Grassland	12	1205	190	1729	5194	59	5705	6506	14 810	151	345	35 906
Semi-improved Grassland	1	451	89	13	391	23	1562	4091	34 979	680	1223	43 503
Imp. Grassland & Arable	31						16	79	4137	71	75	4409
Built	4	8	4		21		123	159	243	1407	125	2094
Miscellaneous		83	2	64	416	58	436	250	465	95	1543	3412
Total	1005	50 252	944	14 916	51 620	415	23 152	16 531	62 041	2505	4368	227 749

(Rows correspond to the 1940s feature groups.)

Note: These figures are preliminary and may be subject to minor change.

vegetation, as may occur with changing moorland management (Miles, 1985; Thompson and Brown, 1992). The overall result was that heather, which had covered 45% of the area in the 1940s, was reduced to 22% in the 1970s and to only 6% in the 1980s. At the same time the number of heather patches increased from 20 to 51 and 115, respectively, with the mean patch size falling from 0.5 km² in the 1940s to 0.1 km² in the 1970s and to 0.01 km² in the 1980s. This can be seen on the aerial photographs.

- Grass-dominated mire was slightly reduced from the 1940s to the 1970s, from 8% to 5% of the overall area, as a result of afforestation. It then expanded throughout the former heather-dominant mire land, to extend across 21% of the overall area in the late 1980s.

- The extent of mire in the 1940s was 13 km², representing over half of the sample area. The vegetation cover, by area, of the mire at that time was estimated to be 85% heather dominated and 15% grass dominated. By the late 1980s the extent of mire had been reduced to 6.6 km², and the vegetation cover was 78% grass dominated and 22% heather dominated. Whereas isolated grassy patches had existed within a heather setting in the 1940s, the converse was true in the late 1980s.

- The overall grassland/arable extent expanded from 33% of the case study area in the 1940s to 38% in the 1970s, and to 41% in the 1980s. The greatest transformation in vegetation composition was from the 1970s to the 1980s: improved grassland/arable accounted for 6% of the grassland/arable area in the 1940s, 13% in the 1970s, but 61% in the late 1980s.

The overall pattern (Figure 13.6) was a reduction in the area of semi-natural features (broadleaved woodland, scrub, heather, wet ground, unimproved grassland); encroachment of grassy vegetation on former heather-dominated mire; afforestation; grassland improvement; and arable expansion. Habitat fragmentation increased, from a relatively simple, open landscape of 268 patches in the 1940s towards greater complexity, with 442 patches in the 1970s and 883 patches in the 1980s.

THE RELIABILITY OF RESULTS

Inaccuracies are inherent in the data capture elements of such methods, which involve aerial photointerpretation, map drawing and digitising. It is always necessary to qualify such findings with an assessment of interpretational accuracy. The accuracy of the NCMS method across Scotland as a whole is therefore being tested in an independent exercise, from which results are not yet available.

More detailed interchange figures (Tables 13.4, 13.5) can highlight the types of difficulty which may arise. Within the time-frame of the study, the creation of blanket mire (peat forming vegetation) cannot be defended ecologically, except where former drainage ditches are blocked to restore active peat formation. Otherwise, the occurrence of a change to mire can only be due to interpretational and/or boundary placement errors. Minor gains and losses may tend to cancel out in the final results and inaccuracies will be reflected in the standard errors.

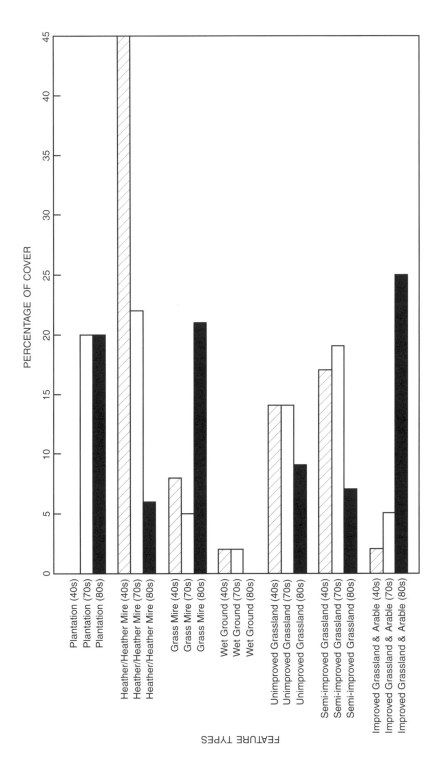

PERCENTAGE OF COVER

FEATURE TYPES

Plantation (40s)
Plantation (70s)
Plantation (80s)

Heather/Heather Mire (40s)
Heather/Heather Mire (70s)
Heather/Heather Mire (80s)

Grass Mire (40s)
Grass Mire (70s)
Grass Mire (80s)

Wet Ground (40s)
Wet Ground (70s)
Wet Ground (80s)

Unimproved Grassland (40s)
Unimproved Grassland (70s)
Unimproved Grassland (80s)

Semi-improved Grassland (40s)
Semi-improved Grassland (70s)
Semi-improved Grassland (80s)

Improved Grassland & Arable (40s)
Improved Grassland & Arable (70s)
Improved Grassland & Arable (80s)

Figure 13.6 Caithness study area: summary of vegetation change

Table 13.4 Caithness interchange matrix for all features: 1940s to 1970s (number of 10 m × 10 m pixels)

1940s \ 1970s	BW	BP	CP	YP	ST	SL	HM	BM	BMG	WG	MI	SNW	SMW	UG	SG	G	A	Q	TC	B	R	BR	BG	RK	Total
BW	1271													76	10	32			45	25					1459
BP		31				1								3	1					2					38
CP																									0
YP																									0
ST					9			30						117	1										157
SL						1130	3	15						328	83	39	2		30						1630
HM							928	16						571	6										1521
BM			29457	10808	27	366	384	45553	9257	1079	27	16	1	10233	2086	57	110	43	483	43					110030
BMG			7565	1180	11	19		4443	2172	60				2802	1046				42	50					19390
WG								3	318	3993	1			38	53										4406
MI										21	100	67		75											263
SNW										24	22	442		50											538
SMW																									0
UG	42	17	1277		9	995	646	1628	289	114		4		16302	13427	807	13		218	144					35932
SG	7	21	407		10	132	360	790	37	41				4316	30161	4050	2237	25	559	396					43549
G	29				6									9	79	2224	287		32	33					2699
A														11		865	802		22	4					1704
Q												7	37	23				17							127
TC	20	3	78		6	17	6		8					146	226	60	45		1441	70					2240
B		2	2			7	3	119	2					130	321	10	13		29	1421					2065
Total	1369	74	38786	11988	78	2667	2330	52760	12083	5332	150	536	38	35230	47500	8144	3509	85	2901	2188					227749

Table 13.5 Caithness interchange matrix for all features: 1970s to 1980s (number of 10 m × 10 m pixels)

1970s \ 1980s	BW	BP	CP	YP	ST	SL	HM	BM	BMG	WG	MI	SNW	SMW	UG	SG	G	A	Q	TC	B	R	BR	BG	RK	Total
BW	916				37	33		10	7					175	100	39	18		16	18		2	2		1373
BP														25	29	16			4	5					79
CP			34738	452	11		128	36	99					3278	2			29	19	13		12	8		38786
YP			10757		7		212	89	361				3	563	1										11989
ST					10				34						1										80
SL						383	10	100	205					1348	243	27	29		25	7		10	23		2667
HM				1981		17	300	520	1106					115	166	289	65			2			29		2326
BM			278					7829	31654					4216	2451	2964			62	1		29	197	2	52760
BMG			278	319	48	12	74	2106	6546	114				1375	540	307	420	15	247	24					12089
WG			6					612	4316	81	1	3		114	138	689	9		63				9		5350
MI								6	78	23	31	29				137	7								144
SNW								11	46	11	49	412	2	21											535
SMW									1	19			22			3									41
UG	26		1377	51	36	280	312	2550	6102	143		55		9277	5926	7209	1336	16	120	103		40	233		35230
SG	7		20	1	1	100	18	2	737	17				2013	6216	30394	6224	1	431	99	125	386	151	3	47501
G	47													129	264	6317	1287		31	39			38		8152
A						1			3					60	25	2033	1261		84	6	43		7		3517
Q			29						15					16	1										85
TC	50		10	7				62	210					343	141	632	59		1267	83	2		16		2882
B	14		13				3		24					103	99	271	6		74	1496	41		16		2163
Total	1060	0	47506	2810	150	826	1057	13933	51544	408	88	499	27	23172	16343	51327	10714	0	2381	2461	211	479	748	5	227749

In the case study area there are occurrences where non-mire features from an earlier period are recorded as mire in a later period. This is especially so in the 1970s–1980s comparison, where $1.73 \, km^2$ of mire creation was recorded, amounting to 27% of the 1980s mire area. This may have arisen from confusion in the interpretation of mire and grassland (54% of the total change), wet ground (28% of the total change) and heather moorland (9% of the total change).

Reference back to the photographs and interpreted maps showed that the 1980s interpreter had noted inconsistencies in an area of formerly drained mire to the south of the conifer block. In such cases the NCMS method requires the earlier interpretation to be checked in this way. The 1980s interpreter seemingly concurred with the 1940s and 1970s interpretations in which this corner of the mire area was classified as wet ground in view of the presence of drainage works, apparently in preparation for afforestation. As the wet ground area had not been planted during the 45-year period, in the 1980s interpretation it appeared to be reverting back to its original mire state through the dereliction of these old drainage ditches. The change, although unusual, was therefore considered by the interpreter to be valid. Complicating factors are that the quality of the aerial photography was poorer in the 1980s, and the interpreters were not the same for the periods under comparison. There were also plausible minor occurrences of unimproved grassland reverting back to mire in pockets along the SW–NE grass/mire margin.

The remaining $0.21 \, km^2$ of apparent mire creation is attributed to:

- Problems in accurately differentiating and drawing boundaries in upland areas where mosaics of plant communities often merge into one another.
- The difficulty of re-drawing boundaries consistently between time periods for areas where no change has occurred.
- The difficulty of digitising boundaries in exactly the same place in each time period. As linework is gridded on a 10 m pixel basis, any matching pairs of lines which are offset by more than half a pixel width will not match exactly when overlaid between two time periods.

The only way of verifying results is to compare them with the true vegetation cover on the ground, a check which can only be accomplished for relatively recent periods of study. A general lesson for future studies is that, within the bounds of affordability, the best possible quality of data source should be chosen, including the use of large-scale colour photography.

A CASE STUDY OF LINEAR FEATURES: EAST LOTHIAN

The above example illustrated a reduction and fragmentation of semi-natural areal features, which might affect the quality of habitat for wildlife. The European Habitats Directive (Council of the European Communities, 1992) and the UK Biodiversity Action Plan (Department of the Environment, 1994a) recognise that linear features, such as hedgerows, streams and ditches, can be important for wildlife. Kirby (1995) states that 'linear features that link up habitat patches have a conservation value in their own right whether or not they act as wildlife corridors'.

While the NCMS recorded five linear features, its most useful information is for hedgerows and lines of trees (termed 'treeline'). These are distinctive features in many parts of lowland Scotland. The hedges, predominantly hawthorn, and the windbreaks of East Lothian commonly date back to the eighteenth century and they are of considerable interest both visually and historically in the cultural landscape.

A case study area in East Lothian (Figure 13.1) has been mapped for the 1940s (Plate 11 (Figure 13.7)) and the 1980s (Plate 12 (Figure 13.8)). Unmanaged hedge may evolve into treeline and former treeline may be cut-back to hedge. It is visually evident that the 1980s landscape became more open, having lost many of the internal hedges which lent a small-scale intricacy to the 1940s rural scene. The length of hedgerow, which was nearly 8 km in the 1940s, was reduced to little over 4 km in the 1980s. The length of treeline, which was nearly 4 km in the 1940s was reduced to less than 3 km in the 1980s. Those features which remained in the 1980s were more dispersed and fragmented.

VEGETATION CHANGE THROUGHOUT SCOTLAND FROM THE 1940s TO THE 1970s

Results from the case studies are consistent with published findings for Scotland as a whole for the 1940s and 1970s (Tudor *et al.*, 1994a). The main gains and losses associated with land cover change in Scotland (Table 13.6) are outlined below:

- Grassland was the most widespread land cover, of which unimproved grassland accounted for 52% of the 1940s grassland area. There was little change in the overall area of grassland from the 1940s to the 1970s, but unimproved grassland was reduced in area by 10% while semi-improved grassland increased in area by 13%. The overall tendency was to grassland improvement, with an apparent increase also in the area of improved grassland.
- Mire was second to grassland in extent and all forms were reduced from the 1940s to the 1970s. It is estimated that heather-dominated blanket mire was reduced by 8%, grass-dominated blanket mire was reduced by 7%, and lowland raised mire was reduced by 23%. The mire group in total was reduced by 8%.
- Heather moorland was estimated to have extended over 20% of Scotland in the 1940s. Managed traditionally by muirburn to suppress woodland regeneration and to maintain a patchwork of varying age for red grouse, red deer and sheep, heather moorland is largely semi-natural in character. It is estimated that the decline in heather cover from the 1940s to the 1970s represented 18% of the 1940s extent.
- Broadleaved woodland was reduced by 14% and coniferous woodland (Caledonian pine forest) was reduced by 51%. Conifer plantation became increasingly evident between the 1940s and the 1970s, with mature plantation increasing by 226% and young plantation increasing by 525%.
- The length of hedgerow was reduced by 37% and the length of treeline was reduced by 9%.

Regions with high semi-natural land cover and low levels of change include Shetland (Scottish Natural Heritage, 1992a) and the Western Isles (Scottish Natural Heritage, 1993a), which were characterised by open landscapes of moorland and

Table 13.6   Feature estimates for Scotland (total area: 77 926 km²)

Feature type	Estimates		% change
	1940s	1970s	
Area groups (km²)			
Grassland			
Unimproved Grassland	12 608	11 414	−9 *
Semi-improved Grassland	3556	4011	13 *
Improved Grassland	8185	8433	3
Group Total	**24 348**	**23 858**	**−2**
Mire			
Blanket Mire – Heather	11 455	10 392	−9 *
Blanket Mire – Grass	8549	7964	−7 *
Lowland Raised Mire	187	144	−23 *
Group Total	**20 191**	**18 500**	**−8**
Heather Moorland			
Heather Moorland	15 377	12 636	−18 *
Group Total	**15 377**	**12 636**	**−18**
Arable			
Arable	8246	8098	−2
Group Total	**8246**	**8098**	**−2**
Woodland & Scrub			
Broadleaved Woodland	1525	1312	−14 *
Coniferous Woodland	187	91	−51 *
Scrub Tall	194	227	17
Scrub Low	465	472	2
Broadleaved Plantation	41	49	20
Coniferous Plantation	945	3080	226 *
Mixed Woodland	580	532	−8
Young Plantation	492	3076	525 *
Felled Woodland	32	64	97
Parkland	83	72	−13
Group Total	**4544**	**8974**	**97**
Water			
Standing Natural Water	1177	1132	−4
Running Natural Water	227	224	−1
Standing Man-made Water	159	246	54
Running Canalised Water	8	10	17 *
Group Total	**1572**	**1612**	**3**

Table 13.6 (*cont.*)

Feature type	Estimates		% change
	1940s	1970s	
Built			
Built	1018	1448	42 *
Transport Corridor	586	618	5 *
Group Total	**1604**	**2066**	**29**
Other			
Bracken	629	693	10
Wet Ground	565	507	−10
Marginal Inundation	48	52	10
Quarry	52	86	63 *
Rock	592	592	0
Bare Ground	87	108	24 *
Recreation	70	145	107 *
Group Total	**2042**	**2183**	**7**
Linear group (km) Hedgerow	42 819	26 874	−37 *
Treeline	19 334	17 600	−9 *
Running Natural Water	123 904	122 728	−1
Running Canalised Water	56 784	71 718	26 *
Unsurfaced Track	31 344	39 240	25 *

All areas are given to the nearest square kilometre, all lengths are given to the nearest kilometre.
* Indicates that the change is statistically significant at the 5% level. This means that the likelihood of finding the observed change, under the hypothesis that no change has taken place, is less than 1 in 20.

heath, and an abundance of open water, rock and cliff. In the 1940s moorland and heath extended over 87% of the Shetland area and over 85% of the Western Isles area. In both cases the area of arable land accounted for 2% or less and woodland cover was insignificant. While successional interchanges were evident, with a tendency to grassland improvement at the expense particularly of mire, the 1970s picture was not greatly different from that of the 1940s.

Highland Region was also relatively little changed, having 83% heath and moorland cover in the 1940s. This was reduced to 80% cover by the early 1970s, with a threefold increase in afforestation. Reduction of hedgerow in places, land drainage and the expansion of unsurfaced track access were also indicative of emerging change (Scottish Natural Heritage, 1993b).

Regions with moderately expansive semi-natural land cover and high levels of change were characterised by landscapes of open hill and lowland agriculture, in which agriculture intensified over the period and afforestation expanded onto moorland. This category encompassed Borders (Nature Conservancy Council and Countryside Commission for Scotland, 1991a), Central (Scottish Natural Heritage, 1993c), Dumfries & Galloway (Scottish Natural Heritage, 1992b), Grampian (Nature Conservancy Council and Countryside Commission for Scotland, 1988), Strathclyde (Scottish Natural Heritage, 1993d) and Tayside (Scottish Natural Heritage, 1993e) Regions.

Regions with little semi-natural land cover and low levels of change included the agricultural lowland regions of Fife (Scottish Natural Heritage, 1993f), Lothian (Nature Conservancy Council and Countryside Commission for Scotland, 1991b) and Orkney (Scottish Natural Heritage, 1992a). Fife and Lothian differed from Orkney in having relatively large and expanding proportions of built land.

THE INTERCHANGE OF VEGETATION GAINS AND LOSSES

The dynamics of change among areal features can be represented for Scotland as a whole by the interchange of gains and losses among feature types (Table 13.7, Figure 13.9), indicating the following main trends:

* A large interchange between improved grassland and arable cover was characteristic of crop rotation, with a small net reduction in area mainly due to an expansion in built land.
* There were interchanges between unimproved grassland, semi-improved grassland and improved grassland, with a tendency to grassland improvement.
* Among the semi-natural heaths and moorland features, there was interchange between heather moorland and unimproved grassland, with a net reduction of heather moorland.
* There was a net reduction in blanket mire to heather moorland as well as to unimproved grassland, which was associated with drainage for afforestation.
* Large areas of unimproved grassland, blanket mire and heather moorland were afforested. The overall extent of heaths and moorland declined by 12%. Afforestation accounted for 62% of this, and conversion to semi-improved grassland accounted for a further 20%.

Among the linear features there was an extension of unsurfaced tracks (largely for access onto moorland and to new conifer plantations) and drainage ditches (associated mainly with land preparation for afforestation), and a reduction in lines of trees and hedges (associated with technological change in agriculture).

RECENT DEVELOPMENTS

Land use practices continue to change. In recent years there has been a reduction in pressures to expand agricultural production and commercial forestry. Agri-environmental schemes and Environmentally Sensitive Areas, which now cover 19% of the agricultural land of Scotland, offer incentives to farm in less intensive ways. Agricultural and forestry policies are drawing back from intensification, with a growing regard for wildlife and landscape (Scottish Natural Heritage, 1995).

The Government has published a UK Action Plan for Biodiversity (Department of the Environment, 1994a), with a view to ratifying the Convention on Biological Diversity which was signed at the United Nations Conference on Environment and Development in Rio de Janeiro in 1992. The programme for its implementation, which is being prepared in 1995, is expected to contain specific action plans for habitats and

Table 13.7 Interchange matrix for Scotland: 1940s–1970s (km²)

(Columns = 1940s features; rows = 1970s features. A blank indicates no interchange. A zero indicates an interchange of less than 0.5 km².)

Feature	UG	SG	G	BM	BMG	LRM	HM	A	BW	CW	ST	SL	BP	CP	MW	YP	FW	P	SNW	RNW	SMW	RCW	B	TC	BR	WG	MI	Q	RK	BG	R	Total
UG	8111	329	118	218	230		9	1718	79	76	12	18	72	2	80	25	27	0	4	3	6	2	0	21	26	93	120	3	4	2	4	0	11 414
SG	745	1683	634	45	23	1	175	446	35	0	7	45	0	11	18	6	1	10	2	3	0	0	17	15	23	55	4	2	0	5	1	4011
G	315	850	4266	8	5	4	98	2607	45	1	8	45	1	20	13	2	3	19	0	3	3	0	42	28	19	24	1	3	0	1	1	8433
BM	36	2	0	9471	795		55	1	2	2	0	1		3	0	1			1	1	1		0	0	3	1	9	0	1	6	1	10 392
BMG	34	1	0	762	7122		18	0	3	0	0	0		0	0	0								1	2	9	1	0	0	5		7964
LRM	5				.1	133	2																									144
HM	870	27	5	314	61	1	11 260	2	14	10	5	8	0	10	5	4	1	0	1	2	0	0	1	1	19	8	1	0	4	1	1	12 636
A	105	347	2684	2	1		19	4806	21		5	22	0	7	7	2	0	5	0	1	0		34	18	3	8	1	1		0	1	8098
BW	59	16	28	2	2		29	121	1028		33	25	1	10	20	2	3	3	1	5	0	0	7	5	15	1	2	1	1	1	0	1312
CW	3	1	1	2	0		2	0	2	75	0	1		2	1	1		0							0		0	0				91
ST	32	9	10	2	0	3	26	5	37		43	15	0	7	20	1	0	1					2	4	7	0	0	0	0	0	0	227
SL	103	32	27	4	2	1	55	34	3		14	117	1	18	16	3	0	1	0	1	0	0	4	2	12	4	1	2	0	2	0	472
BP	6	1	2				1	1	6		2	3	17	3	4	3	1	0					0	0	0	0						49
CP	769	54	35	85	29	14	724	15	68	29	29	67	8	642	60	362	19	3	1	2	1		4	3	51	9	1	0	1	0	0	3080
MW	27	10	13	0	0	0	25	4	50	1	12	8	2	18	340	6	1	2	0	0	0		2	3	7	0	0	0	0	0	0	532
YP	998	51	25	475	249	19	955	18	33	31	6	18	6	68	29	54	2	1	0	0	0		2	1	20	15	1	0	0	0	0	3076
FW	6	0	3	0	0		0	0	2	19	0	3	0	27	2	6	1		0	0			1	0								64
P	2	2	3				10		1	2	1	1	1	7	4	6	0	29	0	0			0	0	1	0					2	72
SNW	2	1	0	2	3		2	0	0		0	1	0	0	0			0	1115	1	0	1		0	0	1	1			0		1132
RNW	5	3	4	1			2	1	5			1	0	1	1			0	0	194	1	0	0	0	1	0	1	0		0	0	224
SMW	13	0	1	0	0		35	1	2		0	1		2	1			1	42	1	142		0	0		1	1					246
RCW	0		0	0	0		0	0	0		0	0	0	0	0		7			1	0	7										10
B	34	47	223	13	0	0	19	166	22	0	4	5	1	7	7	4	2	0	1	1	0	1	865	21	1	0	1	2	2	2	3	1448
TC	25	34	34	12	3	0	11	35	7	1	5	2	0	3	4	2	0	0	0	1	0	0	10	452	1	1	0	0	2	0	0	618
BR	159	13	11	21	5	0	95	3	18	0	3	5	1	3	3	3	1	1	0	0	1	0	0	0	347	3	3	0	0	0	0	693
WG	110	30	8	9	14	0	23	6	1	0	1	2	0	0	0	0	0	0	3	0	0	0	0	0	5	292	1	0	0	0	0	507
MI	3	1	1	1	1		2	0	0		0	0	0	1	0	1	0	0			11		1	1	0	2	26	0	0	0	0	52
Q	9	6	16	1	0		4	7	1	0	0	1	1	0	0	1			0	0	0		1	0	0	0	0	32	0	0	0	86
RK	2	0	0	2	0	0	4				1	0		1	0					1	0		0	1	0	0	0	0	584		1	592
BG	7	7	6	5	1	0	7	8	1		1	1		3	0	2			1				1	1	0	0	0	1	0	54	0	108
R	1	19	30				1	14	1		1	1	0	0	1	2	0	2	0	0	0		2	1	0	0	0	0	0	0	63	145
Total	12 608	3556	8185	11 455	8549	187	15 377	8246	1525		194	465	41	945	580	492	32	83	1177	227	159	8	1018	586	629	565	48	52	592	87	70	77 926

A blank indicates no interchange.
A zero indicates an interchange of less than 0.5 km².

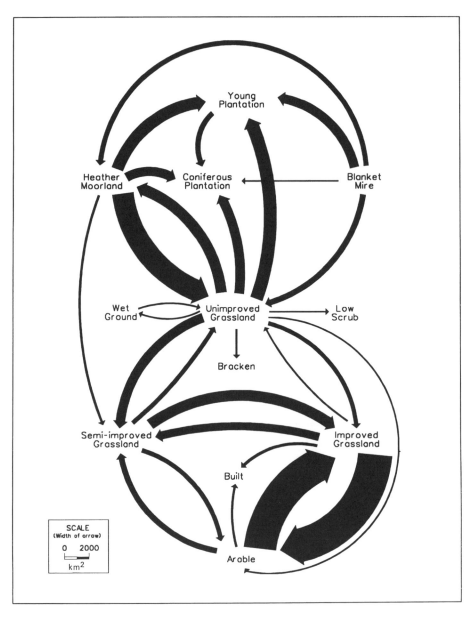

Figure 13.9 Interchanges of more than 100 km² for the whole of Scotland over the 1940s–1970s time period

species. In conjunction with related initiatives (Department of the Environment, 1994b) the Biodiversity Action Plan provides a national policy framework for considering data collection requirements on the biodiversity resource at the local, national and international levels.

Relevant also to the planning of future vegetation studies are requirements to implement the European Directive 92/43/EEC on the Conservation of Natural Habitats and Wild Flora and Fauna, and Directive 79/409/EEC on the Conservation of Wild Birds (Scottish Office, 1995). These include requirements to undertake surveillance of the conservation status of listed natural habitats and species, with particular regard to priority natural habitat types and priority species.

THE INTEGRATION OF LAND COVER STUDIES

A great many methodologies and classifications have been advanced in recent years for land cover mapping (Wyatt *et al.*, 1994). Two which are commonly employed in Scotland are 'Phase I' habitat mapping (Nature Conservancy Council, 1990) at a regional scale and more detailed vegetation community mapping according to the National Vegetation Classification (Rodwell, 1991) at a local scale.

At the national scale, in addition to the National Countryside Monitoring Scheme, the land cover of Scotland has been mapped from an aerial photographic census (Macaulay Land Use Research Institute, 1993) and from satellite imagery (Barr *et al.*, 1993). These provide further types of information, at different scales and resolutions. Among the 59 actions towards meeting objectives in the Biodiversity Action Plan is one to 'improve the database of countryside surveys of Great Britain and Northern Ireland, while further developing the Scottish Office Land Cover of Scotland survey'.

Although there are no equivalent plans at present to extend the NCMS into a further phase of study, knowledge gained and technological advances which have taken place since it was initiated in 1983 might make that a practical proposition if another aerial photographic survey were flown. The priority, however, is to make the most effective possible use of existing NCMS data.

A pilot study has therefore been undertaken, based on NCMS Central Region data, to explore the feasibility of integrating the NCMS sample-based data set with the Land Cover of Scotland census (Brooker, 1997). In this it is planned to combine the two surveys fully in a GIS environment to:

- provide classification comparisons;
- develop the database of land cover in order to analyse habitat fragmentation and countryside change in Scotland;
- provide a possible basis for recompiling NCMS results independently of the regional stratification in its sampling design (i.e. to use the land cover census as a dynamic carrying surface for the NCMS sample data); and to
- develop an understanding of different class definitions and so to move towards standardised land cover descriptions and statistics for Scotland.

A European classification has also been developed for vegetation mapping (Commission of the European Communities, 1991). A pilot study has been undertaken to test the feasibility of converting the land cover map of Great Britain to the CORINE classification (Fuller and Brown, 1994). Conversion from one classification to another is invariably problematic, but in the absence of a universal standard it is likely to remain an important consideration for vegetation mapping and environmental reporting.

CONCLUSIONS

The NCMS was established in June 1983 in response to a demand at that time for information about how the countryside was changing in the face of fiscal and strategic policies aimed at increasing agricultural and forestry productivity. The NCMS has demonstrated that semi-natural habitats have been lost over a period when agricultural and forestry developments were bringing about large-scale changes in the land cover of Scotland. The quantification of these effects will continue to assist in informing the development of policy initiatives for conservation and enhancement of the natural heritage of Scotland. Technical lessons learned from the method will also help to inform future studies of land cover in Scotland.

ACKNOWLEDGEMENTS

The authors would like to thank members of the Environmental Audit Branch of Scottish Natural Heritage for assistance in the preparation of this chapter, and, more especially, the members of the now-dispersed National Countryside Monitoring Scheme team who have, over the years, assembled the data upon which it is based.

REFERENCES

Barr, C.J., Bunce, R.G.H., Clarke, R.T., Fuller, R.M., Furse, M.T., Gillespie, M.K., Groom, G.B., Hallam, C.J., Hornung, M., Howard, D.C. and Ness, M.J. (1993) *Countryside Survey 1990: Main Report.* Department of the Environment, London.
Brooker, N. (1997) The integration of land cover data from NCMS and LCS88. Report for Central Region. *Scottish Natural Heritage Research Survey and Monitoring Series*, Report No. 71 Perth.
Commission of the European Communities (1991) *CORINE Biotopes Manual: Methodology.* Office for Official Publications of the European Communities, Luxembourg.
Council of the European Communities (1992) Council Directive 92/43/EEC of 21 May 1992 on the conservation of natural habitats and of wild fauna and flora. *Official Journal of the European Communities*, No. L 206/7. Office for Official Publications of the European Communities, Luxembourg.
Department of the Environment (1994a) *Biodiversity: The UK Action Plan.* CM2428. HMSO, London.
Department of the Environment (1994b) *Sustainable Development: The UK Strategy.* CM2426. HMSO, London.
Fuller, R.M. and Brown, N.J. (1994) *CORINE Land Cover Map: Pilot Study.* Countryside 1990 Series Vol. 5, Department of the Environment, London.
Kershaw, C.D. (1988) *A Review of the Statistical Aspects of the National Countryside Monitoring*

Scheme. Nature Conservancy Council (internal report).

Kirby, K. (1995) *Rebuilding the English Countryside: Habitat Fragmentation and Wildlife Corridors as Issues in Practical Conservation.* English Nature Science, vol. 10, Peterborough.

Langdale-Brown, I., Jennings, S., Crawford, C.L., Jolly, C.M. and Muscott, J. (1980) *Lowland Agricultural Habitats (Scotland): Air-Photo Analysis of Change.* Chief Scientist's Team Report No. 332, NCC, Peterborough.

Macaulay Land Use Research Institute (1993) *The Land Cover of Scotland 1988: Final Report.* MLURI, Aberdeen.

Marshall, G.F. (1985) *LANDSCAPE: Operating Instructions.* SNH internal report.

Miles, J. (1985) The pedogenic effects of different species and vegetation types and the implications of succession. *Journal of Soil Science,* **36,** 571–574.

Nature Conservancy Council (1987) Changes in the Cumbrian Countryside: First Report of the National Countryside Monitoring Scheme. *Research and Survey in Nature Conservation,* **6,** NCC, Peterborough.

Nature Conservancy Council (1990) *Handbook for Phase 1 Habitat Survey: A Technique for Environmental Audit.* Joint Nature Conservation Committee, Peterborough.

Nature Conservancy Council and Countryside Commission for Scotland (1988) *National Countryside Monitoring Scheme, Scotland: Grampian.* CCS and NCC, Perth.

Nature Conservancy Council and Countryside Commission for Scotland (1991a) *National Countryside Monitoring Scheme, Scotland: Borders.* CCS and NCC, Perth.

Nature Conservancy Council and Countryside Commission for Scotland (1991b) *National Countryside Monitoring Scheme, Scotland: Lothian.* CCS and NCC, Perth.

Ratcliffe, D.A. and Thompson, D.B.A. (1988) The British uplands: their ecological character and international significance. In: M.B. Usher and D.B.A. Thompson (eds) *Ecological Change in the Uplands.* Blackwell Scientific Publications, Oxford, pp. 9–36.

Rodwell, J.S. (ed.) (1991) *British Plant Communities. Volume 1: Woodlands and Scrub* (and further volumes published and in print). Cambridge University Press, Cambridge.

Scottish Natural Heritage (1992a) *National Countryside Monitoring Scheme Scotland: Northern Isles.* SNH, Perth.

Scottish Natural Heritage (1992b) *National Countryside Monitoring Scheme Scotland: Dumfries and Galloway.* SNH, Perth.

Scottish Natural Heritage (1993a) *National Countryside Monitoring Scheme Scotland: Western Isles.* SNH, Perth.

Scottish Natural Heritage (1993b) *National Countryside Monitoring Scheme Scotland: Highland.* SNH, Perth.

Scottish Natural Heritage (1993c) *National Countryside Monitoring Scheme Scotland: Central.* SNH, Perth.

Scottish Natural Heritage (1993d) *National Countryside Monitoring Scheme Scotland: Strathclyde.* SNH, Perth.

Scottish Natural Heritage (1993e) *National Countryside Monitoring Scheme Scotland: Tayside.* SNH, Perth.

Scottish Natural Heritage (1993f) *National Countryside Monitoring Scheme Scotland: Fife.* SNH, Perth.

Scottish Natural Heritage (1995) *The Natural Heritage of Scotland: An Overview* (Ed. E.C. Mackey). SNH, Perth.

Scottish Office (1992) *The Scottish Environment – Statistics. No. 3.* HMSO, Edinburgh.

Scottish Office (1995) *Habitats and Birds Directives.* Circular No. 6/1995, Edinburgh.

Thompson, D.B.A. and Brown, A. (1992) Biodiversity in montane Britain: habitat variation, vegetation diversity and some objectives for conservation. *Biodiversity and Conservation,* **1,** 179–208.

Thompson, D.B.A., Hester, A.J. and Usher, M.B. (eds) (1995) *Heaths and Moorland: Cultural Landscapes.* HMSO, Edinburgh.

Tudor, G.J., Mackey, E.C. and Underwood, F.M. (1994a) *The National Countryside Monitoring Scheme: The Changing Face of Scotland, 1940s to 1970s. Main Report.* Scottish Natural Heritage, Perth.

Tudor, G.J., Mackey, E.C. and Underwood, F.M. (1994b) *The National Countryside Monitoring Scheme: The Changing Face of Scotland, 1940s to 1970s. Technical Report*. Scottish Natural Heritage, Perth.

Tudor, G.J. and Mackey, E.C. (1995) Upland land cover change in post-war Scotland. In: D.B.A. Thompson, A.J. Hester. and M.B. Usher (eds) *Heaths and Moorland: Cultural Landscapes*. HMSO, Edinburgh.

Wyatt, B.K., Greatorex Davies, N., Bunce, R.G.H., Fuller, R.M. and Hill, M.O. (1994) *Countryside Survey 1990: Comparison of Land Cover Definitions: Dictionary of Surveys and Classifications of Land Cover and Land Use*. Institute of Terrestrial Ecology, Monks Wood.

14 Application of CORINE Land Cover Mapping to Estimate Surface Carbon Pools in Northern Ireland

M.M. CRUICKSHANK,[1] **R.W. TOMLINSON**[1] **and R. MILNE**[2]
[1] *School of Geography, Queen's University, Belfast, UK*
[2] *Institute of Terrestrial Ecology, Edinburgh Research Station, Penicuik, UK*

The land cover of Northern Ireland was mapped between 1992 and 1994 following the CORINE methodology of the European Union. This allowed an inventory and map of carbon (C) pools in vegetation to be made similar to those for GB. In GB the method of land cover mapping differed so that links were required between the two sets of land cover classes. Comparable carbon densities for the Northern Ireland land cover classes also had to be derived. The C stored in vegetation in Northern Ireland was estimated as 4.6 M tonnes; forests accounted for 53% of the pool (store) although they occupied only 4.7% of the land area. In contrast, agricultural land accounted for 29% of the C pool but 74% of the land area. Peatland vegetation occupied 10% of the land area and stored 5.7% of the carbon. The carbon pool was mapped by 10×10 km grid squares and those in the highest class ($8 \, t \, C \, ha^{-1}$ or more) were associated with the distribution of conifer forests; less extensive conifer forests explain most squares with 6–$7.9 \, t \, C \, ha^{-1}$, although demesnes are also important. Peatland was significant in accounting for squares with 2–$3.9 \, t \, C \, ha^{-1}$. The Greater Belfast region had relatively high C pools caused by trees in public parks, golf courses and gardens of large houses. The CORINE land cover methodology has been applied throughout Europe and would allow estimation of European C pools following the approach adopted in this chapter.

INTRODUCTION

The land cover of Ireland, including Northern Ireland, was mapped between 1992 and 1994 from satellite images following the European Union CORINE methodology (O'Sullivan, 1994). The polygons of land cover types derived from this were stored in a database which allows the mapping to be applied, for example to conservation and land management issues (Cruickshank and Tomlinson, 1994). Another application is in the estimation of carbon pools and fluxes, one step towards possibly reducing greenhouse gases in the atmosphere (Cruickshank, *et al.*, 1995).

Under the Framework Convention on Climate Change (1992) the United Kingdom (UK) Government agreed to aim to return emissions of greenhouse gases to 1990 levels by the year 2000. It was committed also to establish a national inventory of greenhouse gas pools in forests and other vegetation on the land surface, and in soils and peat below the surface. Policies to protect and enhance these pools, taking account of

Vegetation Mapping: From Patch to Planet. Edited by Roy Alexander and Andrew C. Millington.
© 2000 John Wiley & Sons Ltd.

possible land use changes, were to be developed so that more carbon might be sequestered from the atmosphere. A first estimate of the carbon pools in vegetation and soils covered Great Britain only (Milne, 1994; Howard *et al.*, 1995) but the current inventory includes Northern Ireland (NI) to give national, United Kingdom (UK) coverage (Milne and Brown, Chapter 15, this volume). However, land cover data for NI are different from those used in Great Britain (GB).

This chapter aims to show how the CORINE land cover data for NI were used to produce an inventory and map of carbon pools in vegetation comparable with that of GB. This required links to be found between the two sets of land cover classes and then comparable carbon density factors for the CORINE classes. Carbon densities were multiplied by the areas of the classes to yield vegetation carbon pools. As in GB, the spatial pattern of these pools was shown by 10 × 10 km grid squares, although the grid used was the Irish National Grid.

METHODOLOGY

CORINE land cover

The CORINE methodology required visual interpretation of hard copy images at 1:100 000, derived from Landsat TM data in the red, near-infrared and middle infrared wavebands of May 1990 (some small cloud-covered areas were mapped from May 1989 data). The land cover polygons interpreted at Level 3 of the CORINE classification (Table 14.1) were traced onto acetate overlays from which they were digitised (by scanning) and labelled by the Ordnance Survey (NI). Two classes, pasture and peat bogs, were subdivided at Level 4 outside the EC contract and these subdivisions were used in the carbon study. The CORINE map database in vector format is in ARC/INFO.

At Level 3, CORINE has 44 classes, but only those applicable to Ireland are included in Table 14.1. *Marine waters* were excluded from the carbon study but the two *Continental water* classes (5.1.1 and 5.1.2) were included as part of the land area. The two classes subdivided at Level 4 brought the total number of CORINE classes used for the carbon study to 34. The minimum polygon size mapped is 25 ha, so that even polygons of 'pure' classes contain small patches of others; for example, an arable area (2.1.1) may have small scattered fields of pasture so long as their total does not cross the threshold percentage beyond which it would be classed as 2.4.2 (O'Sullivan, 1994). In other cases 'pure' classes will not be uniform in land cover; for example, *discontinuous urban fabric* (1.1.2), which is used for suburban areas, includes trees, grass and built surfaces. Some agriculture and tree classes are mixed by definition (e.g. 2.4.3) and the proportions of the mix vary. All these points were considered in devising carbon density factors for CORINE classes.

Accuracy of the CORINE land cover mapping was verified independently as part of the EC contract; for the north of Ireland, the interpretation was approximately 80% accurate in the east and 70% in the west. Additional spot checks by the EC CORINE technical team found the interpretation to be good, but did not report a percentage value (O'Sullivan, 1994).

Table 14.1   CORINE classification of land cover used in Ireland

Level 1	Level 2		Level 3
1. Artificial surfaces	1.1 Urban fabric	1.1.1	Continuous urban fabric
		1.1.2	Discontinuous urban fabric
	1.2 Industrial, commercial, transport	1.2.1	Industrial or commercial units
		1.2.2	Roads/railways and associated land
		1.2.3	Sea ports
		1.2.4	Airports
	1.3 Mines, dumps and construction sites	1.3.1	Mineral extraction sites
		1.3.2	Dumps
	1.4 Non-agricultural vegetated areas	1.4.1	Green urban areas
		1.4.2	Sport and leisure facilities
2. Agriculture	2.1 Arable land	2.1.1	Arable land (crops and bare soil)
	2.3 Pastures	2.3.1	Pastures
	2.4 Heterogeneous agriculture	2.4.1	Annual crops with permanent crops
		2.4.2	Arable–pasture mix
		2.4.3	Principally occupied by agriculture, with significant natural vegetation
3. Forest and semi-natural areas	3.1 Forests	3.1.1	Broadleaved forest
		3.1.2	Coniferous forest
		3.1.3	Mixed forest
	3.2 Scrub or herbaceous vegetation	3.2.1	Natural grassland
		3.2.2	Moors and heathlands
		3.2.4	Transitional woodland–scrub
	3.3 Open spaces; little or no vegetation	3.3.1	Beaches and dunes
		3.3.2	Bare rocks
		3.3.3	Sparsely vegetated areas
		3.3.4	Burnt areas
4. Wetlands	4.1 Inland wetlands	4.1.1	Inland marshes
		4.1.2	Peat bogs
	4.2 Coastal wetlands	4.2.1	Salt marshes
		4.2.3	Intertidal flats
5. Water bodies	5.1 Continental waters	5.1.1	Rivers
		5.1.2	Lakes
	5.2 Marine waters	5.2.1	Coastal lagoons
		5.2.2	Estuaries
		5.2.3	Sea and ocean

In Ireland, pasture was subdivided:
2.3.1.1   High productivity – strong NIR on image from healthy plants, good growth. Much would be cut for silage two to three times a year.
2.3.1.2   Low productivity. Low management, usually light grazing. In a minority of cases can be due to recent cutting or heavy grazing.
2.3.1.3   Patchwork of two previous, in units of less than 25 ha.
Peat bogs were subdivided on image evidence: 4.1.2.1 not exploited; 4.1.2.2 exploited.

Links Between CORINE Classes in Northern Ireland and Land Cover Classes Used in Great Britain

In GB 17 key cover types were identified and used to estimate carbon pools (Cannell and Milne, 1995); therefore the 34 land cover classes of CORINE had to be connected to these. Links were established by discussion between the Institute of Terrestrial Ecology (ITE) and NI researchers and guided by those given in Wyatt *et al.* (1994). For CORINE, class definitions in the CORINE Technical Guide (EC, 1992), as extended for Ireland (O'Sullivan, 1994), were used together with personal knowledge of NI land cover. The links established are shown in Table 14.2.

Some classes in the two land cover systems have similar definitions and, in other cases, several CORINE classes can be combined into one GB class, but problems remain:

1. A number of CORINE Level 3 classes are mixed by definition (2.4.1, 2.4.2, 2.4.3, 3.1.3 and 3.2.4) and also the Level 4 subdivision of pasture includes the mixed 2.3.1.3 class. All mixed classes contain components of two or more GB classes.
2. Some 'pure' classes (1.1.2, 1.2.4, 1.4.1 and 1.4.2) include components of land cover which will have different carbon densities.
3. In the CORINE classification all semi-natural vegetation is allocated to 'pure' classes, e.g. natural grassland (3.2.1); there are no defined mixed ones, unlike for agriculture and forest. Classes of semi-natural vegetation therefore are less pure and less true to their definition, than agriculture and forest classes.

Calculation of Carbon Density for CORINE Classes

Where an acceptable link was established between CORINE and GB classes, the carbon density used for the GB class was applied to the NI CORINE class, but about a third of the CORINE classes required additional work to find a carbon density.

Carbon Density of Forests in Northern Ireland

Work on forests as carbon pools (Cannell *et al.*, 1996) revealed that the species and age structures of forests differed from those in GB, so that the carbon densities estimated by Milne (1994) for GB were inappropriate. Mean carbon densities for broadleaved and conifer forests in NI therefore were calculated and to a base date of 1993 rather than 1980 as in GB. The calculations, which followed Milne (1994), were based on broadleaves in the private sector, where most are found, and conifers in the state sector – again where most occur.

To calculate carbon density for broadleaved forest requires area and volume data for species by age class. The *Private Woodland Inventory, 1975–79* (Graham, 1981) includes volume and area tables for major species by age groups, although small amounts are omitted. Annual Reports of Forest Service, Department of Agriculture for Northern Ireland (DANI) give the total area of grant-aided private planting in each year since 1975 and the Forest Service can estimate the percentage likely to be broadleaved. Woodland has been lost as well as planted through the 1975–93 period.

Table 14.2 Links between GB key cover types for summary data and CORINE classes

Key cover types (17 classes)	CORINE classes
Continuous urban	1.1.1, 1.2.1, 1.2.2, 1.2.3
Suburban	1.1.2, 1.2.4
Tilled land	2.1.1, half 2.4.2
Managed grassland	1.4.1, 1.4.2, 2.3.1.1, half 2.3.1.3, half 2.4.2
Bracken	Included in 3.2.2
Rough grass/marsh	2.3.1.2, half 2.3.1.3, 3.3.1
Heath/moor grass	3.2.1
Open shrub heath/moor	3.2.2
Dense shrub heath/moor	
Bog	4.1.1, 4.1.2.1
Deciduous/mixed wood	3.1.1 broadleaved, 3.1.3 mixed
Coniferous woodland	3.1.2
Inland bare	1.3.1, 1.3.2, 3.3.2, 3.3.3, 3.3.4, 4.1.2.2
Salt marsh	4.2.1
Coastal bare	4.2.3
Inland water	5.1.1, 5.1.2
Sea/estuary	5.2.1, 5.2.2, 5.2.3

Note, no links found for CORINE classes 2.4.1, 2.4.3 and 3.2.4

During the Private Woodland Inventory, annual losses of 160 ha per annum were noted, and as two-thirds was broadleaved, an approximate rate of loss of broadleaved forest was 100 ha per annum. More recently, enhanced grants for farm woodland maintenance and planting are thought to have reduced the rate of loss (Guyer and Edwards, 1988). For the period 1979–93, rates of loss were taken as 100 ha per annum for the first half and 50 ha per annum for the second half. It was assumed that losses would be from the oldest age class. Using the volume and area data from the Private Woodland Inventory and Forest Service, the assumed losses, and applying conversion factors (Milne, 1994), a weighted average carbon density for broadleaved forest in 1993 was calculated as 62.3 t ha^{-1}.

Calculation of carbon density for conifer forests was based on data supplied by the Biometrics Division of DANI. Data on area and volume in the present forest estate were given for six groups by age bands: Sitka spruce, Norway spruce, Douglas fir, all pines, all larches, and other conifers. Volume data were based on mean volume per hectare from inventory plots of continuous tree planting located at random. Forests include roads, rides and buildings which may total up to 15% of their area, so that for the volume/area ratios required for carbon density, the areas were increased to allow for these features. Volumes were estimated to a minimum top diameter of 7 cm, therefore some timber in the tips of older trees and the whole of very young trees was omitted. In fact, volume data were not supplied for post-1980 plantings (21% of planted area), nor for two groups in the previous decade even though area data showed trees were planted. To fill these gaps, timber volume/unit area was estimated at half that of the preceding decade. Volume/unit area for different ages and species then was converted into carbon density using calculations as for GB (Milne, 1994).

Carbon conversion factors used for GB by Milne included a stemwood biomass factor of 2 for trees up to 20 years old and 1.5 for older trees. In NI the age ranges provided by DANI were 0–13 years (1993–81) and then by 10-year groups to 64–73 (1930–21) and those planted before 1921. For the youngest group a factor of 2 was used, for the next 1.75 and 1.5 for all older groups. From these calculations the average conifer carbon density in 1993, weighted by species and age, was estimated at $34\,t\,ha^{-1}$.

Mixed forests, in the absence of evidence on the proportions of broadleaves and conifers, were given a carbon density equal to the mean of the values obtained for each, that is $48.2\,t\,ha^{-1}$.

Calculation of Carbon Density for CORINE Classes which Include Components of Several Land Cover Types

Some CORINE classes include components of two or more GB classes, and where these have the same carbon density as each other there is no problem in finding a carbon density for the mixed class. For example, the arable land and good pasture in class 2.4.2 (Table 14.1) both have a carbon density of $1\,t\,ha^{-1}$. However, other CORINE classes contain components which belong to GB classes with different carbon densities and therefore an estimate of the proportions of cover types present in these mixed classes was made before carbon densities for the components were applied.

Suburban areas are classified in CORINE as *discontinuous urban fabric* (1.1.2). Even after town parks, playing fields and golf courses exceeding 25 ha are taken out as separate classes (1.4.1 and 1.4.2), there remains a variable mix of built-over surfaces, grass and trees in gardens, verges, play areas etc. Two sources were used to establish the proportions of the cover types: (i) a small selection for south Belfast of vertical colour air photographs of 1990 at 1:3000 scale, and (ii) OS(NI) maps at 1:2500 scale from the late 1970s. Using a transparent grid, the land cover type at 1 cm intersections was noted. The percentage of intersections falling on each type was calculated. The sample areas included differing sizes and dates of houses, and public and private housing. Non-vegetated surfaces occupied 60% of the area, gardens etc. 35% and trees (predominantly broadleaves) 5%. Using carbon density values of 0, 1.0 and $62.3\,t\,ha^{-1}$, respectively, the carbon density for class 1.1.2 was estimated as $3.5\,t\,ha^{-1}$. For specific $10 \times 10\,km$ grid squares (see Results section), with distinctive image appearance and following work on their surface types, particular carbon density values were calculated.

Airports (1.2.4) have built-over surfaces as runways, aprons, car parks and buildings (carbon density $0\,t\,ha^{-1}$), and large areas of grass (carbon density $1\,t\,ha^{-1}$). Study of the satellite image suggested a 50% cover of each of non-vegetated surface and grass so that a carbon density of $0.5\,t\,ha^{-1}$ was estimated.

The appearance of *green urban areas* (1.4.1) on the images showed that these were not generally mixed and the class had been used mainly for school and other playing fields greater than 25 ha. Trees are rare and most land cover is short grass; a carbon density of $1.0\,t\,ha^{-1}$ was given, equal to that of good pasture.

Sport and leisure facilities (1.4.2) included public parks and golf courses. The surface cover of areas classed as 1.4.2 was examined using the vertical colour air photographs and a transparent grid as for class 1.1.2 above. Additionally, for a large area of 1.4.2 in south Belfast the hectarage of trees was known from the Lagan Valley Regional Park (LVRP) Tree Survey (Tomlinson, 1988). Also for the LVRP, the extent of lakes, buildings and roads, and in grass were estimated from 6 inch OS maps. These, combined with the tree hectarage, allowed a carbon density for the LVRP to be obtained. Using this, together with the carbon densities found for areas of 1.4.2 covered by air photographs, a carbon density for this class was estimated as $12 \, t \, ha^{-1}$.

Scattered orchards, among mainly pasture and some arable fields, in a compact area south of Lough Neagh, were classed as *annual crops associated with permanent crops* (2.4.1). Orchards on a 1:20 000 topographic map of 1975 covered 27% of the area mapped as 2.4.1. DANI statistics and local sources indicated that between 1975 and 1990, the date of the satellite image, the orchard area decreased by 36%. Applying this reduction, 17.3% of the mapped area was taken as orchard in 1990. At the time of this study there was no available carbon density for orchards and it was decided to take the value from one of the poorer woodland types derived from the Private Woodland Inventory (carbon density of $36 \, t \, ha^{-1}$). Using the 17.3% figure for the area of orchard, and giving a carbon density of $1.5 \, t \, ha^{-1}$ for the mixed quality grassland beneath and between the orchards, a carbon density for class 2.4.1 was estimated at $8 \, t \, ha^{-1}$.

Assigning a carbon density value to class 2.4.3, *principally agriculture with significant areas of natural vegetation*, is difficult because the components of the land cover and their proportions vary. To obtain a carbon density, all $10 \times 10 \, km$ squares with 2.4.3 in them were extracted from the CORINE database (152 of the 182 grid squares). A stratified sample included large and small polygons spread across NI. In each of 12 selected grid squares, 2.4.3 polygons were viewed on the satellite image and the components listed, e.g. pasture with bog, pasture with trees. No attempt was made to estimate proportions of the components in polygons because of their intricate mixture. The most common and extensive mixtures were natural grassland with pasture or natural grassland with bog and pasture. This finding results from the predominant location of 2.4.3 in upland margins or around lowland bogs. Mixtures including trees were found, for example in demesnes, alongside incised rivers and around Upper and Lower Lough Erne. However, from the samples, more general inspection of the images and field knowledge, a carbon density of $2.0 \, t \, ha^{-1}$ seemed reasonable and gave weight to the dominant pasture, natural grassland and bog mix.

Transitional woodland-scrub (3.2.4) was used for discontinuous trees, as when colonising or regressing or in a more stable mix maintained by grazing. Trees are normally found within rough grass and it was estimated that each occupied 50% of the class. Trees were given the carbon density ($36 \, t \, ha^{-1}$) for the sycamore–ash–birch grouping as derived from the Private Woodland Inventory, and the rough grass a carbon density of 2.0, leading to an overall carbon density for the class of $19 \, t \, ha^{-1}$.

Pastures (2.3.1) were divided in Ireland to Level 4: 2.3.1.1 are pastures with high biomass/good growth, 2.3.1.2 with low biomass/poor growth and 2.3.1.3 is a patchwork of each of those in units of less than 25 ha and the mix was taken to be 50% of each. 2.3.1.1 was equated with managed grass in the GB classification and given a carbon density of $1 \, t \, ha^{-1}$; 2.3.1.2 was similar to the rough grass in GB and a carbon density of $2.0 \, t \, ha^{-1}$ was assigned. The mixed class 2.3.1.3, therefore, was given a carbon density of $1.5 \, t \, ha^{-1}$.

The derived carbon densities for CORINE classes are summarised in Table 14.3.

Densities were combined with class areas to calculate total carbon pools and their spatial pattern.

Distribution of Carbon by 10 × 10 km Grid Squares

Using ARC/INFO, the CORINE database coverage was intersected with a grid coverage (Irish National Grid) to divide the map into 10 × 10 km squares. ARC routines enabled the total hectarage in each CORINE class to be recorded for every square. Multiplication of these areas by their carbon density factors (Table 14.3) allowed the mean carbon pool per hectare for the land area to be calculated for each square. These values were allocated to the same class ranges as used for GB and the grid map was produced with the boundary of NI superimposed (Figure 14.1).

RESULTS

Carbon Pools by CORINE Classes

The CORINE data suggest that forests (3.1.1, 3.1.2, 3.1.3) cover only 4.7% of the land area of NI (5.8% following DANI figures), but because of their high carbon densities, account for 53% of the vegetation carbon pool (Table 14.4). Conifer forest alone accounts for 39% of the pool. Agriculture (2.1.1, 2.3.1, 2.4.1, 2.4.2, 2.4.3) occupies a much larger percentage of the land area (74%) and accounts for 29% of the carbon pool. However, it should be noted that agricultural land could be redefined, since in CORINE it includes some natural vegetation (in 2.4.3), while some natural grassland (3.2.1) and moorland (3.2.2), both classed as semi-natural vegetation, are used for extensive grazing. Among other classes, the contribution of areas of transitional woodland (3.2.4) may be noted (5% of the carbon pool, 1% of the land area). Peatland vegetation (4.1.2.1) accounts for 5.7% of the carbon pool and 10% of the land area, but its real significance will emerge once sub-surface carbon pools are calculated, i.e. from the peat itself (Howard *et al.*, 1995). It is noted that suburban areas (1.1.2) contribute 2.4% of the carbon pool of NI, following the calculations outlined earlier. Such areas were given a carbon density of zero in GB, a point which is being re-assessed.

Carbon Stored by Forests in Northern Ireland

According to CORINE-based estimates, the total vegetation carbon pool in NI is 4.6 Mt, of which 2.4 Mt is in forest. However, CORINE missed some young forests and treed areas less than 25 ha which are included in the DANI statistics. In CORINE these were mapped as poor pasture, natural grassland and peatland, all of which have lower carbon densities than forests. If these missing forest hectares were assumed to include both conifers and broadleaves and young and old trees, then the carbon density for mixed forests could be substituted – the forest pool is then increased to 3 Mt. This increases the vegetation carbon pool for NI to 5.2 Mt. However, as most of the missed forest is of young trees, the vegetation pool is likely to be towards the lower

Table 14.3   Carbon densities (t ha^{-1}) for CORINE classes

Carbon density	CORINE class
0	1.1.1, 1.2.1, 1.2.2, 1.2.3, 1.3.1, 1.3.2, 3.3.2, 3.3.3, 3.3.4
	4.1.2.2, 4.2.3, 5.1.1, 5.1.2
0.5	1.2.4
1.0	1.4.1, 2.1.1, 2.3.1.1, 2.4.2
1.5	2.3.1.3, 3.2.1, 3.3.1
2.0	2.3.1.2, 2.4.3, 3.2.2, 4.1.1, 4.1.2.1, 4.2.1
3.5	1.1.2
8.0	2.4.1
12.0	1.4.2
19.0	3.2.4
34.0	3.1.2
48.2	3.1.3
62.3	3.1.1

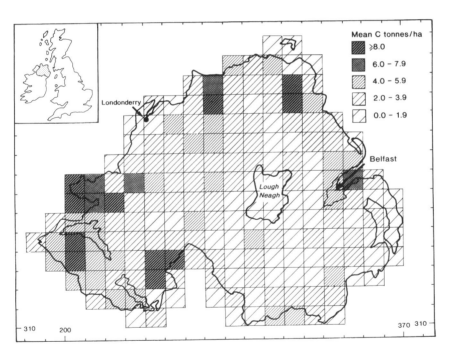

Figure 14.1   Spatial pattern of vegetation carbon pools in Northern Ireland. The two lakes in the west are Upper Lough Erne and Lower Lough Erne (the more northerly)

Table 14.4 Vegetation carbon pools (NI) by CORINE classes

CORINE class	Area (ha)	%/NI	C t ha^{-1}	Carbon (tonnes)	% C
1.1.1	5 327.9	0.38	0	0	
1.1.2	30 649.9	2.17	3.5	107 274.7	2.35
1.2.1	3 476.3	0.25	0	0	
1.2.2	642.3	0.05	0	0	
1.2.3	352.8	0.02	0	0	
1.2.4	1 201.1	0.08	0.5	600.6	0.01
1.3.1	2 465	0.17	0	0	
1.3.2	146.6	0.01	0	0	
1.4.1	897.6	0.06	1	897.6	0.02
1.4.2	4 132.5	0.29	12	49 590.0	1.09
2.1.1	32 176.9	2.28	1	32 176.9	0.71
2.3.1.1	427 323.8	30.22	1	427 323.8	9.37
2.3.1.2	72 692.9	5.14	2	145 385.8	3.19
2.3.1.3	262 491.1	18.56	1.5	393 736.7	8.63
2.4.1	6 363.0	0.45	8	50 904.0	1.12
2.4.2	142 960.3	10.11	1	142 960.3	3.13
2.4.3	62 357.7	4.41	2	124 715.4	2.73
3.1.1	8 489.2	0.60	62.3	528 877.2	11.59
3.1.2	52 182.9	3.69	34	1 774 218.6	38.89
3.1.3	2 676	0.19	48.2	128 983.2	2.83
3.2.1	53 054.9	3.75	1.5	79 582.4	1.74
3.2.2	32 468.4	2.30	2	64 936.8	1.42
3.2.4	12 925.6	0.91	19	245 586.4	5.38
3.3.1	1 248.6	0.09	1.5	1 872.9	0.04
3.3.2	0		0	0	
3.3.3	181.4	0.01	0	0	
3.3.4	0		0	0	
4.1.1	1 830.0	0.13	2	3 660.0	0.08
4.1.2.1	129 117.3	9.13	2	258 234.6	5.66
4.1.2.2	3 369.8	0.24	0	0	
4.2.1	78.3	0.01	2	156.6	
4.2.3	2 464	0.17	0	0	
5.1.1	1 195.6	0.08	0	0	
5.1.2	57 084.3	4.04	0	0	
Total	**1 414 024.0**	**100**		**4 561 674.5**	**100**

C t ha^{-1}: carbon density in tonnes per hectare.
%C: percentage of carbon store in each land cover class.

end of the 4.6 to 5.2 Mt range. The higher estimate for forest carbon (3.8 Mt) by Cannell *et al.* (1996), recognised the age structure but not the species composition and it also included litter.

Conifers in Northern Ireland

The carbon pool in conifer forests in NI has been calculated as 2.16 Mt using DANI area figures (1.77 Mt according to CORINE – Table 14.4). The dominant species in the pool is Sitka spruce; although it has a relatively low carbon density, because most planting post-dates 1960, it accounts for 72% of the conifer area. In contrast, Douglas fir, which has a higher carbon density, contributes little to the pool, because it is relatively rare. Norway spruce trees are generally old and contribute a large amount of carbon in relation to their forest area. In general conifer forests in NI are younger than in GB (Cannell *et al.*, 1996) and therefore their carbon density in 1993 was lower. Nevertheless, the carbon pool still appears relatively large, but this is qualified by the knowledge that over half is planted on drained peatland. Such pre-treatment may reduce the carbon pool in the peat itself, possibly by more than is gained by the growth of trees.

Broadleaves in Northern Ireland

Broadleaved woodlands are less common in NI than in GB despite a marked increase in planting in recent years, and a higher percentage is in the private sector (Cannell *et al.*, 1996). Limitations of detail in the Private Woodland Inventory and in the record for grant-aided planting since 1979, made it impossible to calculate carbon densities for species by age groups in as much detail as for conifers. However, an estimate was made of the average carbon density for broadleaves in 1980, for comparison with GB (Milne, 1994), which indicated the carbon density to be $52.4 \, t \, ha^{-1}$ in NI compared with 61.9 for GB. With an additional 13 years of growth and recent planting, and despite likely losses of older trees since 1979, the carbon store of broadleaved forests appears to have increased from 0.64 Mt in 1979 to 0.79 Mt in 1993 (using DANI areas). This is based on the different age-weighted carbon densities and areas of trees at these dates.

Spatial Distribution of Carbon Pools in Vegetation per 10 × 10 km Grid Square

Comparison of the distribution shown in Figure 14.1 with topographic and thematic maps and images, reveals that squares where the mean carbon pool is equal to or exceeds $8 \, t \, ha^{-1}$ have extensive planted conifer forests. Patchy, or less extensive conifer forests account for those squares with a mean value of $6-7.9 \, t \, ha^{-1}$. An exception is square 340380 where the land area has several demesnes and parks with mature broadleaved woodland; these raise the carbon pool even though much of the square is sea. A mean value of $4-5.9 \, t \, ha^{-1}$ arises from several causes; in the north-west some squares contain small conifer forests, but around Lough Erne there are patches of broadleaved woodland. The Greater Belfast region has a relatively high value caused by trees in public parks, golf courses and gardens of large houses. South of Lough Neagh, the one square in this category results from the apple orchards. Squares with a mean value of $2-3.9 \, t \, ha^{-1}$ are mainly in marginal lands and uplands where woodland fragments, forest patches and extensive peatlands occur. The pasture-dominated lowlands account for many of the squares with a mean value of $0-1.9 \, t \, ha^{-1}$. However, it should be noted that in the present UK inventory of carbon pools, inland water is given a carbon density of 0. As NI has many lakes, including the large Lough Neagh and Upper and Lower Lough Erne, the carbon density of some 10 × 10 km grid squares was reduced. More work is required to obtain a carbon density for lakes, possibly integrating material from lake inventories.

DISCUSSION

CORINE land cover mapping has been applied successfully to estimate vegetation carbon pools in Northern Ireland. Links had to be found between CORINE and GB land cover classes and carbon density factors devised for the CORINE classes, but otherwise the same methods of calculating carbon pools were used as for GB. On this basis, estimation of vegetation carbon pools could be extended into other European countries where CORINE land cover mapping is available. This could form part of a Europe-wide response to global warming.

Carbon densities were related to those used by Milne for GB (1994) and Milne *et al.* (1995). Higher densities have been proposed by Adger *et al.* (1991) based on experimental sites; more recently, densities have been based on MAFF statistics (Adger and Subak, 1995). Additional work is required on the carbon densities to be used in NI, where crop and grass growth may differ from other parts of the UK. This will involve further data integration with DANI statistics. However, any such work needs to take into account the UK-wide basis of the study and continue the land cover class links already established.

Different estimates have been made for total forest carbon pools in NI, depending on the total areas taken and the species/age mixes. The importance of integrating available data from different sources needs to be stressed, but the basis of the data needs to be understood. It is also necessary to know the methodology used to calculate carbon densities and pools before comparisons can be made between NI and GB or elsewhere.

NI was not included in the first attempt to estimate and map vegetation carbon pools. Consequently there was considerable catching up to be done; nevertheless estimates and a map of their distribution have been achieved. These require refinement, not only of the carbon densities and their links to GB values, but of the mapping. The 10 × 10 km grid is too coarse and results in a very generalised pattern; note, for example, the way in which squares in the Greater Belfast area are increased in their carbon pool by higher values deriving from some suburbs with vegetation cover of relatively high carbon densities. Similarly, the influence of forests and upland communities can be seen at only the most general of levels. A decision to map the UK carbon pools on a 1 × 1 km grid would be supported and not only because of the increased detail. Such a grid is used widely by many organisations involved in land management, facilitating detailed data integration, e.g. by the Soil Surveys in GB and by ITE.

The class intervals used for the map impose another level of generalisation. For example, a frequency distribution of vegetation carbon pools by 10 × 10 km grid squares for NI, with a classs interval of 1, shows that 73% of the squares have values between 1 and 4 t ha^{-1}, very few are below 1 and only 9% are above 6. However, NI has to be included within the overall map for the UK (Milne and Brown, Chapter 15, this volume) where there are already problems of legibility for a density shaded map printed at a small scale – increasing the number of classes would compound this.

The values for carbon pools and their distribution presented in this chapter must be regarded as first estimates which may be improved as further work is done on crop and grass growth in Northern Ireland, on refining forest statistics, and on mapping with greater precision. However, considering the global scale of the greenhouse effect, such improvements in methodology and results may be of relatively minor significance.

ACKNOWLEDGEMENT

This work was funded by the UK Department of the Environment's Climate Change Research Programme under contract No. EPG/1/1/3.

NOTE

Since this paper was originally presented many of the improvements to the methodology, highlighted in the discussion, have been employed – see Cruickshank, M.M., Tomlinson, R.W., Devine, P.M. and Milne, R. (1998). Carbon in the vegetation and soils of Northern Ireland. *Biology and Environment: Proceedings of the Royal Irish Academy*, 98B, 9–21.

REFERENCES

Adger, N. and Subak, S. (1995) Carbon fluxes resulting from land use change; land use data and policy. In: *Carbon Sequestration in Vegetation and Soils, Report, March 1995*. Department of the Environment, Global Atmosphere Division.

Adger, N., Brown, K., Sheil, R. and Whitby, M. (1991) *Dynamics of Land Use Change and the Carbon Balance*. Working Paper No. 15, Countryside Change Unit, University of Newcastle upon Tyne.

Cannell, M.G.R. and Milne, R. (1995) Carbon pools and sequestration in forest ecosystems in Britain. *Forestry*, **68**, 361–378.

Cannell, M.G.R., Cruickshank, M.M. and Mobbs, D.C. (1996) *Carbon Storage and Sequestration in the Forests of Northern Ireland*. Forestry, 69, 155–165.

Cruickshank, M.M. and Tomlinson, R.W. (1994) First analysis of CORINE land cover as an indicative framework for environment policy in the north of Ireland. In: D. Bond, J. Reid, M. Stevens and L. Worrall (eds) *GIS, Spatial Analysis and Public Policy 94*. University of Ulster, Coleraine, pp. 81–94.

Cruickshank, M.M., Tomlinson, R.W., Devine, P.M. and Milne, R. (1995) Estimation of carbon stores and production of carbon density maps for Northern Ireland: progress in the use of data based on CORINE land cover, Soil Survey and the Peatland Survey. In: *Carbon Sequestration in Vegetation and Soils, Report March 1995*. Department of the Environment, Global Atmosphere Division.

EC (1992) *CORINE Land Cover Technical Guide, Part 1*. European Commission, Directorate-General Environment, Nuclear Safety and Civil Protection, Brussels.

Graham, T. (1981) *Private Woodland Inventory of Northern Ireland 1975–79*. Forest Service, Department of Agriculture for Northern Ireland.

Guyer, C.F. and Edwards, C.J.W. (1988) The role of farm woodland in Northern Ireland. *Irish Geography*, **22**, 79–85.

Howard, P.J.A., Loveland, P.J., Bradley, R.I., Dry, F.T., Howard, D.M. and Howard, D.C. (1995) The carbon content of soil and its geographical distribution in Great Britain. *Soil Use and Management*, **11**, 9–15.

Milne, R. (1994) *The Carbon Content of Vegetation, and its Geographical Distribution in Great Britain*. Contract Report to Department of Environment.

Milne, R., Brown, T.A.W. and Howard, D.C. (1995) The size, and rate of change, of the pools of carbon in the vegetation and soils of Great Britain. In: *Carbon Sequestration in Vegetation and Soils, Report March 1995*. Department of Environment, Global Atmosphere Division.

O'Sullivan, G. (ed.) (1994) *Final Report, CORINE Land Cover Project (Ireland)*. Ordnance Survey of Ireland, Dublin.

Tomlinson, R.W. (1988) *Lagan Valley Regional Park Tree Survey*. Report to the Department of the Environment for Northern Ireland.

Wyatt, B.K., Greatorex Davies, N., Hill, M.O., Bunce, R.G.H. and Fuller, R.M. (1994) *Comparison of Land Cover Definitions*. Final Report to Department of the Environment. Contract No. PECD 7/2/127, March 1994.

15 Mapping of Carbon Pools in the Vegetation and Soils of Great Britain

R. MILNE and T.A. BROWN

Institute of Terrestrial Ecology, Edinburgh Research Station, Penicuik, UK

Vegetation in Great Britain is estimated to hold 113.8 ($\pm$ 27) million tonnes (Mt) of carbon. Woodlands hold about 80% of this total although occupying only 11% of the rural land area. Broadleaved woodland is the individual vegetation type holding most carbon (47 Mt). A map of carbon in the vegetation of Great Britain at 1 km $\times$ 1 km spatial resolution is presented. The predominant location of vegetation carbon is the broadleaved woodlands of Southern England.

The effect of vegetation on soil carbon content is assessed using the Institute of Terrestrial Ecology Land Cover Map. The soils of Great Britain are estimated to contain 9838 ($\pm$ 2463) Mt of carbon (71% in Scotland). In Scotland most soil carbon is in blanket peats, whereas in England and Wales most soil carbon is in stagnogley soils. The carbon content of the soils of Great Britain is mapped at 1 km $\times$ 1 km spatial resolution. Scottish peat soils show the greatest density of carbon and in total contain about 4523 Mt of carbon, approximately 46% of the GB total.

INTRODUCTION

Sources in the United Kingdom emit about 150 million tonnes (i.e. 150×10^{12} g) of carbon, as carbon dioxide, into the atmosphere each year. This extra atmospheric carbon dioxide is the main greenhouse gas which may be causing an increase in global temperatures and otherwise altering climate. About one-third of the emission comes from power stations and the rest comes about equally from road transport, industry and commercial/domestic sources. Programmes to limit these emissions are discussed in the publication *Climate Change – United Kingdom's Report under the Framework Convention on Climate Change* (HMSO, 1994) and will form the main method for reducing Britain's contribution to global warming. However, the Framework Convention on Climate Change that was signed at the United Nations conference in Rio de Janeiro in June 1992 also commits the United Kingdom to protecting and enhancing greenhouse gas reservoirs and sinks. Forests and other vegetation are an important part of the reservoir, or pool, of carbon, and changes in size and productivity of the pool may act as a sink or source for carbon dioxide. United Kingdom policy aims to secure an annual increase in the net amount of carbon stored in these natural pools. The amount, and geographical distribution, of carbon in vegetation (especially forests) is of considerable importance to this policy. Soils contain much more carbon than vegetation and, although this pool is only able to change slowly, changes in land use will have a major long-term effect on the carbon held in soils. Although the size of these pools and their

Vegetation Mapping: From Patch to Planet. Edited by Roy Alexander and Andrew C. Millington.
© 2000 John Wiley & Sons Ltd.

uptake rates are small compared to industrial emissions, they are the main component for the policy of enhancing uptake and storage of carbon. The Institute of Terrestrial Ecology has, under a contract from the Department of the Environment, been developing an inventory of carbon in the vegetation and soils of Great Britain in support of these United Kingdom commitments. The Climate Change Report (HMSO, 1994) described initial estimates of the total carbon in vegetation and soils and its geographical distribution. Here the latest version of those estimates is presented.

METHODS AND RESULTS

Vegetation Carbon

The carbon stored in the forest and non-forest vegetation of Great Britain was estimated from published studies of biomass partitioning, a census of forests, ecological surveys of sample areas, a land classification and a remotely sensed map of land cover.

Non-forest Carbon

The Institute of Terrestrial Ecology (ITE) 1990 Countryside Survey of land use in Great Britain (Barr *et al.*, 1993) was stratified by using a land classification system developed by Bunce *et al.* (1981). This allocates each $1 \, km \times 1 \, km$ square in Great Britain to one of 32 land classes on the basis of combinations of environmental data which are already in mapped form, such as geology, climate and topography (Barr *et al.*, 1993). In the 1990 Countryside Survey, 512 $1 \, km \times 1 \, km$ squares were visited in Great Britain and the vegetation features were recorded in detail (Barr *et al.*, 1993). Within this total the number of squares visited in each land class was generally in proportion to the occurrence of that class in Great Britain. The average area of 58 different cover types within each land class (Barr *et al.*, 1993) is available from the Countryside Information System (CIS), a computer database developed by ITE (Merlewood) for the Department of the Environment. The 58 cover types were combined (Milne, 1994) into 14 vegetation groups and a carbon density of 0, 1 or 2 tonnes carbon per hectare ($1 \, t \, ha^{-1} = 0.1 \, kg \, m^{-2}$) was selected for each non-woodland group on the basis of the values in Adger *et al.* (1991) and Olson *et al.* (1985) (Table 15.1).

Carbon Content of Woodlands

Woodlands contain large amounts of carbon, and therefore greater detail was included in estimates of carbon density for different types of woodland. The standing volume of timber classified by species and forest age and the equivalent land area found in a Forestry Commission (FC) census of commercial woodlands in the state and private sectors (Locke, 1987) were used to calculate a British average of timber volume per unit area ($m^3 ha^{-1}$) for each age/species group (Milne, 1992). The carbon density ($t \, ha^{-1}$) of each age and species group was calculated using overall conversion factors k_{sa} ($t \, m^{-3}$) $= \rho \omega \phi$ where ρ is the specific gravity of wood, ω is the ratio of the total volume of wood to the 'standing volume' from the FC Census and ϕ is the fraction of wood

Table 15.1 Carbon density used for each vegetation group

Vegetation group	Vegetation carbon density (t ha^{-1})
Cereal	1
Crops	1
Pasture etc.	1
Fallow	0
Horticulture	1
Unimproved grass	1
Shrub	2
Heath	2
Bogs etc.	2
Maritime	2
Broadleaf woodland	32.3–64.7*
Conifer woodland	12.2–33.8*
Mixed woodland	12.4–58.9*
Non-vegetated	0

(1 t ha^{-1} = 0.1 kg m^{-2})
* Woodland carbon densities depend on land class.

mass that is carbon (Milne, 1992). The carbon densities were then weighted by the area in the country of each age, for each species, to provide an estimate of the Great Britain average of carbon density for each species. The distribution with age of carbon density of British woodlands and the net values for each species are presented in Table 15.2. Sitka spruce has a low net carbon density (14.1 t ha^{-1}) due to the preponderance of young stands. Douglas fir, however, has the greatest estimated carbon density (41.3 t ha^{-1}) for conifers due to older stands being more common. Sweet chestnut has the greatest estimated carbon density for broadleaved woodland (90.6 t ha^{-1}) and poplar the smallest (36.4 t ha^{-1}).

The data from the 1990 Countryside Survey were analysed to produce, as an average for each ITE land class, the area of each tree species relative to total woodland area. Weighted averages of carbon density for broadleaf and conifer woodland consisting of the respective species were hence calculated for each land class. Mixed woodland was assumed to have the carbon density in each land class given by calculating the weighted average for all species (Table 15.1).

Total Mass of Carbon in Vegetation

The total mass of carbon in vegetation on land in each ITE land class was calculated by summing the products of carbon density and total area for each 'vegetation' group. The average density for each land class was then calculated by dividing by the total area of land in each land class. Total mass of carbon in the vegetation of Great Britain was estimated to be 113.8 million tonnes (megatonnes (Mt), 1 Mt = 1 × 10^{12}g) (Table 15.3) by adding the 32 land class totals. Woodland held 80.1% of this, although occupying only 11.1% of the rural land area. Broadleaf woodland accounted for 47.3% and conifer woodlands 24.8% of the total carbon in Great Britain.

Table 15.2(a) Estimated carbon density (t ha^{-1}) of coniferous species in Great Britain. The average (NET), weighted by the area distribution with age, for each species is also presented. (1 t ha^{-1} = 0.1 kg m^{-2})

Species	\multicolumn{11}{c}{Age (years)}
	0–10	10–20	20–30	30–40	40–50	50–60	60–70	70–80	80–120	> 120	NET
Scots pine	3.4	8.1	16.9	31.6	49.1	58.6	64.9	68.9	65.7	71.3	30.3
Corsican pine	3.4	18.4	33.6	51.8	66.8	82.4	96.6	92.3	112.7	127.2	34.9
Lodgepole pine	2.5	4.9	14.6	28.6	37.1	37.6	68.3	122.9			6.5
Sitka spruce	2.9	7.5	22.9	36.9	52.6	67.1	85.3	90.2			14.1
Norway spruce	2.9	7.2	19.7	35.2	47.1	56.9	69.1	70.2	113.7	102.7	24.4
Larch	10.3	21.3	29.3	44.1	52.2	58.9	64.1	61.6	69.2	68.4	32.3
Douglas fir	3.4	19.2	31.8	54.2	78.9	92.9	99.2	113.3	133.4	118.1	41.3
All conifers	4.8	9.2	22.1	37.6	52.1	60.8	64.8	64.5	68.7	69.1	21.1

Table 15.2(b) Estimated carbon density (t ha^{-1}) of broadleaved species in Great Britain. The average (NET), weighted by the area distribution with age, for each species is also presented. SAB refers to sycamore, ash and birch taken together. (1 t ha^{-1} = 0.1 kg m^{-2})

Species	\multicolumn{11}{c}{Age (years)}
	0–10	10–20	20–30	30–40	40–50	50–60	60–70	70–80	80–120	> 120	NET
Oak	1.9	3.9	18.1	45.7	57.8	63.6	68.8	73.2	90.1	88.3	73.9
Beech	1.1	2.1	13.6	30.1	49.1	70.6	82.3	104.3	123.1	124.3	78.2
SAB	7.2	13.8	30.5	43.1	48.5	59.3	64.3	66.7	75.5	93.9	46.6
Poplar	9.5	20.7	43.2	62.1	82.7	74.8	95.8	79.9	111.3	103.5	36.4
Sweet chestnut	3.8	7.3	32.2	74.1	63.3	84.2	82.4	110.8	122.8	173.5	90.6
Elm	3.8	8.7	36.2	55.2	83.7	105.2	76.9	94.1	106.7	105.2	88.5
All broadleaves	5.7	12.9	27.7	44.1	52.9	63.7	68.5	76.8	91.3	98.9	61.9

Table 15.3 Area of vegetation cover groups in Great Britain and associated carbon in vegetation. Based on data on cover and species distribution in 512 1 km × 1 km squares visited in Countryside Survey-1990 and scaled to Great Britain using the ITE land classification. (1 Mt = 1 × 10^{12} g)

	Area (km²)	Area (% of GB)	Carbon (Mt)	Carbon (% of GB)
Cover group				
Agricultural	110 547	49.3	10.77	9.6
Semi-natural	66 912	29.9	11.08	9.8
Woodland	24 965	11.1	91.97	80.1
Non-vegetated	21 586	9.6	0.00	0.0
Total	224 010		113.82	
Woodland type				
Broadleaf	9 100	4.1	53.32	47.3
Conifer	13 646	6.1	29.02	24.8
Mixed	2 220	1.0	9.62	8.5

The estimates of vegetation areas in Great Britain have a standard error of at most 23% of their value (Barr *et al.*, 1993) and for woodlands compare well with other estimates. Bunce *et al.* (1983) estimated total woodland cover in Britain, using the ITE land classification and field surveys carried out in 1978, to be 9.4% compared to 8.6% calculated from data supplied by the Forestry Commission. They also carried out a test of the same land classification estimate by selecting 5234 areas of 1 km^2 on appropriate 1:50 000 Ordnance Survey maps and measuring the area of woodland marked. On scaling these measurements to the land area of Great Britain a figure of 9.3% woodland cover was found, in remarkably good agreement with the land classification estimate.

The accuracy of the estimates of carbon stored is more difficult to gauge but Adger *et al.* (1991) calculated that 'High Forests' in Great Britain contain 30.4 Mt in coniferous stands and 50.5 Mt in broadleaf woodlands. These estimates are similar to those in Table 15.3 but Adger *et al.* (1991) used greater carbon densities than the average derived here and in the case of broadleaved woodlands excluded about 4000 km^2 by limiting their study to 'High Forests'. The estimate of 113.8 Mt calculated here can be considered as the sum of the products of an area and a mean carbon density for each land cover group. Standard errors for the areas were taken from the Countryside Information System. Errors in the mean vegetation carbon densities values were assigned by consideration of the possible variation in carbon densities across the constituent crop and forest conditions. Uncertainty in the woodland carbon estimates is the overall controlling factor. The resulting overall standard error in the total carbon in vegetation estimate of 113.8 Mt was found to be $\pm$ 27 Mt.

The Geographical Distribution of Vegetation Carbon

The value of 113.8 Mt for vegetation carbon based on the ITE land classification and Countryside Survey is an acceptable estimate of the total pool size due to the stratified sampling in the survey. Plotting the land class average of the carbon density does not however give a good geographical representation of where carbon is stored in vegetation. The 25 land cover types ('Target Classes') identified in the remotely sensed ITE Land Cover Map can, however, be used to show a better geographical distribution of the total carbon. Cannell and Milne (1995) used that data to present a map of vegetation carbon density in 10 km $\times$ 10 km blocks and here we extend the approach to produce a 1 km $\times$ 1 km spatial resolution map. The 'Target Classes' of the ITE Land Cover Map are those cover classes into which each individual pixel of a remotely sensed image are classified. Each has an equivalent in one of 17 'Key Cover Types' defined for the Countryside Survey (Barr *et al.*, 1993; Wyatt *et al.*, 1994) and hence can be related back to the vegetation groups used above (Table 15.4). The data in the ITE land cover map used for this study consisted of the percentage of each 'Target Class' occurring in each 1 km $\times$ 1 km square in Great Britain.

However, the ITE land cover map data has significantly different estimates of the total area for Great Britain of several 'Key Cover Types' (Barr *et al.*, 1993) compared to those estimated from the Countryside Survey data scaled up using the ITE land classification. The ratio of the areas from the two methods, for each 'Key Cover Type' for each land class, was therefore used to calculate an adjusted carbon density for each

Table 15.4 Equivalent 'Target Classes' and 'Key Cover Types' for ITE Land Cover Map with 'Key Cover Types' of Countryside Survey-1990 and the selected carbon densities for these types

Target classes	ITE Land Cover Map– Key Cover Type	Field Survey– Key Cover Type	Carbon density (t ha^{-1})
Continuous urban	Continuous urban	Communications	0
Suburban	Suburban	Built up	0
Tilled land	Tilled land	Tilled land	1
Mown/grazed turf	Managed grassland	Managed grass	1
Meadow/verge	Managed grassland	Managed grass	1
Ruderal weed	Rough grass/marsh	Rough grass/marsh	2
Felled forest	Rough grass/marsh	Rough grass/marsh	2
Rough grass/marsh	Rough grass/marsh	Rough grass/marsh	2
Bracken	Bracken	Dense bracken	2
Grass heath	Heath/moor grass	Moorland grass	1
Moorland grass	Heath/moor grass	Moorland grass	1
Open shrub heath	Open shrub heath/moor	Open heath	2
Open shrub moor	Open shrub heath/moor	Open heath	2
Dense shrub heath	Dense shrub heath/moor	Dense heath	2
Dense shrub moor	Dense shrub heath/moor	Dense heath	2
Lowland bog	Bog	Wet heaths and saturated bogs	2
Upland bog	Bog	Wet heaths and saturated bogs	2
Scrub/orchard	Deciduous/mixed woodland	Broadleaved/mixed woodland	10
Deciduous woodland	Deciduous/mixed woodland	Broadleaved/mixed woodland	12.2–33.8*
Coniferous woodland	Coniferous woodland	Coniferous woodland	21.7–64.8*
Inland bare ground	Inland bare	Inland bare	0
Saltmarsh	Saltmarsh	Saltmarsh	0
Coastal bare	Coastal bare	Coastal bare	0
Inland water	Inland water	Inland water	0
Sea/estuary	Sea/estuary		0

* Woodland carbon densities depend on ITE land class.

'Key Cover Type'. The adjustments were such that combining the new densities with areas from the land cover map for each 'Key Cover Type' would have produced a carbon total for each land class equal to that obtained using the Countryside Survey and land class method. To calculate the total vegetation carbon in each 1 km × 1 km square the adjusted densities were multiplied by the area of each 'Key Cover Type' in the square. The carbon density produced in this way for the vegetation in each square in Great Britain is mapped in Plate 13 (Figure 15.1). This approach therefore combines the estimate of total vegetation carbon from the land class/survey approach with the locations of different land types from the ITE land cover map. The realistic location of major pools of vegetation carbon can be seen, for example, by the ease of identifying Kielder Forest on the Scottish/English border.

Soil Carbon

In HMSO (1994) it was estimated that the soils of GB contain 9500 Mt carbon and the geographical distribution of the soil carbon pool was mapped in 10 km × 10 km blocks. These estimates were based on data for each 1 km × 1 km square in Great Britain from several sources. The Soil Survey and Land Research Centre (SSLRC) LandIS database provided the dominant soil series for each square and the bulk density and carbon content for each series from representative soil core data for England and Wales. The Macaulay Land Use Research Institute (MLURI) National Soils Database provided equivalent data for Scotland except in the case of the bulk density of peat where a value of $110\,kg\,m^{-3}$ was assumed. For a given soil series the carbon content of the soil is dependent on the vegetation or other cover. In HMSO (1994) and Howard *et al.* (1994) cover for each square was estimated using data from the ITE Countryside Survey-1990 and the ITE land classification. This implied that the dominant vegetation type in each 1 km × 1 km square was the dominant vegetation in the ITE land class of the square. This did not, however, reflect the geographical variation of vegetation accurately and hence may have introduced error into the values of soil carbon estimated from soil series on a square by square basis. In fact some ITE land classes are dominantly non-vegetation but have significant areas of vegetation containing carbon.

Estimates of the Size of Carbon Pool and its Geographical Distribution

In order to provide a description, for the purpose of calculating soil carbon, of the geographical distribution of vegetation more realistic than that in HMSO (1994) and Howard *et al.* (1994), the ITE land cover map (Barr *et al.*, 1993) was again used. The dominant cover in each 1 km × 1 km square of Great Britain was estimated using the 25 'Target Classes' of this map grouped into six cover types (P – permanent grass, A – other agriculture, S – semi-natural, T – woodland, W – inland water, O – no vegetation) (Plate 14 (Figure 15.2)).

   The available database of soil series was that used for HMSO (1994) and Howard *et al.* (1994) but now some squares would have a different dominant vegetation. Fortunately many of the soil series/vegetation combinations had already been assigned a carbon content in the original work. However, this did not apply to woodlands, as no ITE land class has woodland as a dominant vegetation. Where data were not available from the existing database the following main rules were adopted rather than looking for specific carbon content data from soil core databases:

1. Soil under woodlands will have same carbon content as the same soil series under semi-natural vegetation.
2. The carbon content of a given soil/vegetation combination will be equal to the mean value in Scotland of all the available data from the same soil type (i.e. peaty podzol, brown forest soil etc.) under the same vegetation, and in England and Wales the mean of all of the available data for the same Avery Soil Group under the same vegetation.

This approach was taken as it had been found that the range of soil carbon content within any Avery Group or Scottish soil type was about 20% of the mean for the group which was an acceptable level of accuracy. Values for woodland soils were substituted by data from semi-natural areas because, apparently, few measurements under trees have been made by any soil survey organisation and where measurements have been made little difference is observed to similar semi-natural sites.

Scottish peats contain a large proportion of the carbon in British soils and yet, because of lack of data on bulk density, this amount was least well estimated by Howard *et al.* (1994). They used a value for the bulk density of the surface layer of Scottish peats of $350 \, kg \, m^{-3}$ based on data for lowland English peats (Burton and Hodgson, 1987). This value is large compared to other published values for the bulk density of peat, which tend to be about $100 \, kg \, m^{-3}$ (e.g. Miller *et al.*, 1973; Clymo, 1983; Hulme, 1986). However, the Forestry Authority Research Division carried out a peat forest soils survey between 1974 and 1981 which included 302 cores of peat, mostly in Scotland under semi-natural conditions (Pyatt *et al.*, 1979; Pyatt and Anderson, pers. comm.). The data relevant to bulk density for each of these cores were obtained from the Forestry Authority for use in the determination of a reliable estimate of carbon in Scottish peats. Of the 302 cores, 202 were identified as Blanket Peat and 22 cores were Basin Peat. The bulk densities, expressed as dry mass per fresh volume, of the cores were averaged and the mean bulk density for Blanket Peat was found to be $100 \, kg \, m^{-3}$ and for Basin Peat $90 \, kg \, m^{-3}$. These values were therefore adopted for the surface 1 m of peat but other factors affecting the carbon content, e.g. variation of bulk density with depth, variation of total depth with location and carbon fraction of dried peat, were retained from the work of Howard *et al.* (1994).

The results of adding up the soil carbon contents for each square are summarised in Table 15.5. The soil carbon pool for Scotland is 6948 Mt, and for England and Wales 2890 Mt, giving a total for Great Britain of 9838 Mt. Scottish peats hold 46% of this total. These values are similar to those presented in the United Kingdom's *Report under the Framework Convention on Climate Change* (HMSO, 1994). In Scotland most carbon is in blanket peats whereas in England and Wales stagnogley soils cover the largest area and have the largest proportion of carbon. In England and Wales peat soils have about the same carbon total as brown earths.

The error in the estimate of total soil carbon was investigated by similar methods to that used for vegetation. It was assumed that there could be a 20% error in the area estimated for the Scottish peat group. This group contains most carbon and the areas of the other Scottish soils and the area of England and Wales would have negligible error, as these could be estimated from geographical data. A standard error of $\pm \, 20\%$ of the mean soil carbon density was also assumed for the non-peat Scottish soils and the soils of England and Wales. A greater standard error of $\pm \, 800 \, t \, ha^{-1}$ (about 50% of mean) was assumed for the carbon density of Scottish peats. On these assumptions a standard error of $\pm \, 2463$ Mt was calculated in the total soil carbon in Great Britain of 9838 Mt.

The carbon content of the dominant soil in each $1 \, km \times 1 \, km$ square of Great Britain is mapped in Plate 15 (Figure 15.3). The importance of Scottish peatlands to the soil carbon pool is easily seen from this map and the main features are similar to those in the map published in UK's *Report under the Framework Convention on Climate Change* (HMSO, 1994).

Table 15.5 Total soil carbon in Great Britain based on soil series and carbon content of each 1 km × 1 km square. (1Mt = 1 × 10^{12} g)

	Soil carbon (Mt)	Soil carbon (%)
Scotland (peat)	4523	46
Scotland (non-peat)	2425	25
Scotland (total)	6948	71
England and Wales	2890	29
Great Britain	9838	

DISCUSSION

Existing databases have proved to be a valuable data source in calculating the size and distribution of the carbon pools in the vegetation and soils of Great Britain. The estimates of total carbon in the vegetation (113.8 Mt) and soils (9838 Mt) have standard errors of about ± 25%. The distribution maps usefully show the broad pattern of variation in carbon density, but because of the method of preparation and the scale of error they cannot be used to investigate localised situations. Similarly temporal change in stored carbon will only be identifiable where it is large and is recorded by future surveys. However, existing databases do not always have all the required information and further work is often necessary to provide this.

In carbon sequestration studies relevant to the United Kingdom's commitment to the Framework Convention on Climate Change the next requirement is for information on the present geographical distribution of sources and sinks of carbon in vegetation and soils. Forests are the main sink of carbon and their density, age and productivity varies throughout the country. Changes in land use, due to new government policies and schemes, e.g. set aside, may also be significant sources or sinks for carbon. The challenge now is to extend the work on the pool size to include rates of change of stored carbon. This is a complex questions and will only be amenable to solution by limiting the changes examined to four of five major processes and by accepting geographical disaggregation at a scale larger than 1 km × 1 km.

ACKNOWLEDGEMENTS

This research was funded under the UK Department of the Environment's Climate Change Programme (Contract EPG 1/1/3). The soil series and carbon content database for England and Wales is leased from the Soil Survey and Land Research Centre (SSLRC), Silsoe, and the equivalent database for Scotland from the Macaulay Land Use Research Institute (MLURI), Aberdeen, for use specifically within the above contract at the Institute of Terrestrial Ecology (ITE). The Environmental Information Centre at the Institute of Terrestrial Ecology (Monks Wood) supplied the ITE Land Cover Map. The Scottish peat data were provided by the Research Division of the Forestry Authority. The advice of Peter Howard and David Howard at ITE

(Merlewood), Peter Loveland and Iain Bradley at SSLRC, Frank Dry at MLURI, and Graham Pyatt and Russell Anderson at the Forestry Authority is acknowledged.

REFERENCES

Adger, N., Brown, K., Sheil, R. and Whitby, M. (1991) *Dynamics of Land Use Change and the Carbon Balance.* Working Paper No. 15, Countryside Change Unit, University of Newcastle upon Tyne.

Barr, C.J., Bunce, R.G.H., Clarke, R.T., Fuller, R.M., Furse, M.T., Gillespie, M.K., Groom, G.B., Hallam, C.J., Hornung, M., Howard, D.C. and Ness, M.J. (1993) *Countryside Survey 1990, Main Report.* Department of the Environment, London.

Bunce, R.G.H., Barr, C.J. and Whittaker, H.A. (1981) An integrated system of land classification. *Annual Report, Institute of Terrestrial Ecology 1980,* pp. 28–33.

Bunce, R.G.H., Barr, C.J., Whittaker, H.A., Mitchell, C.P. and Pearce, M.L. (1983) Estimates of present forest cover in Great Britain. In: A. Strub, P. Chartier and G. Schleser (eds) *Energy from Biomass, Berlin 1982.* Applied Science, London, pp. 186–189.

Burton, R.G.O. and Hodgson, J.M. (1987) *Lowland Peat in England and Wales* (Special Survey No. 15). Soil Survey of England and Wales, Harpenden.

Cannell, M.G.R. and Milne, R. (1995) Carbon pools and sequestration in forest ecosystems in Britain. *Forestry,* **68,** 362–378.

Clymo, R.S. (1983) Peat. In: A.J.P. Gore (ed.) *Mires, Swamp, Bog, Fen and Moor. Ecosystems of the World, 4A.* Elsevier, Amsterdam, pp. 159–224.

HMSO (1994) *Climate Change – United Kingdom's Report under the Framework Convention on Climate Change.* London, HMSO.

Howard, P.J.A., Loveland, P.J., Bradley, R.I., Dry, F.T., Howard, D.M. and Howard, D.C. (1994) The carbon content of soil and its geographical distribution in Great Britain. *Soil Use and Management,* **11,** 9–15.

Hulme, P.D. (1986) The origin and development of wet hollows and pools on Craigeazle Mire, South West Scotland, *International Peat Journal,* **1,** 15–28.

Locke, G.M.L. (1987) *Census of Woodlands and Trees 1979–82.* Forestry Commission Bulletin 63. London, HMSO.

Miller, H.G., Robertson, R.A. and Williams, B.L. (1973) Evaluation of peatland sites according to their physical and chemical characteristics. In: *Proceedings of the Symposium on Peatland Forestry, Edinburgh, 1968.* Swindon, Natural Environment Research Council, pp. 165–175.

Milne, R. (1992) The carbon content of vegetation and its geographical distribution in Great Britain. In: *Carbon Sequestration by Vegetation in the UK.* Interim Report to the Department of the Environment, Contract No. PECD 7/12/79, March 1992.

Milne, R. (1994) The carbon content of vegetation and its geographical distribution in Great Britain. In: *Carbon Sequestration by Vegetation in the UK.* Final Report to the Department of the Environment, Contact No. PECD 7/12/79, March 1994.

Olson, J.S., Watts, J.A. and Allison, L.J. (1985) *Carbon in Live Vegetation of Major World Ecosystems.* Oak Ridge National Laboratory Publication ORNL-5862. Report to US Department of Energy, Contract No. W-7405-eng-26.

Pyatt, D.G., Craven, M.M. and Williams, B.L. (1979) Peatland classification for forestry in Great Britain. In: *Proceedings of the International Symposium on Classification of Peat and Peatlands, Finland, September 1979.* International Peat Society, pp. 351–366.

Wyatt, B.K., Greatorex Davies, N., Hill, M.O., Bunce, R.G.H. and Fuller, R.M. (1994) *Comparison of Land Cover Definitions.* Final Report to Department of the Environment. Contract No. PECD 7/2/127, March 1994.

16 Mapping the Forests of Cameroon: Scales and Scaling-up

P.M. ATKINSON
Department of Geography, University of Southampton, UK

It is important when measuring environmental properties that the scales of measurement are chosen in relation to the scales of the underlying variation of interest. Further, where two properties are to be compared it is important that the scales of measurement of one property match those of the other. An extensive field survey of several biophysical properties of the forests of central Cameroon was undertaken in January and February 1994 as part of the UK NERC's Terrestrial Initiative in Global Environmental Research (TIGER) 1.4 project, 'Remote Sensing of Tropical Vegetation'. A hierarchical sampling strategy was adopted with 66 sparsely placed plots, each comprising three sub-plots of 40 m by 250 m. These data were obtained in support of Advanced Very High Resolution Radiometer (AVHRR) imagery with a nominal spatial resolution of 1.1 km by 1.1 km. The biophysical variables were examined for spatial dependence by computing the variogram and two major scales of spatial variation were identified: that between sub-plots (at distances less than 1 km) and that between plots (at about 35 km). The results imply that the spatial resolution of the AVHRR is appropriate for mapping spatial variation acting at lags of around 35 km. Once the scales of spatial variation were identified it was possible to evaluate the sampling strategy adopted for scaling-up the biophysical variables (i.e. for averaging the spatial variation within plots). It was found that three systematically placed sub-plots reduced the error variance to a 'tolerable' level.

INTRODUCTION

Scale has become an increasingly important concept in environmental remote sensing in recent years (e.g. Lam and Quottrochi, 1992; McGwire et al., 1993; Foody and Curran, 1994a; Raffy, 1994a, 1994b, 1994c; Atkinson, 1997), mainly as a result of a shift in focus from local scale to regional and global-scale investigations. This change in the scale of investigation is, in part, a result of an increasing awareness of the finite nature of the environment in which we live, and a desire to monitor changes (and in particular anthropogenic changes) to our environment as a whole. To study regional and global-scale phenomena researchers have used remotely sensed imagery with moderate spatial resolutions (around 1 km by 1 km) and coarse spatial resolutions (around 8 km by 8 km) (e.g. Justice et al., 1985, 1991; Tucker et al., 1985; Townshend et al., 1991).

Strictly speaking, scale is the difference in size between two representations of the same phenomenon. An example of this strict use of scale is cartographic scale. For example, a 1:50 000 scale map implies that the real world is 50 000 times larger (in one-dimension) than the cartographic representation of it. However, this use of scale has often caused confusion (e.g. 1:50 000 is a smaller scale than 1:25 000 even though a

Vegetation Mapping: From Patch to Planet. Edited by Roy Alexander and Andrew C. Millington.
© 2000 John Wiley & Sons Ltd.

larger area is covered on the map). To avoid such confusion the meaning of scale used widely in physics (Raffy, 1994a) is adopted throughout, that is, scale is used to mean size. This may seem at first inappropriate or even incorrect, but we commonly use scale with this meaning. For example, a 'large scale investigation' is simply one of large size, and it is implicit that we mean in relation to something else.

With the above definition of scale in mind, two types of scale are of interest. The first are the scales of the underlying process, form or, more generally, spatial variation that we are interested in. The second are the scales of measurement. The interaction between these two types of scale determines the scales of variation that are detectable from and embedded in both environmental data (Hill *et al.*, 1994; Stoms, 1994) and remotely sensed imagery (Belward and Lambein, 1990; Clark, 1990; Sèze and Rossow, 1991).

The scales of measurement may be further subdivided into the support and the spatial coverage. The support is the size, geometry and orientation of the space over which a measurement is made or datum is defined. Most environmental properties are defined in two dimensions so that the support is the area that each individual observation represents. In remote sensing, the support is approximated by the spatial resolution, although strictly the support is given by the point spread function (PSF) of the sensor. The PSF means that the true support is centre weighted so that variation close to the centre of the support receives more weight than that at the edges.

Whatever the geometry, the support of a measurement is always (except where location is the property of interest) of positive size (i.e. it cannot be zero). Consequently, all measurements of spatially varying properties are integrals or averages of the underlying spatial variation over the support (Jupp *et al.*, 1988; Marceau *et al.*, 1994a; Raffy, 1994b, 1994c). The effect of integration is to produce a filter through which we see the underlying variation. The scales of measurement associated with the support determine the size and shape of this filter, and consequently the scales of spatial variation that we are able to detect. Given that to observe the real world we must measure, it follows that we always see reality through a filter determined by the support.

The spatial coverage refers to the set of lags (vector distances and directions) between pairs of observations that comprise the sample. The maximum lag between any pair of observations defines the spatial extent of the sample, and we cannot observe scales of variation that are larger than this. In fact, the only scales of variation that may be captured in sample data are those that occur between the limits given by the support (fine scale) and the spatial extent (large scale): that is, the scales given by the spatial coverage.

Where the objective is to map spatial variation in some environmental property the scales of measurement should be chosen carefully to match the scales of spatial variation of interest (Woodcock and Strahler, 1987; Townshend and Justice, 1988; Marceau *et al.*, 1994b; Atkinson and Curran, 1995). Clearly, if the support is too large it may preclude the detection of certain fine scales of spatial variation, but if it is too small unwanted variation may be introduced and effort may be wasted. Further, the interaction of the scales of variation and of measurement can lead to uncertainties in the data such as those caused by the mixed pixel problem in remote sensing (Jasinski, 1990; Settle and Drake, 1993). Finally, many techniques and methods that may be applied to sample data such as kriging (an optimal technique for linear unbiased estimation that may be used for interpolation) (e.g. Journel and Huijbregts, 1978) depend on knowledge of the scales of spatial variation. In particular, the errors of estimation associated with such techniques depend on the scales of spatial variation.

Scaling-up is the process of transforming the sampling frame from a small scale to a large scale, and may be necessary when measurements made at one scale are to be related to those made at a larger scale (Foody and Curran, 1994a; Atkinson, 1997). To scale-up successfully both the support and the spatial coverage of the small-scale sampling frame must be increased to match those of the large-scale sampling frame. The latter amounts to ensuring that the coverage of the small-scale sample is equal to that of the large-scale sample. The former, increasing the size of support, is more problematic and requires careful sample design.

Scaling-up is often necessary in remote sensing where observations of properties made at the ground are to be compared with remotely sensed image pixels (Dozier and Strahler, 1983). Remotely sensed images are usually calibrated to radiance or reflectance. However, generally, we are interested in some property at the ground to which the remotely sensed imagery is related. To use the imagery to map the property of interest it is usually necessary to measure the property at several places within ground resolution elements (GREs) which correspond to certain image pixels. These observations may then be related to the corresponding image pixels, for example through a regression (or other) relation, and this relation applied to the remainder of the image to map the property of interest.

For regional-scale investigations one is likely to require a moderate spatial resolution such as the 1.1 km by 1.1 km provided by the National Oceanographic and Atmospheric Administration (NOAA) Advanced Very High Resolution Radiometer (AVHRR). Problems of scale arise with such imagery because measurements made of properties at the ground may be very small in relation to the GRE that each remotely sensed image pixel represents. The task then is to 'scale-up' the measurements at the ground so that they match the support of the image pixels. In certain circumstances, the problems of increasing the support may to some extent be circumvented with remotely sensed imagery of intermediate and fine spatial resolutions to bridge the gap between the properties at the ground and the remotely sensed imagery. In this chapter, it is assumed that such intermediate spatial resolution imagery is unavailable.

In the analyses presented below the scales of variation in three biophysical properties (mean diameter at breast height, mean diameter at first leafing branch, and tree density) of the tropical and dry forests of Cameroon are measured and recommendations made for sampling and scaling-up the spatial variation in these properties.

SCALE AND SPATIAL DEPENDENCE

In this section, spatial dependence and a function used to represent it known as the variogram are introduced and the effect of the support on the variogram is explained.

Spatial Dependence and the Variogram

Spatial dependence may be thought of as the tendency for close observations to be more alike than those further apart. It is useful in understanding scale because (i) it simplifies our view of spatial variation, (ii) it identifies scales of spatial variation and (iii) it provides a link between spatial variation and the sampling framework (and, in particular, the support).

Most environmental properties distributed spatially over the surface of the Earth are spatially dependent, at least at some scale. Spatial dependence at a variety of scales has been measured in geology (Journel and Huijbregts, 1978), soil survey (Webster, 1985), ecology (Rossi et al., 1992) and most recently in remote sensing (Yoder et al., 1987; Ramstein and Raffy, 1989; Wald, 1989; Clark, 1990; Cohen et al., 1990; Bian and Walsh, 1993; Gohin and Langlois, 1993; McGwire et al., 1993). If properties were spatially independent at all scales, all values of a given property at all places would be equal and there would be no form or structure.

We may represent spatial dependence with the variogram. A random function, RF, is a set of n random variables, RVs, distributed spatially. We may treat a sample of a spatially varying property Z as a single realization $z(\mathbf{x})$ of a RF $Z(\mathbf{x})$ (the term realization refers, in this context, to a set of values drawn from a set of RVs). The semivariance is defined as half the expected squared difference between the RFs $Z(\mathbf{x})$ and $Z(\mathbf{x} + \mathbf{h})$ at a particular lag $\mathbf{h}$ (Matheron, 1965). The variogram $\gamma(\mathbf{h})$ is then the function that relates semivariance to lag (equation 16.1).

$$\hat{\gamma}(\mathbf{h}) = 1/2 \, E[\{Z(\mathbf{x}) - Z(\mathbf{x} + \mathbf{h})\}^2] \qquad (16.1)$$

The experimental variogram may be estimated for supports of size $|v|$ by equation 16.2.

$$\hat{\gamma}_v(\mathbf{h}) = 1/2P(\mathbf{h}) \sum_{\substack{(i,j) \\ D(\mathbf{h})}}^{P(\mathbf{h})} \{z_v(\mathbf{x}_i) - z_v(\mathbf{x}_j)\}^2 \qquad (16.2)$$

where $D(\mathbf{h})$ is defined as $\{(i, j) \; \mathbf{x}_i - \mathbf{x}_j = \mathbf{h}\}$ (ensuring that the lag distance between $\mathbf{x}_i$ and $\mathbf{x}_j$ is $\mathbf{h}$), and $P(\mathbf{h})$ is the cardinality of $D(\mathbf{h})$.

To use the variogram in geostatistical analyses it is usually necessary to represent the scatter of points that constitute the experimental variogram with a mathematical model. This model must be such that the variances that result from linear combinations of the RF must be positive. It would not make sense to have negative uncertainty associated with an estimate. For covariances this is ensured by models that are positive definite or 'authorised'. For the variogram, which is essentially the inverse of the covariance, the model must be Conditional Negative Semi Definite (CNSD). Checking for the CNSD criterion is time-consuming and, therefore, models are usually selected from several that are already known to be CNSD for the dimensions of the space over which the RF is defined. Some common models are presented by Webster and Oliver (1990).

The fitted model often approaches and intercepts the ordinate at some positive value of semivariance, known as the nugget variance. The term derives from the gold mining industry where displacements of very short distances sometimes lead to variations of large magnitude. The nugget variance encompasses microscale variation, the uncertainty inherent in sampling, measurement errors, and the errors in fitting a mathematical model to sample data.

The question is 'how can the variogram be used to detect scales of spatial variation?'. Here, scales of spatial variation are taken to be the distances over which spatial forms exist and spatial processes operate. This is conceptualised simply in terms of objects, but less so for continuous fields. In one dimension, an object (line) has a single scale: its length. In two or more dimensions, an object may have many scales (measured in terms of lag). For example, a disc represents a range of scales from zero at its edge to the diameter across its centre.

The variogram may be used to detect scales of variation as defined above because it presents semivariance as a function of lag. In one dimension, a line results in a variogram which increases linearly to a maximum, known as the sill, at a finite lag, known as the range. The range is determined by and equal to the scale or length of the line. For two-dimensional objects such as discs placed on a background, the set of scales of variation are translated into a set of ranges which are superimposed to give a variogram which is convex upwards from the origin. The convexity of the variogram increases as the range of spatial scales encountered increases (e.g. a distribution of disc sizes will increase the convexity). This model extends to the description of continuous fields. Therefore, the variogram may be used to determine the scales of spatial variation present in the property of interest. It is worth noting that spatial variation may be scale invariant or fractal (Mandelbrot, 1982; Barnsley, 1989) in which case there are no predominant scales of variation. Interestingly, the variogram may also be used to detect fractal variation (e.g. Rees, 1992).

The experimental variogram, like the sample data from which it is derived, is defined for a support of given positive size. Therefore, the experimental variogram represents the underlying scales of spatial variation as seen through a filter. It is important, therefore, to know the support for which the variogram is defined. The effect of the support on the scales of spatial variation that are detectable from the variogram is considered next.

The Support and Regularisation

The effect of imposing the support of the sampling frame on the underlying variation of interest can be modelled through spatial dependence. Moreover, the effect of increasing the size of support (known as regularisation) can be modelled. This is important for comparing one scale with another.

Equation 16.3 (Journel and Huijbregts, 1978) describes the relation between the punctual semivariance and the regularised semivariance at a given lag.

$$\gamma_v(\mathbf{h}) = \bar{\gamma}(v,v_{\mathbf{h}}) - \bar{\gamma}(v,v) \tag{16.3}$$

Here $\bar{\gamma}(v,v_{\mathbf{h}})$ is the integral of the punctual semivariance between two supports of size v whose centroids are separated by $\mathbf{h}$, given formally by equation 16.4,

$$\bar{\gamma}(v,v_{\mathbf{h}}) = \frac{1}{v^2} \int_v \int_{v(\mathbf{h})} \gamma(\mathbf{y},\mathbf{y}')d\mathbf{y}d\mathbf{y}' \tag{16.4}$$

where $\mathbf{y}$ describes an observation of size v and $\mathbf{y}'$ describes independently another observation of equal size and shape at lag $\mathbf{h}$ away. The quantity $\bar{\gamma}(v,v)$ is the average punctual semivariance within an observation of size v and is written formally as in equation 16.5,

$$\bar{\gamma}(v,v) = \frac{1}{v^2} \int_v \int_v \gamma(\mathbf{y},\mathbf{y}')d\mathbf{y}d\mathbf{y}' \tag{16.5}$$

where $\mathbf{y}$ and $\mathbf{y}'$ now sweep the same pixel independently.

Equation 16.3 provides a means by which to assess the effect of the support on the nature and scale of measured spatial variation (Clark, 1977; Jupp et al., 1988, 1989; Isaaks and Srivastava, 1989; Zhang et al., 1990; Atkinson, 1993). It also provides an analytical tool for investigating the effect of changing the scale of measurement on detectable scales of spatial variation. The only spatial variation detectable from the sample values (defined on a support v) is that described by the term on the left-hand side of equation 16.3. The variation described by the second term on the right-hand side of equation 16.3 (the within-block variance) is completely obscured from analysis by integration over the support. Therefore, the support acts like a filter removing small-scale spatial variation in favour of larger-scale variation occurring within the spatial coverage of the sampling frame.

KRIGING AND OPTIMAL SAMPLING

To scale-up observations made at the ground so that they are comparable with remotely sensed image pixels it may be necessary to sample within several GREs. To do this efficiently we might use kriging (a linear technique for optimal and unbiased estimation) to design optimal sampling strategies (Matheron, 1965; Journel and Huijbregts, 1978; Isaaks and Srivastava, 1989; Cressie, 1991).

Kriging

The kriging estimator $\hat{z}_V(\mathbf{x}_0)$ of the average value over a cell V is a linear weighted sum of n sample observations (equation 16.6),

$$\hat{z}_V(\mathbf{x}_0) = \sum_{i=1}^{n} \lambda_i z(\mathbf{x}_i) \tag{16.6}$$

where the n observations $\{z(\mathbf{x}_i), i = 1,2, ..., n\}$ are defined on quasi-point supports, $\mathbf{x}_i$. The n weights, λ_i, are chosen such that the estimate is unbiased and has minimum estimation or kriging variance. For ordinary kriging the unbiasedness is ensured by equation 16.7.

$$\sum_{i=1}^{n} \lambda_i = 1 \tag{16.7}$$

The estimation variance is minimised (with the constraint that the weights sum to one) by standard Lagrangian techniques resulting in a system of $(n + 1)$ linear equations in

$(n + 1)$ unknowns, the kriging system. The minimum estimation variance or kriging variance σ_k^2 is then given by equation 16.8.

$$\sigma_k^2 = \sum_{i=1}^{n} \lambda_i \, \bar{\gamma}(\mathbf{x}_i, V(\mathbf{x}_0)) + \mu - \bar{\gamma}(V(\mathbf{x}_0), V(\mathbf{x}_0)) \tag{16.8}$$

where μ is the Lagrange parameter, $\bar{\gamma}(\mathbf{x}_i, V(\mathbf{x}_0))$ is the integral semivariance between the point $\mathbf{x}_i$ and the block to be estimated $V(\mathbf{x}_0)$ and $\bar{\gamma}(V(\mathbf{x}_0), V(\mathbf{x}_0))$ is the integral semivariance between $V(\mathbf{x}_0)$ and itself; the within-block variance. For a fuller description of kriging in this form see Burgess and Webster (1980).

Optimal Sampling

Kriging estimates with minimum estimation or kriging variance, σ_k^2, and also produces an estimate of this kriging variance for every estimated value. The kriging variance depends only on the geometry of the domain or support $V(\mathbf{x}_0)$ to be estimated, the distances between $V(\mathbf{x}_0)$ and the n data points $\mathbf{x}_i$, the geometry of the n data, and finally the variogram. The values of the sample observations have no influence (equation 16.8). Therefore, if the variogram is known, the kriging variance can be computed for any proposed sampling strategy prior to the actual survey. Kriging may, therefore, be used to design optimal sampling strategies.

Kriging as a means of optimising sampling is described with application to soil survey by Burgess *et al.* (1981), and McBratney and Webster (1981). Kriging has been applied to design optimal strategies for sampling reflectance with field radiometers (Webster *et al.*, 1989), and for sampling variables at the ground for remote sensing investigations (Atkinson, 1991).

The optimal sampling intensity for a given sampling scheme can be chosen by solving the kriging equations for several sampling intensities and plotting σ_k^2 against sample spacing (McBratney and Webster, 1981). If the budget for the survey is limited then so too is the maximum precision attainable. For block kriging to square cells, σ_k^2 can occur either when the blocks are centred on the grid cells or when they are centred on the grid nodes. It depends on the size of the blocks in relation to the grid mesh. Therefore, the equations must be solved twice for block kriging.

If the survey is not limited by funding and the investigators can define a maximum tolerable error, then the optimal sampling strategy is the one that just achieves the desired precision. Greater precision would be wasteful. The optimal strategy is found as before by reading the required sample spacing from the graph of σ_k^2 against sample spacing.

FIELD SITE AND EXPERIMENTAL DATA

The TIGER Project

The study region encompasses a rectangular (100 km north–south and 500 km east–west) area of rain forest in Cameroon, Africa, south-east of the capital Yaoundé (Figure 16.1). In January and February 1994 a field campaign was undertaken in this area in

Figure 16.1   Location of study area

support of the Natural Environment Research Council's (NERC) Terrestrial Initiative in Global Environmental Research (TIGER) 1.4 project 'Remote Sensing of Tropical Vegetation'.

One of the aims of the TIGER project was to map the extent of different stages of forest regeneration. Tropical forests are most often thought of as sources of carbon (e.g. from forest clearing). However, where the forest is regenerating it may act as a sink of carbon (e.g. Foody and Curran, 1994b). Maps of regenerative stage are one means of 'scaling-up' point estimates of carbon flux. In the context of this broader aim, the objective is to map regenerative stage through the combination of NOAA AVHRR imagery (with a spatial resolution of 1.1 km by 1.1 km) and extensive field data.

Field Campaign

Radiometrically corrected and geometrically registered AVHRR imagery for three dates (30 December 1992, 7 January 1993 and 9 January 1993) were used to select the sample sites. The Normalised Difference Vegetation Index (NDVI) was computed as NDVI = (Near-infrared − Red)/(Near-infrared + Red) for each image and the maximum NDVI for each pixel over three dates taken to produce a Maximum Value Composite (MVC) of NDVI. This image was used to select 66 sites of known latitude and longitude covering a range of MVC NDVI between 0.25 and greater than 0.45.

The 66 sites were located in the field using a Magellan Global Positioning System (GPS). Each site was accessed from the road by cutting an access line through the forest (Figure 16.2). From the end of this access line a further 100 m was cut and then a 40 m by 250 m (i.e. 1 ha) sub-plot was sampled. Two further 100 m lines were cut at angles of 120° and 240° to the initial 100 m line, and two further sub-plots sampled. This process was repeated for all 66 sites giving a total of 198 sub-plots and a two-stage hierarchical sampling design (Figure 16.3). The clusters of sub-plots all fell within latitudes of 3° N and 4° N, and longitudes of 11.4°E and 13°E.

Three biophysical properties were measured: mean diameter at breast height, mean diameter at first leafing branch (of size greater than 10 cm) (both to the nearest 10 cm), and tree density (i.e. number of trees per hectare). These variables were obtained by measuring *all* the trees within each sub-plot. All trees were, in addition, identified to species. In the future, forest biomass will be estimated using the relations between the three properties and species, and biomass will be related to indices derived from the AVHRR imagery to map biomass and regenerative stage over the study area.

Clearly, the data set represents an extensive source of information on biophysical properties of the forests of Cameroon. Further, the hierarchical sampling design makes it possible to (i) determine the scales of spatial variation in the rain forest properties using the variogram, and (ii) assess the sampling strategy in terms of the ability to scale-up the field measurements so that they may be interpreted effectively with NOAA AVHRR imagery.

ANALYSIS

Of the 66 sample sites eight were unsuitable for analysis because one part or another of the data for each of the eight sites (e.g. full location details) was incomplete. Therefore, the analysis was undertaken for each of the three biophysical variables based on 58 times 3 (i.e. 174) observations.

Experimental Variograms

The locations of the sub-plots were provided in latitude and longitude as shown in Figure 16.3. The absolute distance (in say kilometres) corresponding to a given longitude depends on latitude. However, at the range of latitudes indicated variation in absolute distance is small. Therefore, degrees of longitude and latitude were multiplied by a scalar to obtain approximate distance in kilometre units.

Experimental variograms were computed for the three biophysical properties using the GSLIB software (Deutsch and Journel, 1992) and are plotted (as a series of discrete data points) in Figure 16.4a–c. These results should be interpreted with some caution

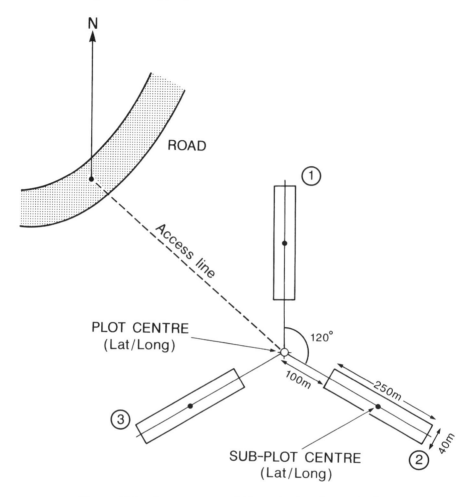

Figure 16.2   Arrangement of 1 ha sub-plots for each plot

as the number of pairs of data used in computing the experimental semivariance was limited by the spatial array of data available, particularly for the larger lags. Unreliable data points representing lags of 2, 3, 5, 24 and 34 km were removed from the graphs. Several mathematical models were fitted to the experimental values of semivariance and the spherical model given by equation 16.9 was found to provide the best fit.

$$\gamma(h) = c_0 + c_1\{3h/2a - 1/2(h/a)^3\} \qquad \text{for } 0 < h < a$$
$$\gamma(h) = c_0 + c_1 \qquad\qquad\qquad \text{for } h > a$$
$$\gamma(0) = 0 \qquad\qquad\qquad\qquad\qquad (16.9)$$

where a is the range, c_0 is the nugget variance and c_1 is the spatially dependent or structured component. The three variables were found to be highly inter-correlated, and so a single range was chosen for the three variograms as 35 km, and the remaining coefficients were obtained by fitting the models up to lags of 35 km using least squares approximation.

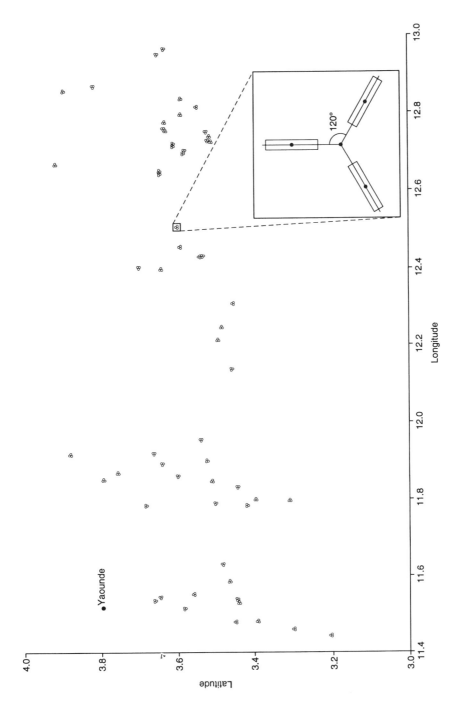

Figure 16.3 Location of sample plots

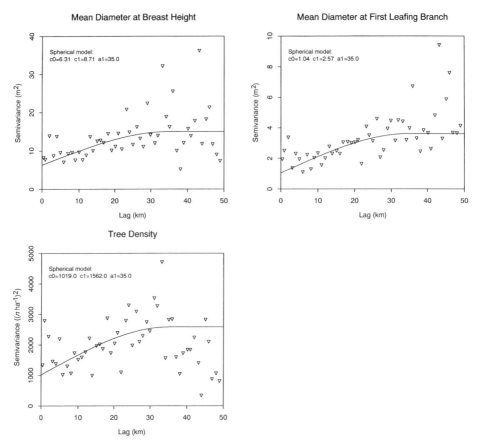

Figure 16.4 Experimental variograms of mean diameter at breast height (MDBH), mean diameter at first leafing branch (MDFLB), and tree density defined on a support of 40 m by 250 m

The spherical models fitted and their coefficients are shown in Figure 16.4a–c. The form of these variograms reveals something of the scales of spatial dependence present in the biophysical variables. For example, the fact that the experimental variograms appear to reach a sill at a fixed range indicates that there may be a maximum scale of spatial variation (around 35 km) operating.

The nugget variance of all three experimental variograms is around 20% to 40% of the overall amount of variation present. This implies one of two things. Either, that there is a sizeable amount of measurement error, or that there is very fine-scale or microscale variation present in the data (analogous to the gold nuggets of the mining industry). Since the sampling is known to have been exhaustive within each 1 ha sub-plot, it is likely that at least some and probably most of the nugget variance of each variogram is attributable to fine-scale spatial variation.

The results demonstrate that at least two scales of spatial variation appear to be present in the tropical forests of central Cameroon. One must account for these scales of variation when sampling and mapping. In the present TIGER study the objective was to map the biophysical properties of the Cameroonian forests using NOAA AVHRR imagery with a spatial resolution of 1.1 km by 1.1 km. From equation 16.3 this spatial resolution will ensure that the spatial variation at a scale of 35 km is captured in the imagery at the expense of the spatial variation between sub-plots of 1 ha. The latter spatial variation will be averaged over the 1.1 km by 1.1 km support. Providing that the investigators are interested in the larger scale of spatial variation the results indicate that the choice of spatial resolution is appropriate.

The effect on the variogram of averaging three 1 ha sub-plots to obtain a set of estimates representative of the 1.21 km^2 GRE of the AVHRR imagery can be seen in Figure 16.5a–c. The regularised variograms are shown as a set of experimental values of semivariance at a set of discrete lags. The models plotted in Figure 16.5a–c are the same as those fitted to the variograms computed for the smaller support of 1 ha and are shown for comparison. As equation 16.3 suggests, the experimental values of semivariance have decreased at lags larger than $|v|$ by an amount approximately equal to the within-block variance. These experimental variograms describe the spatial variation that is potentially correlated with the AVHRR imagery.

Scaling-up

It is possible to compare directly the measurements made at the ground over 1 ha sub-plots with the AVHRR image pixels. However, the simple correlation between the two variables may be low (and, therefore, the accuracy of both the regression relations and the predictions made from the imagery may be low also). Rather, it is recommended that some attempt be made to scale-up the measurements made at the ground. To scale-up the ground data it is necessary to estimate the mean value for GREs that correspond to selected image pixels. Given that the support of the biophysical variables at the ground is 40 m by 250 m and that the support or spatial resolution of the imagery is 1.1 km by 1.1 km it is clear that complete coverage of a GRE is impractical (requiring some 121 observations). Therefore, to estimate the mean value for each of several GREs one must sample *within* the GREs.

For the TIGER field campaign three 1 ha sub-plots were grouped together to allow natural averaging within each of the 66 1.1 km by 1.1 km supports as shown in Figure 16.3. The question in relation to the TIGER project is 'are three observations sufficient to estimate the mean of each variable?'. The answer depends on the spatial dependence and scales of spatial variation present in the variables. In particular, kriging may be used to determine from the variogram the kriging variance associated with this and other potential sampling strategies.

Each variogram was input in turn to the kriging equations and the kriging variances computed using equation 16.8 for an equilateral triangular grid and a range of sample sizes (Figure 16.6a–c). In the present case the sample size was three. The kriging variances $\sigma_k^2(V \mid 3v)$ of estimating the mean within a support V from three observations v set out on an equilateral triangular grid are given in Table 16.1. If the investigators had not chosen to sample hierarchically, that is, they had sampled with isolated 1 ha sub-plots, the kriging variances $\sigma_k^2(V \mid 1v)$ would have been much larger as shown in Table 16.1.

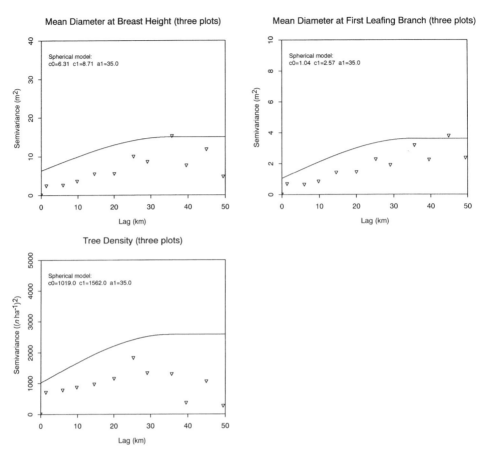

Figure 16.5 Experimental variograms of mean diameter at breast height (MDBH), mean diameter at first leafing branch (MDFLB), and tree density computed for clusters of three 40 m by 250 m sub-plots. The solid curves are the models fitted in Figure 16.4

Where the objective is solely to estimate the mean the above results stand alone. However, where the estimated values are to be used in further analyses it may be necessary to investigate the results further to aid interpretation. In the present case, the objective is to regress the biophysical variables on AVHRR image pixels. Then the kriging variances σ_k^2, as set out in Table 16.1, may be interpreted in relation to the dispersion or sample variances $D^2(V/R)$ of each variable defined on a 1.21 km^2 support V within the 500 km by 100 km region of interest R. The dispersion variance $D^2(V/R)$ describes the variation between supports of 1.21 km^2, and it is this variation that may be correlated with the AVHRR image pixels.

The dispersion variance may be computed from the variogram as a discrete approximation of the integral semivariances defined (in general terms) by equation 16.5. Since the experimental variograms in Figure 16.4a–c represent spatial variation defined on a support v of 1 ha one may compute from them the dispersion variances $D^2(v/V)$ within a support of 1.21 km^2 and $D^2(v/R)$ within a region R of 50 000 km^2. The dispersion

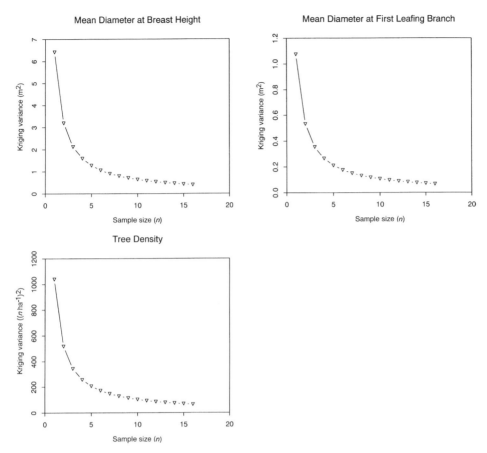

Figure 16.6 Kriging variance plotted against sample size for mean diameter at breast height (MDBH), mean diameter at first leafing branch (MDFLB), and tree density

variance required $(D^2(V/R))$ may then be computed from Krige's relation which is given by

$$D^2(V/R) = D^2(v/R) - D^2(v/V) \qquad (16.10)$$

Using Krige's relation the dispersion variances $D^2(V/R)$ were computed for each of the three biophysical variables and these are given in Table 16.2. By comparing the kriging variances σ_k^2 to the dispersion variances $D^2(V/R)$ one can see the amount of variation that is attributable to variation *between* plots of 1.21 km^2 (i.e. variation that is potentially correlated with the AVHRR imagery) and the amount of variation that is due to the error in estimating the mean in those plots (i.e. variation that is orthogonal to or unrelated to the AVHRR imagery). The ratio of dispersion variance to kriging variance (which may be seen as a measure of signal to noise) is shown for each of the three variables in Table 16.1. Clearly, averaging three sub-plots together makes a large difference to this ratio. The investigator should strive to increase this ratio, thereby increasing the likelihood that relations in the data are detected and represented

Table 16.1 Comparison of the kriging variances (noise) for samples of one observation, $\sigma_k^2(V\mid 1v)$, and three observations, $\sigma_k^2(V\mid 3v)$, to the dispersion variance (signal), $D^2(V/R)$, defined for a support of 1.21 km², V, within a region of 50 000 km², R

	$\sigma_k^2(V\mid 1v)$	$D^2(V/R)$	D^2/σ_k^2	$\sigma_k^2(V\mid 3v)$	$D^2(V/R)$	D^2/σ_k^2
MDBH	6.42	8.34	**1.3**	2.12	8.34	**3.93**
MDFLB	1.08	2.46	**2.28**	0.354	2.46	**6.95**
Tree Density	1039	1496	**1.44**	344	1496	**4.35**

Table 16.2 Three dispersion variances for mean diameter at breast height (MDBH), mean diameter at first leafing branch (MDFLB), and tree density

	$D^2(v/V)$	$D^2(v/R)$	$D^2(V/R)$
MDBH	6.54	14.88	**8.34**
MDFLB	1.11	3.57	**2.46**
Tree density	1061	2557	**1496**

accurately.

The second prerequisite to scaling-up is that the spatial coverage and spatial extent of the sample at the ground match that of the remotely sensed imagery. The reason for this is that the full dynamic range of variation in both properties must be measured so that the bivariate distribution function between both variables is adequately characterised. Strictly, this does not necessitate an even spatial coverage. If the full range of spatial variation in both variables is encountered in a small region then it is sufficient that only that region be sampled. However, experience has shown that variation tends to increase as the spatial extent increases, and in some cases there is no identifiable limit to this variation.

In the present study the full dynamic range of spatial variation was ensured by the sampling framework set out at the beginning. In other words, the 66 pixels were chosen from imagery such that the full range of MVC NDVI from 0.25 to greater than 0.45 was included.

DISCUSSION

In the analysis above the sample size was investigated and recommendations made for ensuring tolerable errors. However, the sample size is not all that may be varied. One might wonder if efficiency can be increased by altering the sampling scheme (an equilateral triangular grid), the sampling intensity, and the support (40 m by 250 m cells) chosen by the TIGER investigators.

Both theoretical and empirical studies have shown that systematic sampling schemes are more efficient than more random alternatives (Yates, 1948; Matérn, 1960; Burgess et al., 1981). Most environmental variables are spatially dependent at some scale so that observations close together tend to be alike. Random schemes inevitably place some observations close together so that information is duplicated. Systematic schemes ensure that the observations are as far apart and as statistically independent as possible, maximising information. The only drawbacks of systematic sampling are that it may not be possible to estimate the precision with which the mean is estimated (Dunn and Harrison, 1993) and that there is some risk of bias due to periodicity (Finney, 1950). Generally, the equilateral triangular grid is the most efficient scheme.

In the analysis above it was assumed that the three sub-plots are evenly distributed over the 1.21 km^2 GREs. The most even spatial distribution (i.e. with the sub-plots placed at the centres of Dirichlet tiles) represents the optimal sampling intensity (Burgess et al., 1981). The sub-plots were actually placed closer together than this (with a sample spacing between the sub-plots of 225 m). Consequently, the actual kriging variances may be slightly less than those indicated in Figure 16.6a–c.

In practice, the support of each AVHRR image pixel is centre-weighted due to the PSF of the sensor. Further, the spatial extent of each pixel far exceeds that of the nominal GRE. The positioning of the sub-plots (i.e. with more observations at the centre of the GRE and less towards the edges) actually reflects the centre-weighting of the PSF. Further analysis would be necessary to account for the PSF in computing the kriging variances.

As for the sampling intensity, the support chosen may not be optimal. Specifically, the size, geometry and orientation of the sub-plots may each be sub-optimal. Research suggests that many smaller supports represents a more efficient strategy than fewer larger ones (e.g. Atkinson and Curran, 1995). The reason is that, total areal coverage remaining constant, the smaller the support the greater the spatial independence between data. Therefore, field techniques permitting, future investigators might consider sampling with more supports of smaller size. Similarly, elongated sub-plots tend to be more efficient than square sub-plots, again because this increases spatial independence. In this respect, the sampling strategy adopted *is* efficient. Orientation may be important where the spatial variation is anisotropic or periodic. In the present study it was not feasible to search for anisotropy or periodicity because the sample size was limited. However, where anisotropy is detected, the procedures for dealing with it are straightforward (Burgess et al., 1981).

Finally, it has been assumed throughout that the mean of a given cell is equal to the linear average of the components of that cell. That is, the relation between spatial variation and spatial resolution is linear. This is not necessarily the case. For example, for remotely sensed imagery a very bright 10 m by 10 m object may cause the signal of a pixel to be saturated whether the spatial resolution is 100 m by 100 m or 1 km by 1 km. Clearly, the average of all 100 m by 100 m pixels within a 1 km by 1 km pixel will be less than the saturated value for the larger cell. Such non-linear scaling is beyond the scope of the present chapter, but has important implications, particularly for scaling-up remotely sensed imagery (as opposed to biophysical variables).

CONCLUSIONS

It is important when measuring environmental properties that the scale of measurement is chosen in relation to the scales of spatial variation of interest. In the examples presented in this chapter two scales of spatial variation were identified; one at less than 1 km (i.e. between the 1 ha sub-plots within each 1.21 km² plot) and the other at 35 km (i.e. between the 1.21 km² plots). If the investigator is interested in spatial variation occurring over distances of up to 35 km then a spatial resolution of 1.1 km by 1.1 km should be considered appropriate. The spatial variation between 1.21 km² plots will be captured in data at the expense of the spatial variation within plots (equation 16.3).

To scale-up from 1 ha sub-plots to the 1.1 km by 1.1 km cells one must sample and average several 1 ha sub-plots to estimate values for the larger cells. In the examples given, three sub-plots were measured within each larger cell. By computing the kriging variances for different sample sizes it was possible to plot the relation between precision and sample size for each of the three biophysical variables. It was found that for a sample size of three the kriging variances were 2.12 m (mean diameter at breast height), 0.354 m (mean diameter at first leafing branch) and 344 trees per hectare (tree density).

To interpret the above results it is necessary to consider the dispersion or sample variances of the biophysical variables defined for 1.21 km² cells within a 50 000 km² area. The dispersion variances were computed as 8.34 m (mean diameter at breast height), 2.46 m (mean diameter at first leafing branch) and 1496 trees per hectare (tree density). Clearly, the errors in estimating the biophysical properties over plots of 1.21 km² are around four to seven times smaller than the amount of spatial variation between the plots. This implies that if a relation exists between the biophysical variables and the AVHRR data, a simple bivariate distribution function based on these data (with a sample size of three sub-plots per cluster) should reveal it.

ACKNOWLEDGEMENTS

The author acknowledges the Natural Environment Research Council for funding the TIGER 1.4 programme of research and thanks Dr Richard Lucas of the University College of Swansea for providing the data and background information on the TIGER project. Phillip Bailey and Doreen Boyd are thanked for help with data preparation.

REFERENCES

Atkinson, P.M. (1991) Optimal ground-based sampling for remote sensing investigations: estimating the regional mean. *International Journal of Remote Sensing*, **12**, 559–567.
Atkinson, P.M. (1993) The effect of spatial resolution on the experimental variogram of airborne MSS imagery. *International Journal of Remote Sensing*, **14**, 1005–1011.
Atkinson, P.M. (1997) Scale and spatial dependence. In: P.R. van Gardingen, G.M. Foody and P.J. Curran (eds) *Scaling-up*. Cambridge University Press, Cambridge, pp. 35–60.

Atkinson, P.M. and Curran, P.J. (1995) Defining an optimal size of support for remote sensing investigations. *IEEE Transactions on Geoscience and Remote Sensing*, **33**, 1–9.

Barnsley, M.F. (1989) *Fractals Everywhere*. Freeman, San Francisco.

Belward, A.S. and Lambein, E. (1990) Limitations to the identification of spatial structures from AVHRR data. *International Journal of Remote Sensing*, **11**, 921–927.

Bian, L. and Walsh, S.J. (1993) Scale dependencies of vegetation and topography in a mountainous environment in Montana. *The Professional Geographer*, **45**, 1–11.

Burgess, T.M. and Webster, R. (1980) Optimal interpolation and isarithmic mapping of soil properties II. Block kriging. *Journal of Soil Science*, **31**, 333–341.

Burgess, T.M., Webster, R. and McBratney, A.B. (1981) Optimal interpolation and isarithmic mapping of soil properties IV. Sampling strategy. *Journal of Soil Science*, **32**, 643–659.

Clark, C.D. (1990) Remote sensing scales related to the frequency of natural variation: an example from paleo-ice-flow in Canada. *IEEE Transactions on Geoscience and Remote Sensing*, **28**, 503–515.

Clark, I. (1977) Regularization of a semi-variogram. *Computers and Geosciences*, **3**, 341–346.

Cohen, W.B., Spies, T.A. and Bradshaw, G.A. (1990) Semivariograms of digital imagery for analysis of conifer canopy structure. *Remote Sensing of Environment*, **34**, 167–178.

Cressie, N.A.C. (1991) *Statistics for Spatial Data*. Wiley, New York.

Deutsch, C.V. and Journel, A.G. (1992) *GSLIB Geostatistical Software Library User's Guide*. Oxford University Press, Oxford.

Dozier, J. and Strahler, A.H. (1983) Ground investigations in support of remote sensing. In: R.N. Colwell (ed.) *Manual of Remote Sensing*, 2nd edition. American Society of Photogrammetry, Falls Church, Virginia, pp. 959–986.

Dunn, R. and Harrison, A.R. (1993) Two dimensional systematic sampling of land use. *Journal of the Royal Statistical Society, Series C: Applied Statistics*, **42**, 585–601.

Finney, D.J. (1950) An example of periodic variation in forest sampling. *Forestry*, **23**, 96–111.

Foody, G.M. and Curran, P.J. (1994a) Scale and environmental remote sensing. In: G.M. Foody and P.J. Curran (eds) *Environmental Remote Sensing from Regional to Global Scales*. Wiley, Chichester, pp. 223–232.

Foody, G.M. and Curran, P.J. (1994b) Estimation of tropical forest extent and regenerative stage using remotely sensed data. *Journal of Biogeography*, **21**, 223–244.

Gohin, F. and Langlois, G. (1993) Using geostatistics to merge *in situ* measurements and remotely sensed observations of sea surface temperature. *International Journal of Remote Sensing*, **14**, 9–19.

Hill, J.L., Curran, P.J. and Foody, G.M. (1994) The effect of sampling on the species–area curve. *Global Ecology and Biogeography Letters*, **4**, 97–106.

Isaaks, E.H. and Srivastava, R.M. (1989) *Applied Geostatistics*. Oxford University Press, Oxford.

Jasinski, M.F. (1990) Sensitivity of the normalized difference vegetation index to sub-pixel canopy cover, soil albedo and pixel scale. *Remote Sensing of Environment*, **32**, 169–187.

Journel, A.G. and Huijbregts, C.J. (1978) *Mining Geostatistics*. Academic Press, London.

Jupp, D.L.B., Strahler, A.H. and Woodcock, C.E. (1988) Autocorrelation and regularization in digital images I. Basic theory. *IEEE Transactions on Geoscience and Remote Sensing*, **26**, 463–473.

Jupp, D.L.B., Strahler, A.H. and Woodcock, C.E. (1989) Autocorrelation and regularization in digital images II. Simple image models. *IEEE Transactions on Geoscience and Remote Sensing*, **27**, 247–258.

Justice, C.O., Townshend, J.R.G., Holben, B.N., and Tucker, C.J. (1985) Analysis of the phenology of global vegetation using meteorological satellite data. *International Journal of Remote Sensing*, **6**, 1271–1318.

Justice, C.O., Dugdale, G., Townshend, J.R.G., Narracott, A.S. and Kumar, M. (1991) Synergism between NOAA-AVHRR and Meteosat data for studying vegetation development in semi-arid West Africa. *International Journal of Remote Sensing*, **12**, 1349–1368.

Lam, N.S.-N. and Quottrochi, D.A. (1992) On the issues of scale, resolution and fractal analysis in the mapping sciences. *Professional Geographer*, **44**, 88–98.

Laporte, N., Justice, C. and Kendall, J. (1995) Mapping the dense humid forest of Cameroon and Zaire using AVHRR satellite data. *International Journal of Remote Sensing*, **16**, 1127–1145.

McBratney, A.B. and Webster, R. (1981) The design of optimal sampling schemes for local estimation and mapping of regionalized variables II. Program and examples. *Computers and Geosciences*, **7**, 335–365.

McGwire, K., Friedl, M. and Estes, J.E. (1993) Spatial structure, sampling design and scale in remotely sensed imagery of a California savanna woodland. *International Journal of Remote Sensing*, **14**, 2137–2164.

Mandelbrot, B. (1982) *The Fractal Geometry of Nature*. Freeman, New York.

Marceau, D.J., Howarth, P.J. and Gratton, D.J. (1994a) Remote sensing and the measurement of geographical entities in a forested environment. 1. The scale and spatial aggregation problem. *Remote Sensing of Environment*, **49**, 93–104.

Marceau, D.J., Gratton, D.J., Fournier, R.A. and Fortin, J.-P. (1994b) Remote sensing and the measurement of geographical entities in a forested environment. 2. The optimal spatial resolution. *Remote Sensing of Environment*, **49**, 105–117.

Matérn, B. (1960) Spatial variation. *Meddelanden Från Statens Skogsforskningsinstitut*, **49**, 1–144.

Matheron, G. (1965) *Les Variables Régionalisées et Leur Estimation*. Masson, Paris.

Raffy, M. (1994a) Change of scale theory: a capital challenge for space observation of earth. *International Journal of Remote Sensing*, **15**, 2353–2357.

Raffy, M. (1994b) Heterogeneity and change of scale in models of remote sensing: spatialization of multi-spectral models. *International Journal of Remote Sensing*, **15**, 2359–2380.

Raffy, M. (1994c) The role of spatial resolution in quantification problems: spatialization method. *International Journal of Remote Sensing*, **15**, 2381–2392.

Ramstein, G. and Raffy, M. (1989) Analysis of the structure of radiometric remotely sensed images. *International Journal of Remote Sensing*, **10**, 1049–1074.

Rees, W.G. (1992) Measurement of the fractal dimension of ice-sheet surfaces using Landsat data. *International Journal of Remote Sensing*, **13**, 663–671.

Rossi, R.E., Mulla, D.J., Journel, A.G. and Franz, E.H. (1992) Geostatistical tools for modeling and interpreting ecological spatial dependence. *Ecological Monographs*, **62**, 277–314.

Settle, J.J. and Drake, N.A. (1993) Linear mixing and the estimation of ground cover proportions. *International Journal of Remote Sensing*, **14**, 1159–1177.

Sèze, G. and Rossow, W.B. (1991) Effects of satellite data resolution on measuring the space/time variations of surfaces and clouds. *International Journal of Remote Sensing*, **12**, 921–952.

Stoms, D.M. (1994) Scale dependence of species richness maps. *The Professional Geographer*, **46**, 346–358.

Townshend, J.R.G. and Justice, C.O. (1988) Selecting the spatial resolution of satellite sensors required for global monitoring of land transformations. *International Journal of Remote Sensing*, **9**, 187–236.

Townshend, J., Justice, C., Lei, W., Gurney, C. and McManus, J. (1991) Global land cover classification by remote sensing: present capabilities and future possibilities. *Remote Sensing of Environment*, **35**, 243–255.

Tucker, C.J., Vanpraet, C.L., Sharman, M.J. and Van Ittersum, G. (1985) Total herbaceous biomass production in the Senegalese Sahel: 1980–1984. *Remote Sensing of Environment*, **17**, 233–249.

Wald, L. (1989) Some examples of the use of structure functions in the analysis of satellite images of the ocean. *Photogrammetric Engineering and Remote Sensing*, **55**, 1487–1490.

Webster, R. (1985) Quantitative spatial analysis of soil in the field. *Advances in Soil Science*, **3**, 1–70.

Webster, R. and Oliver, M.A. (1990) *Statistical Methods for Soil and Land Resources Survey*, Oxford University Press, Oxford.

Webster, R., Curran, P.J. and Munden, J.W. (1989) Spatial correlation in reflected radiation from the ground and its implications for sampling and mapping by ground-based radiometry. *Remote Sensing of Environment*, **29**, 67–78.

Woodcock, C.E. and Strahler, A.H. (1987) The factor of scale in remote sensing. *Remote Sensing of Environment*, **21**, 311–322.

Yates, F. (1948) Systematic sampling. *Philosophical Transactions of the Royal Society*, **A241**, 345–377.

Yoder, J.A., McClain, C.R., Blanton, J.O. and Oey, L.-Y. (1987) Spatial scales in CZCS-chlorophyll imagery of the southeastern U.S. continental shelf. *Limnology and Oceanography*, **32**, 929–941.

Zhang, R., Warrick, A.W. and Myers, D.E. (1990) Variance as a function of sample support size. *Mathematical Geology*, **22**, 107–122.

17 Mapping the World's Tropical Moist Forests

C. BILLINGTON
World Conservation Monitoring Centre, Cambridge, UK

There is great uncertainty attached to area estimates of forest extent and rates of deforestation for tropical moist forests. The World Conservation Monitoring Centre (WCMC) has developed a programme of activities designed to quantify the extent of forest ecosystems, make assessments on their protection, identify their importance for the conservation of biodiversity and highlight the most pressing threats. Since 1989, WCMC has gathered spatial data on the world's tropical moist forests. National-level data, for all three tropical regions, have been aggregated into broad forest classes to produce the first global map of moist tropical forest extent. These digital data are housed in WCMC's geographic information system, the Biodiversity Map Library, and have been published in *The Conservation Atlas of Tropical Forests.* By using a Geographic Information System (GIS) it is possible to calculate forest area and estimate forest area under protection. This chapter addresses the methodology of tropical moist forest mapping at the World Conservation Monitoring Centre.

INTRODUCTION

Accurate information on forest extent, condition and management is essential to understand the scale of forest loss and degradation. This was emphasised at the United Nations Conference on Environment and Development (UNCED) in Rio de Janeiro in 1992, and was stated in the non-legally-binding Forest Principles: 'The provision of timely, reliable and accurate information on forests and forest ecosystems is essential for public understanding and informed decision-making'. However, baseline information on the status and distribution of species and ecosystems is often not available, and ways of storing, sharing and disseminating information must be determined in order to assist analysis, planning and implementation of conservation measures. In response to these needs the World Conservation Monitoring Centre's mission is 'to provide information services on the conservation and sustainable use of species and ecosystems, and to support others in the development of their own information systems'. WCMC has developed a programme designed to quantify the extent of forests, make assessments on their protection and evaluate their condition.

WCMC is an independent, non-governmental organisation providing information on the status, protection and management of the Earth's living natural resources. WCMC has been collecting and analysing data on the world's tropical moist forests for a number of years, as part of a broad information management programme covering other ecosystem types, protected areas and threatened species. WCMC is well

Vegetation Mapping: From Patch to Planet. Edited by Roy Alexander and Andrew C. Millington.
© 2000 John Wiley & Sons Ltd.

placed to disseminate data for biodiversity decision-makers at national and international levels by collaborating with expert national and international institutions, maintaining these data in global spatial databases, and developing information management standards.

When WCMC began collating information on tropical forests in 1989, no *global* map of actual forest extent existed. Vegetation maps were mostly available at the national and sub-national level but the continental and global perspectives on actual extent were lacking. Since 1989 WCMC, in close collaboration with IUCN, has been gathering mapped data on tropical moist forests, collecting national information to incorporate into regional and global compilations. Through a number of connected projects within its Forest Programme, WCMC will continue to gather and evaluate forest data, to improve currency of data holdings and to monitor area change. WCMC has focused on mapping the closed moist forests of the tropics because of their importance for biodiversity.

Figure 17.1 shows the extent of tropical forests according to the WCMC forest dataset. These data have been published in *The Conservation Atlas of Tropical Forests* in three volumes: *Asia and the Pacific, Africa* and *The Americas* (Collins *et al.*, 1991; Sayer and Harcourt, 1992; Harcourt and Sayer, 1996). The digital data are maintained in the WCMC Biodiversity Map Library (BML), a computer interface designed in-house to allow easy access to a large library of ARC/INFO Geographic Information System (GIS) files. Spatial data on temperate and boreal forests also form a part of the global data set.

METHODS

Data Gathering

WCMC is not in a position to compile primary data or to undertake field analyses but manages data that are already in existence. WCMC communicates with a large number of data gatherers and data providers through collaboration with a network of governments, environmental groups, scientific institutions and individuals working at the species, site, national and international levels. For example,

- UNEP/GRID Geneva supplied classified AVHRR imagery showing the extent of West Africa's moist forests;
- the Natural Resources Institute (NRI) provided paper maps of the forests of Indonesia, compiled under the Regional Physical Planning Programme for Trans-migration (RePPProT) (RePPProT, 1990), and
- the Conservation Data Centre–Bolivia, a non-governmental organisation, provided digital data on Bolivia's moist forests.

Data Types

Data cover a wide range of types, derived from remote sensed imagery (satellite imagery and aerial photography) or ground survey data compiled by forest depart-

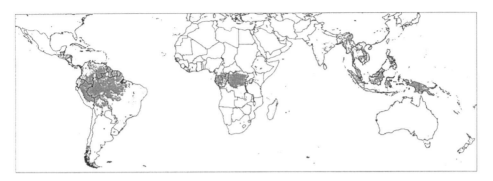

Figure 17.1   Extent of forests mapped in the WCMC Tropical Forest Global Dataset
Note: Forests include moist forests for all three tropical regions. Also mapped are the
closed tropical dry forest and coniferous forests of South and Central America. See
The Conservation Atlas of Tropical Forests (Collins *et al.*, 1991; Sayer and Harcourt,
1992; Harcourt and Sayer, 1996)

ments, vegetation or land cover mapping projects. Data are mainly collected at the
national level, although sub-national information is often available. National level
data were generally gathered at a scale of 1:1 million. Often data were available at
higher resolutions, for example for Costa Rica (1:500 000) and some at lower resolutions, for example for Indonesia (1:2 million) and Brazil (1:3 million). Spatial data
are derived from various sources including published and unpublished maps, in
paper and digital form. WCMC does not have the capacity to classify satellite
imagery in-house, but liaises closely with organisations collecting and classifying
satellite data, such as the EC-JRC TREES project (Malingreau, 1991), UNEP/EAP
(Giri and Shrestha, 1995), FAO (FAO, 1993, 1994) and NASA (Lawrence, 1992).

WCMC obtained the most recent available data as the Tropical Forest Conservation Atlases were being prepared. For some countries it was not possible to obtain
recent data, for example 1971 data for Cambodia and 1979 data for Myanmar were
used. This has resulted in a global data set comprising a moderately wide range of
different ages of data, from the late 1970s to the early 1990s. However, on average,
most data gathered were compiled in the late 1980s and later. Data were located for
60 tropical countries; 81% of those comprised data compiled in the late 1980s and
early 1990s. If WCMC is to attempt to keep pace with area change it is essential that
the spatial dataholdings are maintained and, for example, WCMC now holds data
for Cambodia which were compiled in 1990 (FAO/UNDP, 1994).

Once digitised, the national forest cover data were harmonised and fitted to national boundary data. The supporting political and topographical information, such
as national boundaries, waterbodies and contours, were derived from the 1:1 million
Operational Navigation Charts (ONCs) from the digital database, Mundocart, compiled by Petroconsultants.

Forest Types

The objective was to produce a synoptic view of *tropical moist forests*, by combining forest formations into major groups. Tropical moist forests, defined by Schimper (1903) as *tropical rain forest* and *tropical seasonal (monsoon) forests*, were included. Tropical rain forests occur in perhumid climates where the rainfall is well distributed throughout the year, there is no regular dry season, or a dry season of only one or a few month's duration and no month with rainfall less than 60 mm. Tropical seasonal forests occur where there is a regular and longer dry season, usually more than three months with less than 60 mm rainfall (Collins *et al.*, 1991). In areas with long dry seasons, seasonal forests grade into open canopy deciduous woodlands, then into closed canopy thorn forests, scrub and eventually grasslands. None of these drier open habitats has been mapped by WCMC.

Tropical forests can be divided into closed and open forest. Tropical closed forest has a closed canopy, where the crowns of the trees are interlocking, and includes forest in both the humid tropics (tropical moist forests) and the dry tropics. Tropical open forest includes a diversity of open woodland, mainly in drier tropical areas, where trees are widely scattered and there is no continuous closed canopy (Grainger, 1993). All moist forests have a closed canopy and the term 'closed forests' is also often used to denote rain and monsoon forests. FAO's definition of closed forest is predominantly woody formations with a minimum crown-cover of 40% (FAO/UNEP, 1981). Vegetation types with less than 40% tree canopy cover are technically called woodlands or open forest.

Harmonisation of Forest Classes

When compiling continental and global coverages from higher resolution national data, data harmonisation, whereby detailed vegetation types are aggregated into broader classes, is unavoidable. One difficulty experienced at the start of the project was the lack of a standard vegetation classification, needed to aid data harmonisation and standard reporting.

Land cover information is required in various forms, and methods for collecting information are diverse and purpose driven, often resulting in large disparities in the type of data collected (Young, 1993). Most national bodies have their own individual classifications, often not directly comparable with other national schemes. A vegetation classification for Papua New Guinea, Paijmans (1975), comprises a hierarchy of major vegetation types subdivided into finer classes where structure, floristics and environmental attributes are described in great detail; the map, based mainly on interpretation of aerial photographs, is intended to give the user information both on the various vegetation types and on their ecology and habitats. However, the ReP-PProT map of Indonesia (RePPProT, 1990), compiled mainly from satellite imagery, is used for land use planning and comprises broad forest classes with no detailed descriptions.

Different nomenclatures result in similar vegetation types being classified differently, making comparisons difficult. Therefore, data classified according to different schemes are usually not compatible and at present there is no single, recognised and accepted classification which can be applied across the globe at all scales. To address this problem, the Institute of Terrestrial Ecology (ITE), WCMC and the International Institute for Aerospace Survey and Earth Sciences (ITC) are currently working on a UNEP/FAO project to determine a General Global Nomenclature for Land Cover and Land Use. The project will attempt to find a means of translating between nomenclatures, enabling data collected under one application to be compared with another.

For this project a coarse classification of forest classes, which could be pragmatically applied at regional and global levels, was compiled from more detailed national forest classes. Although floristic variation is not shown in the global dataset, the detailed national vegetation classes were examined and incorporated into the broad classes. Inevitably there will be compromise with vegetation mapping, at any scale. This is especially true when mapping transitional and mosaic formations where a single boundary line is not representative of the true situation on the ground. It has not been possible, for example, to map ecotones between monsoon forests and open canopy deciduous woodlands. In addition, aggregation of vegetation types to broader, general classes results in loss of detail. It is difficult to recognise forest that has been locally disturbed. The areas mapped as single units of forest are therefore frequently mosaics of undisturbed and disturbed patches.

A hierarchy of three main forest types, comprising aseasonal dryland tropical moist forests, aseasonal wetland tropical moist forests and seasonal tropical forests was determined. These main forest types were then further subdivided by lowland and montane formations and mangrove and inland swamp. Where appropriate national classifications did not exist, for example where only 'dense forest' was depicted on the source map, then either potential vegetation boundaries were used to delimit classes, or an altitude cut-off of 1000 m was used to differentiate lowland and montane forest. This results in contrived boundaries, which can be accepted in low resolution continental and global datasets, but are not realistic at the local level.

Potential vegetation continental schemes were used broadly to define tropical moist forest limits for each continent; several continental frameworks were consulted. MacKinnon's classification (IUCN, 1986) based on Schimper (1903), Udvardy (1975) and Whitmore (1984) was used for South-East Asia, and White's *Vegetation Map of Africa* (White, 1983) was used for tropical Africa. The Americas proved more problematic, as no common scheme covering the region, could be found. Holdridge (1967) for Central America, based on climate and altitude, and UNESCO (1981) for South America, based on climatic and floristic elements, were referred to.

Table 17.1 outlines the resultant forest categories used in the WCMC dataset, their component classes and example descriptions of the more detailed national categories. The categories in italics represent those comprising the broad forest types in the global dataset.

The mapping project developed over a period of three years. The methodology adopted evolved over this time, in part in response to the differing problems posed by mapping tropical forests in the three major regions. In particular it was felt that the very significant areas of coniferous and closed dry forests in the Neotropics should be taken into account, although these forest types had not been dealt with in the other regions. This has led to some inconsistency in global coverage which is currently being addressed.

Table 17.1 Broad forest categories mapped by WCMC, their component classes and example descriptions from national maps

Major forest groupings	Main mapped classes	Component classes	Sample descriptions
Aseasonal dryland tropical moist forests	*Tropical Rain Forest*	Tropical evergreen rainforest Tropical humid forest Tropical perhumid forest Riverine/gallery forest	• Forest dense, rather thin stemmed, canopy 25 to over 30 m high. Locally dipterocarps, *Intsia*, *Casuarina*, *Campnosperma* (Papua New Guinea) • Dipterocarp and/or other broadleaved forest, closed canopy, mature trees covering > 50% (Philippines) • Dense evergreen forest with *Ocotea–Aningeria–Cassipourea* (Kenya)
	Tropical Montane Rain Forest	Montane wet forest Lower montane rain forest Upper montane rain forest Subalpine rain forest Cloud forest	• Altimontane vegetation in tropical Africa (White, 1983) • Very small-crowned lower montane forest. Above 1400 m on ultramafic rocks and limestone (Papua New Guinea) • Elfin forest (St Lucia)
Aseasonal wetland tropical moist forests	*Mangrove* *Inland Swamp Forest*	Mangrove Peat swamp forest Freshwater swamp forest	• Edaphic swamp forest – mangrove (Trinidad) • Freshwater alluvial swamp forest and peat swamp forest (Malaysia) • Edaphic swamp forest – palm swamp; swamp forest; marsh forest (Trinidad)
Seasonal tropical moist forests	*Monsoon Forest (Lowland and Montane)*	Tropical semi-evergreen forest Tropical semi-deciduous forest	• Mixed deciduous forest (Thailand)

Note:
Riverine or 'gallery' forests were mapped only when they were located within a moist forest block. Riverine forest surrounded by 'non-forest' (scrub, grassland, woodland, cultivated or disturbed land etc.) was not mapped. This was because of practical difficulties in that gallery forests are often too narrow to map at the working scale of 1:1 million and they are not consistently depicted on the source material.

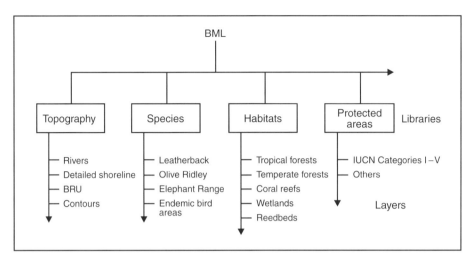

Figure 17.2   Example library and layers within the WCMC Biodiversity Map Library

Data Management and Quality Control

As more spatial data were collected it became important to set up procedures for data management and documentation. A user-friendly interface, the Biodiversity Map Library (BML), was designed in-house to structure data storage and facilitate easy use of GIS datafiles (Rhind, 1993). The BML uses the ARC/INFO librarian module and is based on a hierarchical model where information is organised into thematic libraries. Within each library there are further subdivisions, or layers, of related data (see Figure 17.2). Each layer is spatially indexed to facilitate the management of large datasets. The user can display or query data by a geographical area, state, country, region or continent.

Biodiversity information is continually added to the BML. Other themes include: protected areas (IUCN, 1992–1993), coral reefs, wetlands (Scott, 1989), temperate and boreal forests; species information such as BirdLife International's Endemic Bird Areas (ICBP, 1992), Centres of Plant Diversity (WWF and IUCN, 1994), seabird colonies, turtle nesting sites, topography and socio/economic data such as settlements, roads and railways. The BML is particularly useful as an aid to data dissemination. Access to the various libraries and layers enables inexperienced GIS users to obtain useful information. It is envisaged that in the long term users will be able to query the BML via the Internet.

WCMC does not compile primary data and thus it is not possible to verify the accuracy of the original data compilation. However, WCMC uses expert review (in-house and external) to judge whether a dataset should be used and harmonised into the global database. For instance, during the compilation of the Asian atlas, regional experts were able to comment on draft maps presented at an IUCN Forest Advisory Group Meeting held at Bako National Park, Sarawak. The draft maps of Neotropical countries were reviewed by regional experts attending the 1994 IUCN General

Assembly, Buenos Aires. Information on data sources and harmonisation is maintained and provided to users with the datasets as a matter of routine.

Protected Areas Data

WCMC manages a unique global protected areas database in collaboration with the IUCN Commission on National Parks and Protected Areas. Summary data are maintained in a relational, computerised database (FoxPro), linked to text site sheets (WordPerfect) and linked to the GIS (ARC/INFO), supported by an extensive collection of related and cross-referenced documentation. Data elements can be combined in a variety of ways to provide a wide range of outputs, some being conventional publications (e.g. IUCN, 1992–1993, 1994) others being broadcast via electronic networks such as the Internet. By overlaying management boundary data over the forest extent data using the GIS, it is, for example, possible to calculate the area of forest under protection.

RESULTS

From national-level data WCMC has compiled its first global map of the moist forests of the tropics; this dataset provides a baseline for future monitoring. Data have been gathered for most tropical countries where moist forests occur, although a few gaps, such as for some of the Pacific and Caribbean countries/islands, remain. Some inconsistency currently exists within the pan-tropical dataset as closed dry forests and coniferous forests have been mapped for the Neotropics but not covered in the other regions; however, projects are currently being developed to map the Asian and African dry forests. The forest data collected and analysed by WCMC have been used, mainly for regional priority setting for sustainable development, by a wide range of users, from individual scientists to international research organisations such as the Center for International Forestry Research (CIFOR) located in Bogor, Indonesia.

The use of GIS in the preparation of the maps has allowed an important advance over earlier analyses; it has been possible to estimate areas of forest extent, both statistically and graphically. In the past there have been maps without statistics and statistics without maps. Statistical assessments include the FAO *Forest Resources Assessments* (FAO/UNEP, 1981; FAO, 1988, 1993). From the spatial data gathered under this project, WCMC has compiled area statistics for each country's moist forest cover. Table 17.2 gives estimates of forest extent for tropical Asia, compiled from the digitised maps using GIS analysis. Also included are FAO's 1980 and 1990 figures for closed tropical moist forest (FAO, 1988, 1993) to provide comparison. However, comparisons between sets of statistics may be unreliable, since the dates of available statistics and maps rarely match, and the forest categories employed are usually different. Taking this into consideration, it is interesting to note that the total figure gathered under the WCMC project falls between the totals for FAO's baseline data for 1980 and 1990.

Table 17.3 shows the extent to which moist forest types are represented in protected areas (Murray *et al.*, 1995). These statistics are compiled from the forest extent data

Table 17.2 Original extent of closed canopy moist forests in the Asia region, compared with remaining extent as judged from the WCMC tropical moist forest GIS dataset and FAO statistics for 1980

	Estimated remaining extent of moist forests (km²)					% moist forest remaining		
	Approximate original extent of closed canopy tropical moist forests (km²)	From WCMC dataset; rain monsoon forests	Publication date of maps used in WCMC dataset	FAO (1988) data for 1980, closed broadleaved plus coniferous forests	FAO (1993) data for 1990, closed forest	From WCMC dataset	From FAO (1988) data	From FAO (1993) data
Bangladesh	130 000	9 730	1981–86	9 270	5 920	7	7	5
Brunei	5 000	4 692	1988	3 230	4 580	94	65	92
Cambodia	160 000	113 250	1971	71 680	62 090	71	45	39
India	910 000	228 330	1986	5 040 102†	287 470	25	55†	32
Indonesia	1 700 000	1 179 140	1985–89	1 138 950	863 930	69	67	51
Laos	225 000	124 600	1987	78 100	104 180	55	35	46
Malaysia	320 000	200 450	–	209 960	175 830	63	66	55
Peninsular	(130 000)	(69 780)	1986	–	–	(54)	–	–
Sabah	(70 000)	(36 000)	1984	–	–	(51)	–	–
Sarawak	(120 000)	(94 670)	1979	–	–	(79)	–	–
Myanmar	600 000	311 850	1987	313 090	287 410	52	52	48
Papua New Guinea	450 000	366 750	1975	342 300	318 080	82	76	71
Philippines	295 000	66 020	1988	95 100	76 060	22	32	26
Sri Lanka	26 000	12 760	1988	16 590	13 740	47	64	53
Thailand	250 000	106 900	1985	83 350	82 160	43	33	33
Vietnam	280 000	56 680	1987	75 700	49 460	20	27	18
Totals	5 749 000	2 781 152		2 941 330	2 330 910	48	51	41

* Data for Asian countries are adapted from IUCN (1986).
† Note that for India the FAO figures are not comparable with the WCMC data. FAO have included India's extensive thorn forests in their forest assessment.

Table 17.3 Protection of moist forest types in tropical regions

Forest type	Africa		South and South-East Asia		Insular South-East Asia		South America		Central America and Caribbean		Total	
	Area (km²)	% protected	Area (km²)	% protected	Area (km²)	% protected	Area (km²)	% protected	Area (km²)	% protected	Area (km²)	% protected
Mangrove	34 242	2.04	13 679	3.38	59 937	11.27	31 396	15.55	14 477	12.29	150 730	9.45
Inland swamp	198 994	1.26	12 033	22.53	266 407	6.03	99 411	7.99	15 437	20.09	592 281	5.46
Montane rain forest	159 739	20.19	110 994	8.77	239 352	19.95	268 093	25.24	37 255	17.21	815 432	20.09
Lowland rain forest	1 693 135	6.89	409 924	12.86	1 201 108	9.27	5 356 149	14.37	272 760	20.52	8 933 077	12.39
Sub-montane rainforest							362 573	25.40	10 318	46.62	372 891	25.99
Montane monsoon forest			20 008	12.39	5 254	53.84					25 263	21.01
Lowland monsoon forest			332 509	8.50	35 530	8.63					368 039	8.52
Dry forest							162 895	7.01	194 172	0.66	357 066	3.56
Pine forest							25 457	2.11	388 334	2.07	413 792	2.07
Total	2 086 109	7.29	899 147	10.72	1 804 589	10.39	6 305 973	15.13	932 753	8.72	12 028 571	12.24

and protected areas boundary data that are housed within the BML. According to these data, the lowest overall percentage protection by area of moist forests is found in Africa with 7.3% of forest cover protected. Mangrove, inland swamp and lowland rain forest are all relatively poorly protected, the latter being well below the level of protection found in every other region. This is a matter of some concern, given the importance of this forest type for biodiversity. However, on a positive note, it can be seen that some 20% of montane rain forest is protected in Africa. South America, which has much the largest area of moist forests, has the highest regional protection (15.1%) and this is reflected in above-average protection of all forest types. This particular study, carried out at WCMC once the global forest dataset was complete, also examines protection of ecofloristic zones (Sharma, 1986, 1988), or potential habitats, and illustrates the strategic importance of using GIS datasets to assess the adequacy of protected area networks in tropical countries.

DISCUSSION

This global, harmonised map of tropical moist forest is the first attempt to map digitally the actual extent of forest across all three tropical regions. WCMC and IUCN developed the project as no such map existed, and to understand the scale of forest loss, it was essential first to compile a picture of the remaining forests, to estimate their extent and to provide a baseline for future monitoring of forest area change. The map, which has taken six years to complete, presents a broad pragmatic approach, suitable for regional and global priority setting, but is not applicable at the local scale. The dataset is proving to be valuable for many users, especially in the international conservation and development community; it is also a useful contribution towards the pan-tropical mapping activities being carried out by remote sensing agencies, especially where vegetation mapped from satellite imagery is obscured by cloud.

It is important to emphasise that this is WCMC's first attempt to aggregate and map national forest data into a global coverage. The mapping will be ongoing as every attempt will be made by WCMC to track change in forest cover. New data will be sought to incorporate into the BML and improved methods of data distribution, such as regular output on CD-ROM and the Internet, are being explored. A great deal has been learnt during this mapping project. For example it has not always been possible to obtain up to date or detailed data for all countries and additional effort will be needed to improve the dataset by gathering more accurate data at higher resolutions. This will involve continuing to work closely with collaborating international and national organisations and building up the network of forest/land cover experts. It will also be important to ensure that the dataset is made consistent across all regions, by mapping the closed dry forests of Asia and Africa. As well as improving the tropical coverage, WCMC is building up spatial information on other forest and woody formations, such as temperate and boreal forests, and is developing projects to map open tropical woodlands and savannas.

REFERENCES

Collins, N.M., Sayer, J.A. and Whitmore, T.C. (eds) (1991) *The Conservation Atlas of Tropical Forests. Asia and the Pacific.* Macmillan, London.

FAO (1988) *An interim report on the state of forest resources in the developing countries.* FAO, Rome.

FAO (1993) *Forest Resources Assessment 1990 – Tropical Countries.* FAO Forestry Paper 112. FAO, Rome.

FAO (1994) Africover Project – land cover map and data base of Africa based on satellite remote sensing. Unpublished report on the Technical Consultation and Donor Consultation on the AFRICOVER Project and Annexes. ECA Headquarters, Addis Ababa, 4–11 July 1994. FAO, Rome.

FAO/UNDP (1994) *Cambodia – Land Cover Atlas.* Prepared by the Mekong Secretariat for Project CMB/92/005. FAO, Rome.

FAO/UNEP (1981) *Tropical Forest Resources Assessment Project.* Forest Resources of Tropical Asia, Vol. 3. FAO, Rome.

Giri, C. and Shrestha, S. (1995) Land cover assessment and monitoring at UNEP/EAP-AP: a remote sensing and GIS approach. In: S. Shindo and R. Tateishi (eds) *International Symposium on Vegetation Monitoring, August 29–31, 1995, Chiba University, Japan,* pp. 33–44.

Grainger, A. (1993) Rates of deforestation in the humid tropics: estimates and measurements. *Geographical Journal,* **159**(1), 33–44.

Harcourt, C.S. and Sayer, J.A. (eds) (1996) *The Conservation Atlas of Tropical Forests. The Americas.* Compiled by the World Conservation Monitoring Centre (WCMC) and IUCN – The World Conservation Union. Macmillan, USA.

Holdridge, L.R. (1967) *Life Zone Ecology.* Tropical Science Center, San Jose, Costa Rica.

ICBP (1992) *Putting Biodiversity on the Map: Priority Areas for Global Conservation.* International Council for Bird Preservation, Cambridge, UK.

IUCN (1986) *Review of the Protected Areas System in the Indo-Malayan Realm* (edited by J. MacKinnon and K. MacKinnon). International Union for the Conservation of Nature and Natural Resources, Gland, Switzerland.

IUCN (1992–1993) *Protected Areas of the World: A Review of National Systems.* Compiled by WCMC, 3 volumes. IUCN, Gland, Switzerland, and Cambridge, UK.

IUCN (1994) *1993 United Nations List of National Parks and Protected Areas.* Prepared by WCMC and CNPPA. IUCN, Gland, Switzerland, and Cambridge, UK.

Lawrence, W.T. (1992) *The NASA Landsat Pathfinder Tropical Deforestation Project.* Revista SELPER, June 1992.

Malingreau, J.P. (1991) *Global Tropical Forest Monitoring: Towards the Development of a Methodological Package Using Satellite Data.* EC-JRC/FAO Publication.

Murray, M.G., Green, M.J.B., Bunting, G.C. and Paine, J.R. (1995) Biodiversity conservation in the tropics: gaps in habitat protection and funding priorities. Unpublished report compiled at WCMC for the Overseas Development Administration. WCMC, Cambridge, UK.

Paijmans, K. (1975) *Explanatory Notes to the Vegetation of Papua New Guinea.* Land Research Series No. 35, Commonwealth and Industrial Research Organization, Australia.

RePPProT (1990) *The land resources of Indonesia. A national overview.* Natural Resources Institute, London.

Rhind, J. (1993) Managing environmental data. *Mapping Awareness and GIS in Europe,* **7**(2), 3–7.

Sayer, J.A. and Harcourt, C.S. (eds) (1992) *The Conservation Atlas of Tropical Forests. Africa.* Compiled by the World Conservation Monitoring Centre and IUCN – The World Conservation Union. Macmillan, London.

Schimper, A.F.W. (1903) *Plant Geography Upon a Physiological Basis* (translated by W.R. Fisher, P. Groom, and I.B. Balfour). Oxford University Press, Oxford.

Scott, D.A. (compiler) (1989) *A Directory of Asian Wetlands.* IUCN, Gland, Switzerland.

Sharma, M.K. (1986) *Eco-floristic Zone and Vegetation Maps of Tropical Continental Asia.* FAO,

Rome.

Sharma, M.K. (1988) *Eco-floristic Zone Map of Africa.* FAO, Rome.

Udvardy, M.D.F. (1975) *A Classification of the Biogeographical Provinces of the World.* Project No. 8, IUCN Occasional Paper No. 18. Prepared as a contribution to UNESCO's Man and the Biosphere Programme, IUCN, Gland, Switzerland.

UNESCO (1981) *Vegetation Map of South America: explanatory notes.* UNESCO, Paris.

White, F. (1983) *The Vegetation of Africa.* Unesco/AETFAT/UNSO, Paris.

Whitmore, T.C. (1984) Vegetation of Malesia – 1:5,000 000. Commonwealth Forestry Institute, Oxford University. A contribution to Global Environment Monitoring System, United Nations Environment Programme. *Journal of Biogeography,* **11**.

WWF and IUCN (1994) *Centres of Plant Diversity. A Guide and Strategy for their Conservation.* 3 volumes. IUCN Publications Unit, Cambridge, UK.

Young, A. (1993) Land use and land cover classification: a discussion paper. Background paper to UNEP/FAO Expert Meeting on Land Cover and Land Use Classification Harmonisation, Geneva, 23–25 November 1993. Unpublished.

ACRONYMS AND ABBREVIATIONS

AVHRR	Advanced Very High Resolution Radiometer
BML	Biodiversity Map Library (WCMC)
CIFOR	Center for International Forestry Research
EAP	Environment Assessment Programme (UNEP)
EC–JRC	European Commission–Joint Research Centre (Ispra, Northern Italy)
ESRI	Environmental Systems Research Institute
FAO	United Nations Food and Agriculture Organisation
GIS	Geographic Information System
GRID	Global Resource Information Database (UNEP)
ICBP	International Council for Bird Preservation
IGBP	International Geosphere and Biosphere Project
ITC	International Institute for Aerospace Survey and Earth Sciences
ITE	Institute for Terrestrial Ecology
IUCN	The World Conservation Union
NASA	National Aeronautics and Space Administration
NRI	Natural Resources Institute of the Overseas Development Administration (formerly ODNRI – Overseas Development Natural Resources Institute)
ONC	Operational Navigation Chart
RePPProT	Regional Physical Planning Programme for Transmigration (Indonesia)
TREES	Tropical Ecosystem Environment Observations by Satellite
UNCED	United Nations Conference on Environment and Development
UNDP	United Nations Development Programme
UNEP	United Nations Environment Programme
WCMC	World Conservation Monitoring Centre
WWF	World Wide Fund for Nature

Section 4

CONCLUSIONS

18 Vegetation Mapping in the Last Three Decades of the Twentieth Century

A.C. MILLINGTON[1] and R.W. ALEXANDER[2]

[1] Department of Geography, University of Leicester, UK
[2] Environment Research Group, Department of Geography, Chester College, UK

INTRODUCTION

It is a little over three decades since Kuchler's (1967) seminal work on vegetation mapping was published. Since that time there have been many developments in the subject. Perhaps the most influential of these have been:

1. The increasing demands for vegetation information to assess and help manage pressing environmental problems such as deforestation and land degradation, for environmental policy-making and natural resource planning, and to provide vegetation maps for the many protected areas that have been created in the 1980s and 1990s.
2. The need for vegetation and land cover information for predictive modelling of future climate change and its wide range of impacts.
3. The increased availability of satellite remotely sensed data and its use as the prime data source for vegetation and land cover mapping.
4. The development of sophisticated techniques in spatial analysis and Geographical Information Systems (GIS) which have significant influences on map production and interpretation.

Although these developments have proceeded apace, few scientists have attempted to review their impact on vegetation mapping over the past 30 years (notable exceptions are Kuchler and Zonneveld (1988) and Roberts and Cooper (1989)). In this, the concluding chapter of this book, we attempt a brief review by examining trends in vegetation mapping over this period and identifying some elements for a research agenda.

Vegetation Mapping: From Patch to Planet. Edited by Roy Alexander and Andrew C. Millington.
© 2000 John Wiley & Sons Ltd.

WHY MAP VEGETATION?

The answer to this question may, at first, appear trivial. We could answer 'To communicate to a wide audience information about the vegetation of a specific area'. The map in this case would be a visual tool that is used to convey a condensed, and usually simplified, set of information about the vegetation properties in a spatial framework. In the past (say up until the early 1980s), vegetation maps were in essence only used as communication mechanisms. Many maps still are only ways of communicating information. However, an important and encouraging trend during the last three decades has been for vegetation maps to be part of spatially referenced databases which can be analysed in conjunction with other spatially referenced data to meet the specific requirements of a wide variety of users. Indeed, spatially referenced databases of environmental variables can be used to generate predictive vegetation maps (Franklin, 1995). Technological advances in computer cartography and GIS have probably been the main reason why vegetation 'maps' have evolved into analytical tools, rather than because of the demands of end-users or the influence of vegetation mappers. There are then two paradigms for vegetation maps – communication and analysis. DeMers (1991) examines these two paradigms in some detail.

The question 'Why do we map vegetation?' can therefore be answered in two ways. First, to communicate a complex set of information about vegetation in a simplified, spatially referenced form. Secondly, to provide spatially referenced numerical data about vegetation that can be used for analytical purposes. Both answers indicate that mapping vegetation is an applications-research area: the second answer suggesting that vegetation mapping is more overtly applied than the former. Evidence of the breadth of vegetation mapping as an applications-research area is found in the wide variety of uses of vegetation maps. In this book only a limited number of uses of vegetation maps have been illustrated. A more comprehensive (but not exhaustive) list of uses of vegetation maps is provided in Table 18.1.

ISSUES ARISING FROM FITNESS-FOR-PURPOSE OF VEGETATION MAPS

The many uses that vegetation maps are put to give rise to a set of interesting issues about their fitness-for-purpose. First, most vegetation mapping techniques have been developed by ecologists and biogeographers who are trained in vegetation sampling, plant taxonomy and vegetation classification. Vegetation mapping techniques are therefore underpinned by these sub-disciplines. However, the question arises as to whether mapping techniques underpinned in this manner provide maps that are compatible with the expectations of end-users who often have little ecological or biogeographical background. The most obvious manifestation of this has been the shift from vegetation mapping to land cover mapping, which has accelerated with the wider use of satellite remotely sensed data. These terms are often, and incorrectly, used synonymously. Vegetation mapping should, *sensu stricto*, be retained for mapping biological entities (e.g. species, communities, ecosystems and biomes), whereas land

Table 18.1   Examples of uses of vegetation maps
Examples from this book are in italics, those from elsewhere are in normal typeface.

COMMUNICATION OF VEGETATION AND LAND COVER INFORMATION
Vegetation (Paijmans, 1966; Plumb, 1991; Stone et al., 1994) – Land Cover e.g. EU CORINE
(O'Sullivan, 1994)

AS INPUTS TO OTHER APPLICATIONS
Ecology and biogeography
   Vegetation conservation policy and planning (Scott et al., 1993; *Billington*)
   Animal habitat survey and modelling (Spjelkavik and Elvebakk, 1987; *Brookes et al.*)
   Protected area planning and management (Michalik, 1967; Kachhwaha, 1993; White et al.,
   1995; *Wellens et al.*)
   Understanding and modelling vegetation–environment relationships (Poissonet, 1966;
   Butera, 1977; *Coker; Lázaro et al.*)
   Vegetation change (Chaturvedi, 1978; Guillemyn, 1989; Turcotte et al., 1993)
Silviculture
   Forestry applications (Lee, 1966; Steellingwerf, 1966; Howard, 1967)
Earth surface processes (e.g. water and wind erosion)
   Land degradation assessment and modelling, especially wind and water erosion assessment
   (Coude-Gaussen et al., 1993; Alexander et al., 1994)

Hydrology
   Inputs to hydrological models (*Millington and Jehangir*)
Geology
   Geobotanical prospecting
Climatology and Meteorology
   Inputs to global climate change models
Environmental planning
   Countryside change (*Gulliver; Mackey and Tudor*)
   Resource inventory and management (small areas) (Tablet et al., 1976; Jewell, 1988; Furley
   et al., 1994)
   Environmental impact assessment (Westinga and van Wisngaarden, 1985; van Stokkom et
   al., 1988)
Environmental policy formulation
   Biogeochemical cycling and carbon sequestration (Badhwar et al., 1986; *Cruickshank et al.;
   Milne and Brown*)
   Resource inventory (large areas) (Fontes and Guinko, 1991)
Other
   Terrain mobility maps for military purposes

cover mapping is a more general term, the usage of which has mirrored the rise in mapping using remotely sensed data. Land cover mapping units refer to both natural (and semi-natural) vegetation units as well as other types of land cover – e.g. agricultural crops, forest plantations and, of course, non-vegetated land cover types.

The differences between vegetation and land cover mapping noted above have a direct bearing on the second issue concerning fitness-for-purpose. This concerns the increased reliance on the use of satellite remotely sensed data (at a variety of spatial resolutions) as a key source of information, and how far this conflicts with schemes of vegetation classification that are used to provide the mapping legend for vegetation maps (Adams, 1996). Gerard *et al.* (1998) have shown that, in the humid tropical forest zone of South America, many of the vegetation mapping schemes in use cannot be used to produce vegetation maps from satellite imagery. In particular, vegetation mapping schemes based on floristic taxonomy cannot be applied using remotely sensed data, whereas physiognomic vegetation classifications are ideally suited to the use of such data. Has the increased use of satellite imagery as the prime source of data for land cover mapping led to developments in vegetation mapping? A handful of studies have compared satellite and airborne sensor-derived land cover with floristic distributions obtained from field data collection (Weaver, 1987; Nusser and Schickhoff, 1996; Armitage *et al*, Chapter 6). Though interesting, these studies suggest the two sets of mapping criteria are still too far apart to provide an immediate way forward. Perhaps the approach with the most promise is that based on multitemporal analysis of satellite imagery (*remote sensing phenology*) with seasonal changes in vegetation at the species, community, ecosystem or biome level (*vegetation phenology*).

Thirdly, have new priorities emerged in the last 30 years which require vegetation information (in the form of maps) but which are inadequately served by existing methods of mapping? There are two developments that currently seem to bear on vegetation mappers: (i) the production of vegetation and/or land cover information for the global climate modelling community, and (ii) the production of vegetation and/or land cover maps that can be converted into habitat maps for ecological and biodiversity studies. The global climate modelling community require land surface information to couple with climate models. The scale at which global climate models require such information, 0.5° of latitude and longitude, requires the production of simplified biome maps on a global scale. Moreover, the current requirements of this community are for simplified maps with less classes than there are biomes on the land surface. Mapping biomes is a relatively new area for vegetation mappers, despite the concept being well known to the ecological community. Production of habitat maps also poses new problems for vegetation mappers because, although the concept of a habitat is reasonably well accepted, the definition of a habitat in terms of measurable variables is extremely complex.

Simply talking to people who have produced vegetation maps for non-scientists, elicits a wealth of anecdotal evidence suggesting that users often do not fully appreciate the amount and type of information contained in vegetation maps. The many reasons for this range from an inability fully to comprehend relevant botanical and ecological concepts (e.g. vegetation, species, communities, ecosystems and biomes) through to cartographic issues (e.g. the relationship between scale and generalisation). This suggests to us that either the techniques used to produce vegetation maps need to be modified to account more for the requirements of the end-users, or that at least some of the information contained in the maps needs to be communicated in different ways.

Clearly dialogue between mappers and end-users to produce feasible mapping protocols is required at the outset of a mapping exercise. While an easy statement to make, it is somewhat more difficult to put into practice! The stringency of the requirements and specifications of end-users suggests that general-purpose vegetation maps have little credibility in the end-user community, and that, as a minimum, general-purpose vegetation maps need keys which convert vegetation or land cover classes into classes with parameters the end-users understand. Two issues in particular that need to be tackled by the vegetation mapping community are the generalisation of vegetation information at different mapping scales, and ways of providing information

Table 18.2   Some key issues to be considered in vegetation mapping

---

GENERAL ISSUES
Is the mapping paradigm analysis or communication?
Is the purpose of the map defined by the end-user, the vegetation mapper or both?

CARTOGRAPHIC CONSIDERATIONS
Mapping scale?
Symbols, shading and other cartographic techniques to be used?
How will class boundaries be depicted?
Is the map to be printed on paper or stored as a digital product?
How detailed will the legend on the map be and will there be an accompanying memoir?
Is the map part of a series, and if so what constraints does that put on the mappers and
cartographers?

VEGETATION/LAND COVER CONSIDERATIONS
Is the primary objective to provide vegetation or land cover information?
The vegetation or land cover classification system to be used?
Type of biological entities to be mapped (e.g. species, communities etc.)?

---

on the reliability of the vegetation maps. These issues are introduced in this book, to some degree, in the Chapters by Atkinson (Chapter 16) and Green and Hartley (Chapter 7).

The purpose for which a vegetation map is produced has a significant influence on a number of key issues (Table 18.2). It is clear from this table that purpose, the biological entities mapped, the type of vegetation or land cover classification used, scale and mapping symbols are intricately linked.

DATA COLLECTION TECHNIQUES

The way that vegetation data are collected is also intricately linked to the issues of fitness-for-purpose and those issues listed in Table 18.2. Techniques range from field data collection (these encompass techniques which have been reviewed by Kent and Coker (1992) among others) through to the use of remotely sensed data without reference to ground data collection or ground verification. Mapping techniques are scale dependent (Table 18.3) and the techniques used for a particular mapping project depend on the purpose and the scale of the map that is being produced.

Although the methods of data acquisition clearly vary according to the purpose and the scale of mapping, there have been significant shifts in the dominant data collection techniques in vegetation mapping over the last 30 years. There has been an exponential growth of vegetation maps based on satellite data since the early 1980s (Figure 18.1). To a great extent this growth has been at the expense of the use of aerial photography, which three decades ago was the main source of data (Lee, 1966; Howard, 1967). Studies which rely entirely on data collected in the field, though a small proportion of the total number of maps produced, have remained relatively constant over this period. However, it is not simply that satellite data have ousted aerial photography as a major provider of information that needs to be considered. There is clear evidence that the availability of satellite imagery has had positive effects on vegetation mapping

Table 18.3 Mapping techniques and mapping scale

Final mapping scale	Data collection technique
< 1:1 000 000	Compilation of larger-scale maps (with or without reference to reports)
	Multidate analysis of meteorological satellite data
	Reconnaissance-scale field survey
1:250 000–1:1 000 000	Compilation of larger-scale maps (with or without reference to reports)
	Multidate or multispectral analysis of meteorological and Earth-resource satellite data
	Reconnaissance-scale field survey
1:50 000–1:250 000	Multispectral analysis of earth-resource satellite data
	Semi-reconnaisance-scale field survey
> 1:50 000	Multispectral analysis of Earth-resource satellite data
	Mapping from aerial photography
	Semi-reconnaissance to detailed-scale field survey, with vegetation sampling using quadrats or transects

in two ways. First, it has led to an increase in the overall volume of vegetation mapping, though this increase is also linked to demand. Between 1991 and 1995 there were 33 vegetation mapping publications in total in the searches we made while compiling this chapter (29 of which used satellite imagery as their primary source) while there were only 14 mapping publications in total between 1967 and 1970. Secondly, small-scale vegetation (and land cover) maps covering entire continents have been produced. It was generally not possible to produce these with any reasonable level of accuracy prior to developments which led to the use of NOAA AVHRR for mapping African vegetation (Tucker *et al.*, 1985). Notable exceptions were the maps of Africa (White, 1983) and South America (UNESCO, 1981) produced by groups of 'experts' compiling existing maps. In the case of the map of Africa (White, 1983), a map of high accuracy for continental-scale mapping was produced.

The reasons for the changes in the dominant mapping methods over the last three decades may mainly be construed as technology-led, though it is almost impossible to disentangle issues of cost-effectiveness and the ability to acquire different types of information about vegetation from different techniques, from simply the availability of the 'latest new technique'. What is clear, however, is that the technological advances in Earth observation and GIS have had little input from the vegetation mapping community who have either followed or exploited the technological advances of others.

When satellite imagery first became widely available in the early 1970s, with the launch of the first Landsat (then ERTS) satellite, one of the specific objectives in the design of the Multispectral Scanning System (MSS) mounted on-board the satellite was to acquire ecological information. In this instance, vegetation mappers clearly followed where technology led them! However, advances in our understanding of the nature of the information acquired by sensors on Earth Resources satellites in relation to vegetation-related issues since the early years of satellite remote sensing have had significant inputs from ecologists and biogeographers (e.g. the red edge shift – (Filella and Penuelas, 1994)). There are also cases where technological advances that were not

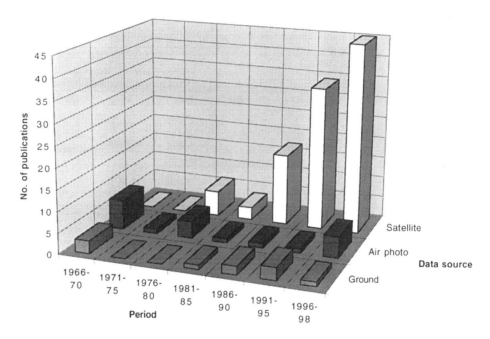

Figure 18.1   Vegetation mapping publications by time period and data source. Note that the most recent period covers only three years (1996–98)

geared towards the vegetation mapping, and the wider biological, community have, nonetheless, been exploited by them. The most obvious example of this is the use of data acquired by the Advanced Very High Resolution Radiometer carried on the NOAA satellites. This satellite–sensor combination was primarily designed to acquire information about cloud patterns for weather forecasting, but because of its high temporal rate of data acquisition and spectral channels located in parts of the electromagnetic spectrum that are sensitive to vegetation growth, it has been extensively exploited for vegetation mapping and monitoring since the early 1980s.

CURRENT RESEARCH AREAS

From the above review of key issues in vegetation mapping and their current status, we identify a number of research areas in vegetation mapping at the present time. This is not meant to be a comprehensive list, rather it represents an initial stab at what vegetation mappers might wish to consider.

The Nature of Vegetation Boundaries

With two exceptions (Armitage *et al.*, Chapter 6, and Wellens *et al.*, Chapter 11), the spatial frameworks for presenting vegetation and land cover information in this volume have used hard or crisp boundaries. Yet it is generally accepted that the boundaries of most vegetation communities are zones of gradual transition – ecotones – which can be termed soft or fuzzy boundaries, e.g. water–land margins, forest– savanna boundaries. In land cover mapping the situation may be more ambiguous because there are many examples of both soft and hard boundaries, the latter mainly arising from human influences. The issue of distinction between hard and soft bound- aries is a pertinent research issue, but one which continues to be ignored by many vegetation mappers despite a long history of references within relevant literature (Burrough, 1989; Foody, 1992, 1998; Atkinson, 1997; Kent *et al.*, 1997; Bastin *et al.*, 1998; Brown, 1998). Some important considerations here are:

(a) Ecological continua. Vegetation should simply be viewed as a response to environ- mental processes and attributes which are, with extremely few exceptions, always continuously variable. Consequently, vegetation must be continuously variable and thus should be represented as such on maps. Indeed, vegetation distribution might be predicted from such variables, and the comparative analysis of predictive and actual vegetation maps represents a potentially fruitful area for research.

(b) Mapping scale. For large-scale maps it is more likely that more hard boundaries to vegetation communities can be identified because of the large volume of detailed data available within the map. As mapping scale decreases, data volumes decrease and become more generalised. This makes the location of hard bound- aries more difficult and soft boundaries should prevail. However, the thickness of the boundary lines themselves means that they cover considerable distances on continental or global maps and thus these boundaries might be considered as representing ecotones in some cases. In such cases this should be made clear on map legends.

(c) The representation of soft boundaries on vegetation maps is both technically difficult for mappers and conceptually difficult for most end-users. Traditional vegetation mapping methods produce maps with hard line boundaries between classes. However, modern mapping methods based on image analysis techniques in a GIS environment can allow ecotonal boundaries to be mapped. Wellens *et al.* (Chapter 11) use a simple approach for mapping savannas in Bolivia. In the region mapped, savanna grassland communities are controlled by soil moisture status which is continuously variable over space and time. To represent variation on a vegetation map, the reflectance values of the savanna communities (which range from short, dry grassland to long, flooded grassland) were mapped along an axis of over 200 Digital Number (DN) values. More complex spatial analysis techniques can be used to present this information and these are considered below.

Mapping Vegetation Classes from Satellite Imagery

There are two active research issues relating to vegetation mapping from satellite imagery: first, matching the type of information received by sensors about vegetation canopies to the different types of vegetation classification; secondly, representing the true nature of vegetation classes or land cover in a landscape given the artificial grid imposed upon it due to the way sensors sample upwelling radiance.

Reflectance values provide information about vegetation canopies and clearly it is information about the elements of the canopy that need to be foremost in vegetation classifications that can be used with satellite data. Most vegetation mapping from satellite imagery relies on simple approaches that reduce the complexity of vegetation canopies, e.g. vegetation indices. An alternative approach in investigating vegetation canopies from remotely sensed data is to develop mathematical models of the interaction of electromagnetic radiation with the canopy. Invertible models, which allow vegetation properties to be obtained from reflectance data obtained at-satellite, offer much in terms of vegetation studies. Until now they have mainly been applied over small areas to estimate specific parameters. However, some parameters that can be predicted are important elements in ecologically based classifications of vegetation (e.g. canopy height, canopy roughness, vegetation phenology) and the application of invertible models over large areas is required to see whether canopy modelling can be used to map vegetation classes.

The second issue relates to mapping the proportional vegetation or land cover in pixels. Proportional mapping allows the artificial grid-like nature that pixels impose on vegetation maps to be broken down. Techniques which are being investigated in this context include spectral mixture modelling, fuzzy classifiers and neural networks. Most of these techniques are used in interpreting and mapping from remotely sensed data, though fuzzy logic has been applied to quadrat data as well where an artificial grid-like nature is imposed on vegetation communities being sampled.

Visualising Map Accuracy and Reliability

Most vegetation maps provide little, if any, information on their accuracy and reliability (see Green and Hartley, Chapter 7). In the GIS community accuracy and reliability are important research issues, and a number of techniques are being investigated to provide information on accuracy and reliability in a general mapping context. These techniques can be applied to vegetation and land cover mapping and a few applications already exist (Millington *et al.*, 1994). Further joint research needs to be undertaken between vegetation mappers and GIS experts to develop the most appropriate methods of visualising accuracy and reliability of vegetation data.

REFERENCES

Adams, J.M. (1996) Towards a better vegetation scheme for global mapping and monitoring. *Global Ecology and Biodiversity Letters*, **5**, 3–6.

Alexander, R.W., Harvey, A.M., Calvo, A., James, P.A. and Cerda, A. (1994) Natural stabilisation mechanisms on Badland Slopes: Tabernas, Almería, Spain. In: A.C. Millington and K. Pye (eds) *Environmental Change in Drylands*. Wiley, Chichester, pp. 85–111.

Atkinson, P.M. (1997) Mapping sub-pixel boundaries from remotely sensed images. In: Z. Kemp (ed.) *Innovations in GIS 4*. Taylor & Francis, London.

Badhwar, G.D., MacDonald, R.B. and Metta, N.C. (1986) Satellite derived leaf-area index and vegetation maps as an input to global carbon cycle models – a hierarchical approach. *International Journal of Remote Sensing*, **7**(2), 265–281.

Bastin, L., Fisher, P., Hughes, M., Millington, A. and Wellens, J. (1998) Characterising change in semi-natural vegetation. In: P.J.A. Burt, C.H. Power and P.M. Zukowskyj (eds) *Developing*

International Connections: Proceedings of the 24th Annual Conference and Exhibition of the Remote Sensing Society. Remote Sensing Society, Nottingham, pp. 619–625.

Brown, D.G. (1998) Classification and boundary vagueness in mapping presettlement forest types. *International Journal of Geographical Information Systems,* **12,** 105–129.

Burrough, P.A. (1989) Fuzzy mathematical models for soil survey and land evaluation. *Journal of Soil Science,* **40,** 477–492.

Butera, M.K. (1977) A technique for the determination of a Louisiana marsh salinity zone from vegetation mapped by multispectral scanner data: a comparison of satellite and aircraft data. NASA Technical Manuscript 58203, JSC-12529. NASA Lyndon B Johnson Space Center, Houston.

Chaturvedi, A.C. (1978) Vegetation damage surveying in India. In: *Symposium on Remote Sensing for Vegetation Damage Assessment,* Seattle, pp. 439–454.

Coude-Gaussen, G., Poncet, Y., Rognon, P. and Davy, M.C. (1993) Evaluation of aeolian soil erosion risk by soil and vegetation mapping in the Sahel: the STARS experiment in Tillaberi, Niger. *Zeitschrift für Geomorphologie,* **37**(4), 403–422.

DeMers, M. (1991) Classification and purpose in automated vegetation maps. *Geographical Review,* **81**(3), 267–280.

Filella, I. and Penuelas, J. (1994) The red edge position and shape as indicators of plant chlorophyll content, biomass and hydric status. *International Journal of Remote Sensing,* **15**(7), 1459–1470.

Fontes, J. and Guinko, S. (1991) Remote detection based mapping inventory of the renewable resources of Burkina Faso. In: *Teledetection applique á la cartographie thematique et topographique. Actes des journees scientifiques, Montreal.* pp. 227–238.

Foody, G.M. (1992) A fuzzy sets approach to the representation of vegetation continua from remotely sensed data: an example from lowland heath. *Photogrammetric Engineering and Remote Sensing,* **55,** 221–225.

Foody, G.M. (1998) Issues in the location and characterisation of inter-class boundaries. In: P.J.A. Burt, C.H. Power and P.M. Zukowskyj (eds) *Developing International Connections: Proceedings of the 24th Annual Conference and Exhibition of the Remote Sensing Society.* Remote Sensing Society, Nottingham, pp. 626–632.

Franklin, J. (1995) Predictive vegetation mapping: geographical modelling of biospatial patterns in relation to environmental gradients. *Progress in Physical Geography,* **19**(4), 474–499.

Furley, P.A., Dargie, T.C.D. and Place, C.J. (1994) Remote sensing and the establishment of a geographic information system for resource management on and around Maraca Island. In: J. Hemming (ed.) *The Rainforest Edge.* Manchester University Press, Manchester, pp. 115–133.

Gerard, F.F.G., Wyatt, B.K., Millington, A.C. and Wellens, J. (1998) The role of data from intensive sample plots in the development of a new method for mapping tropical forest types using satellite imagery. In: F. Dallmeier and J.H. Comiskey (eds) *Forest Biodiversity Research, Monitoring and Modeling: Conceptual Background and Old World Case Studies.* Man and the Biosphere Series, Volume 20. UNESCO, Paris, pp. 141–158.

Guillemyn, D. (1989) Die back of mangrove swamps in the Gambia as seen by SPOT. *Bulletin Societe Francaise de Photogrammetrie et de Teledetection,* **114,** 56–58.

Howard, W.J. (1967) Vegetation surveys of savanna forest reserves in Northern Nigeria based on aerial photographs. *Commonwealth Forestry Review,* **46**(1), 36–50.

Jewell, N. (1988) The operational use of satellite data for vegetation mapping in the North York Moors National Park. *Proceedings EARSeL Workshop on Operational and Classification Problems in the Use of Remote Sensing for Monitoring and Inventory of Protected Landscapes,* Capri. EARSeL Special Publication, SP-1 88.46. Boulonge-Ballencourt, pp. 78–89.

Kachhwaha, T.S. (1993) Temporary and multisensor approach in forest/vegetation mapping and corridor identification for effective management of Rajaji National Park, Uttar Pradesh, India. *International Journal of Remote Sensing,* **14**(17), 3105–3114.

Kent, M. and Coker, P.D. (1992) *Vegetation Description and Analysis – a Practical Approach.* Belhaven Press, London.

Kent, M., Gill, W.J., Weaver, R.E. and Armitage, R.P. (1997) Landscape and plant community boundaries in biogeography. *Progress in Physical Geography,* **23,** 315–353.

Kuchler, A.W. (1967) *Vegetation Mapping*. Ronald Press, New York.

Kuchler, A.W. and Zonneveld, I.S. (1988) *Vegetation Mapping*. Kluwer, Handbook of Vegetation Science, Dordrecht.

Lee, P.C. (1966) The use of aerial photographic interpretation in forest ecology and forest inventory in Malaya. *Malayan Forester*, **29**, 276–281.

Michalik, S. (1967) Vegetation map of the 'Turbacz' nature reserve, Gorce Mountains, West Carpathians. *Ochrona Przyrody*, **32**, 89–131.

Millington, A.C., Douglas, T.D., Critchley, R.W. and Ryan, P. (1994) *Woody Biomass Assessment in Sub-Saharan Africa*. World Bank Publications, Washington, DC.

Nusser, M. and Schickhoff, U. (1996) Traditional methods of vegetation geography in transition: potentials and limitations of the digital processing of vegetation data. *Die Erde*, **127**(2), 93–113 (in German).

O'Sullivan, G. (ed.) (1994) *Final Report, CORINE Land Cover Project (Ireland)*. Ordnance Survey of Ireland, Dublin.

Paijmans, K. (1966) Typing of tropical vegetation by aerial photographs and field sampling in northern Papua. *Photogrammetrica*, **21**(1), 1–25.

Plumb, G.A. (1991) Assessing vegetation types of Big Bend National Park, Texas for image based mapping. *Vegetatio*, **94**(2), 115–124.

Poissonet, P. (1966) The place of photo interpretation in a programme of detailed study of flora, vegetation and environment. *Actes de IIe Symposium International de Photointerpretation, Paris*. Volume IV, 2.51–2.55.

Roberts, D.W. and Cooper, S.V. (1989) *Concepts and Techniques of Vegetation Mapping*. General Technical Report, US Department of Agriculture, Forest Service, INT-257.

Scott, J.M., *et al.* (1993) Gap analysis: a geographic approach to protection of biological diversity. *Wildlife Monographs*, **123**, 41 pp.

Spjelkavik, S. and Elvebakk, A. (1987) Mapping winter grazing areas for reindeer on Svalbard using Landsat Thematic Mapper data. In: T.D. Guyenne and G. Calebresi (eds) *Monitoring the Earth's Environment. Proceedings Workshop on Landsat TM Applications, Frascati*. European Space Agency, ESTEC, Noordwijk, Netherlands, pp. 199–206.

Stellingwerf, D.A. (1966) *Practical Applications of Aerial Photographs in Forestry and other Vegetation Studies*. International Training Centre for Aerial Survey.

Stone, T.A., Schlesinger, P., Houghton, R.A. and Woodwell, G.M. (1994) A map of the vegetation of South America based on satellite imagery. *Photogrammetric Engineering and Remote Sensing*, **60**(5), 541–551.

Tablet, D., Inglis, M., Morain, S., Love, L. and Feldman, S. (1976) *Analysis of LANDSAT B Imagery as a Tool for Evaluating, Developing and Managing the Natural Resources of New Mexico*. New Mexico Bureau of Mines and Mineral Resources, Socorro.

Tucker, C.J., Townshend, J.R.G. and Goff, T.E. (1985) African land cover classification using satellite data. *Science*, **227**, 369–375.

Turcotte, K.M., Lulla, K. and Venugopal, G. (1993) Mapping small scale vegetation changes of Mexico. *Geocarto International*, **8**(4), 73–85.

UNESCO (1981) *Vegetation Map of South America*. UNESCO, Paris.

van Stokkom, H.T.C. and Klostermann, E.H. (1988) Operational vegetation mapping of coastal areas for nature management and monitoring. *Proceedings EARSeL Workshop on Operational and Classification Problems in the Use of Remote Sensing for Monitoring and Inventory of Protected Landscapes, Capri. EARSeL Special Publication, SP-1 88.46, Boulonge-Ballencourt*, pp. 90–98.

Weaver, R.E (1987). Spectral separation of moorland vegetation in airborne Thematic Mapper data. *International Journal of Remote Sensing*, **8**(1), 43–55.

Westinga, E. and van Wisngaarden, W. (1985) A rapid vegetation mapping method for environmental impact assessment in a Netherlands dune area. *ITC Journal*, **4**, 242–245.

White, F. (1983) *The Vegetation of Africa*. Unesco/AETFAT/UNSO, Paris.

White, J.D., Kroh, G.C. and Pinder III, J.E. (1995) Forest mapping of Lassen Volcanic National Park, California, using Landsat TM data and a geographical information system. *Photogrammetric Engineering and Remote Sensing*, **61**(3), 299–305.

Index